MICROBIAL DIVERSITY AND ECOLOGY IN HOTSPOTS

MICROBIAL DIVERSITY AND ECOLOGY IN HOTSPOTS

Edited by

APARNA GUNJAL

Department of Microbiology, Dr. D.Y. Patil, Arts, Commerce & Science College, Pimpri, Pune, India

SONALI SHINDE

Annasaheb Kulkarni Department of Biodiversity, MES Abasaheb Garware College, Pune, India

ELSEVIER

ACADEMIC PRESS

An imprint of Elsevier

Academic Press is an imprint of Elsevier
125 London Wall, London EC2Y 5AS, United Kingdom
525 B Street, Suite 1650, San Diego, CA 92101, United States
50 Hampshire Street, 5th Floor, Cambridge, MA 02139, United States
The Boulevard, Langford Lane, Kidlington, Oxford OX5 1GB, United Kingdom

British Library Cataloguing-in-Publication Data
A catalogue record for this book is available from the British Library

Library of Congress Cataloging-in-Publication Data
A catalog record for this book is available from the Library of Congress

ISBN: 978-0-323-90148-2

For Information on all Academic Press publications
visit our website at https://www.elsevier.com/books-and-journals

Publisher: Stacy Masucci
Acquisitions Editor: Linda Versteeg-Buschman
Editorial Project Manager: Megan Ashdown
Production Project Manager: Maria Bernard
Cover Designer: Christian J. Bilbow

Typeset by MPS Limited, Chennai, India

Contents

List of Contributors

Rashmi Arya CSIR-Unit for Research & Development of Information Products, Pune, India

Ashok Bankar Department of Microbiology, MES Abasaheb Garware College, Savitribai Phule Pune University, Pune, India

Devidas S. Bhagat Department of Forensic Chemistry and Toxicology, Government Institute of Forensic Science, Aurangabad, India

Renuka Bhor Centre for Advanced Studies, Department of Zoology, Savitribai Phule Pune University, Pune, India

Parameswaran Binod Microbial Processes and Technology Division, CSIR-National Institute for Interdisciplinary Science and Technology (CSIR-NIIST), Trivandrum, India

Joyasree Das Microbial Type Culture Collection & Gene Bank (MTCC), CSIR-Institute of Microbial Technology, Chandigarh, India

Julius Eyiuche Nweze Soil and Water Research Infrastructure, Biology Centre CAS, Ceske Budejovice, Czech Republic; Department of Ecosystem Biology, Faculty of Science, University of South Bohemia, Ceske Budejovice, Czech Republic

Aparna Gunjal Department of Microbiology, Dr. D. Y. Patil Arts, Commerce & Science College, Pune, India

Shruti Gupta Soil and Water Research Infrastructure, Biology Centre CAS, Ceske Budejovice, Czech Republic

W.B. Gurnule Department of Chemistry, Kamla Nehru Mahavidyalaya, India

Saima Hamid Centre of Research for Development, University of Kashmir, Hazratbal, India; Department of Environmental Sciences, University of Kashmir, Hazratbal, India

Yogeshwari Hivarkar Annasaheb Kulkarni Department of Biodiversity, MES Abasaheb Garware College, Pune, India

Chieh Chen Huang Department of Life Sciences, National Chung Hsing University, Taichung, Taiwan

Laxmi Jadhav Department of Microbiology, MES Abasaheb Garware College, Savitribai Phule Pune University, Pune, India

Kamlesh Jangid National Centre for Microbial Resource, National Centre for Cell Science, SP Pune University Campus, Pune, India

Chaitanya Kumar Jha Microbiology Department, Gujarat Arts and Science College, Ahmedabad, India

Nishra Joshi Microbiology Department, Gujarat Arts and Science College, Ahmedabad, India

Srinivasan Krishnamurthi Microbial Type Culture Collection & Gene Bank (MTCC), CSIR-Institute of Microbial Technology, Chandigarh, India

Sarita Manandhar Department of Microbiology, Tri-Chandra Multiple College, Tribhuvan University, Kathmandu, Nepal

Amitsinh V. Mangrola Department of Biochemistry, Shri Alpesh. N. Patel P. G. Institute of Science & Research, Charotar Education Society, India

Gincy Marina Mathew Microbial Processes and Technology Division, CSIR-National Institute for Interdisciplinary Science and Technology (CSIR-NIIST), Trivandrum, India

Mohammad Yaseen Mir Centre of Research for Development, University of Kashmir, Hazratbal, India

Sandhya Moghe Department of Biotechnology, Kamla Nehru Mahavidyalaya, India

Pratik Munot MIT-School of Bioengineering Sciences and Research, MIT-ADT University, India

Divya Narayanan Prakash Centre for Advanced Studies, Department of Zoology, Savitribai Phule Pune University, Pune, India

Kalpana Pai Centre for Advanced Studies, Department of Zoology, Savitribai Phule Pune University, Pune, India

Ashok Pandey Center for Innovation and Translational Research, CSIR-Indian Institute of Toxicology Research (CSIR-IITR), Lucknow, India

Suneeta Gireesh Panicker Department of Microbiology, D.Y. Patil Arts Commerce and Science College, Pune, India

Stutee Panse Department of Microbiology, PES Modern College of Arts, Science and Commerce, Savitribai Phule Pune University, Ganeshkhind, Pune, India

Javid A. Parray Department of Environmental Sciences, Government Degree College Eidgah, India

Shrushti Patil Annasaheb Kulkarni Department of Biodiversity, MES Abasaheb Garware College, Pune, India

Smita Patil Department of Microbiology, Camp Education Society's Arvind B. Telang Senior College of Arts, Commerce and Science, Savitribai Phule Pune University, Pune, India

Vikas Patil National Centre for Microbial Resource, National Centre for Cell Science, SP Pune University Campus, Pune, India

Ankur Patwardhan Annasaheb Kulkarni Department of Biodiversity, MES Abasaheb Garware College, Pune, India

Vrushali Phalke Department of Microbiology, MES Abasaheb Garware College, Savitribai Phule Pune University, Pune, India

Aniruddh Rabari Microbiology Department, Gujarat Arts and Science College, Ahmedabad, India

Shaima Rifaie National Centre for Microbial Resource, National Centre for Cell Science, SP Pune University Campus, Pune, India

Chinmayee Bar Routaray Centre for Advanced Studies, Department of Zoology, Savitribai Phule Pune University, Pune, India

Janki Ruparelia Microbiology Department, Gujarat Arts and Science College, Ahmedabad, India

Pradipta Saha Department of Microbiology, The University of Burdwan, Golapbag Campus, Burdwan, India

Jitendra Kumar Saini Department of Microbiology, Central University of Haryana, Mahendergarh, Haryana, India

Himanshu Departmentof Microbiology, Central University of Haryana, Mahendergarh, Haryana, India

Sonali Shinde Annasaheb Kulkarni Department of Biodiversity, MES Abasaheb Garware College, Pune, India

Manju Shree Shakya Hada Department of Microbiology, Tri-Chandra Multiple College, Tribhuvan University, Kathmandu, Nepal

Anima Shrestha Department of Microbiology, Tri-Chandra Multiple College, Tribhuvan University, Kathmandu, Nepal

Raveendran Sindhu Microbial Processes and Technology Division, CSIR-National Institute for Interdisciplinary Science and Technology (CSIR-NIIST), Trivandrum, India

Anjana Singh Central Department of Microbiology, Tribhuvan University, Kirtipur, Kathmandu, Nepal

Sharad Dnyandev Subugade Department of Biotechnology, Vidya Pratishthan's Art, Science & Commerce College, Pune, India

Rebecca S. Thombre Blue Marble Space Institute of Science, Seattle, WA, United States; Department of Biotechnology, Modern College of Arts, Science and Commerce, India

Sarita Tiwari Department of Biotechnology, Kamla Nehru Mahavidyalaya, India

Reshma Tuladhar Central Department of Microbiology, Tribhuvan University, Kirtipur, Kathmandu, Nepal

Preface

Being researchers and enthusiasts of Ecology, contemplating for protracted hours on subject matters of biodiversity, its associated flora and fauna and such is allegorically labor of loves! Having assumed the role of a past master with regards to Intellectual Property and its associated rights at an Institute, this thought struck us—"How would Ecological factors be blended in contemporary times?"

To our consternation, the postulations dealing with the Intellectual Properties associated with microbes with respect to their Ecological Niche was an underenvisaged subject matter for research. Eyeballing further, we decided to get down into venturing the prospects of patenting and legitimately procuring the microbial subjects from their inherent habitats—that too particularly from the Biodiversity Hotspot regions distributed across the world.

Brevity is the soul of wit, they say! For it was a collective effort from all the authors involved in carrying out an intensely vigorous sense of study and research. In a chucklesome sense, our research methodology can be manifested into a computer code with "if, while and for"—loops repeating ad nauseam. Collating and juxtaposing the resulting endeavors of ours was a quenching task—for we received wise counsels from many eminent personalities of this field.

We present this book that covers wide range of topics, namely, microbial diversity and ecology in hotspots; actinobacteria diversity in Western Ghats; habitat-specific microbial community associated with biodiversity hotspots; ecology of saline environments in India; marine microbial hotspots related to corals; phyllosphere microbiome diversity; microbial diversity at the polluted sites; bacterial diversity in termite gut ecosystem. The book has also given importance and covered topics such as bacterial diversity from warm springs; diversity of cellulase degrading organisms from hotspots; diversity and isolation methods for evaluation of extremophiles from Antarctica region. The aspects such as advances in microbial databases, sequencing technology, and various tools for studying microbial diversity are also taken into consideration in this book. Last but not least, the legal protection of microbial diversity is also beautifully highlighted in this book.

This book will serve the purpose of rightfully shepherding aspiring researchers and erudite cerebrals to understand and formulate a legitimate approach toward procuring and wielding subject matters involving the use of microbial life forms, especially from the Biodiversity Hotspot regions. It contains deliberate use of flowcharts and diagrams, quintessentially to simplify the whole understanding of the subject matter. Special efforts have been made to acknowledge the sources from where the information has been collated vis-à-vis through the bibliographical content.

Suggestions for the improvement of the book will be thankfully acknowledged and incorporated in further editions.

Aparna Gunjal and Sonali Shinde

Acknowledgments

- The editors—Aparna Gunjal and Sonali Shinde—are thankful to all the authors for the contribution of their chapters for the book *Microbial Diversity and Ecology in Hotspots*.
- The editors are also thankful to the publisher and their whole team for their management of book publication process.
- The editors are thankful to Vighnesh Shinde, Content Developer and Outreach Person at Arannya Environment Research Organisation, pursuing MSc Biodiversity, Annasaheb Kulkarni Department of Biodiversity, MES Abasaheb Garware College, Pune, Maharashtra, India, for preparing the excellent Cover Page Design for our book *Microbial Diversity and Ecology in Hotspots*. The Cover Page is of bracket fungi, which was taken at Bhagwan Mahaveer Wildlife Sanctuary, Goa, India. The spectrum of colors can be seen by illuminating artificial light. These spores are dispersed during monsoon nights.

****Cover Page Copyright**—Photograph was taken by Omkar Dharwadkar, Naturalist, Partner at Mrugaya Xpeditions and President of Goa, Bird Conservation Network.

- Last but not least, the editors are thankful to their family members for their support and cooperation during this book publication process.

Aparna Gunjal and Sonali Shinde

CHAPTER

1

Exploration of microbial ecology and diversity in hotspots

Sonali Shinde[1], Pratik Munot[2], Yogeshwari Hivarkar[1], Shrushti Patil[1] and Ankur Patwardhan[1]

[1]Annasaheb Kulkarni Department of Biodiversity, MES Abasaheb Garware College, Pune, India, [2]MIT-School of Bioengineering Sciences and Research, MIT-ADT University, India

1.1 Introduction

"Man is a microcosm"—so rightly said by Democritus, an ancient Greek philosopher, as the Alchemy of Nature is boundless, seamless—as if nature talks and whispers to us! As mankind strives toward scientific and technological progresses, it is a mere wonderful sight to watch earth, revealing its hidden gems and secrets in front of us. Delighted, as every being is, we can experience the spread of different forms of life and its supporting elements throughout the span of this planet. Astrophysicists are now even exploring the possibilities of life being present in other regions of the Universe as well (Steven et al., 2015). While these are significant steps toward ensuring an unimaginable world of possibilities, it is ironic to learn that even on our planet, plants and animals on land have yet to be named and cataloged (Mora et al., 2011). The case for microbes is even more surprising as a recent study estimates that a significant portion of the microbial species present on the earth are yet to be discovered (Locey & Lennon, 2016). A series of such serious facts strengthens and encapsulates the importance of studying the biodiversity and natural heritage spanning the earth in a systematic and sustainable way, in order to discover life to its fullest!

Microbial Diversity and Ecology in Hotspots
DOI: https://doi.org/10.1016/B978-0-323-90148-2.00015-8

1.2 Meaning of biodiversity, threats associated and need for its protection

First used by Thomas Lovejoy in the year 1980, the term Biological Diversity or simply put, biodiversity, was meant to quantitatively include the variety of species of living organisms on earth. However, the brief definition of biodiversity can be obtained from the charter of the Convention on Biodiversity (CBD). Article 2 of the CBD describes biological diversity as "the variability among living organisms from all sources including inter alia, terrestrial, marine and other aquatic ecosystems and the ecological complexes of which they are part; this includes diversity within species, between species and of ecosystems" (refer to Chapter 16: Legal Protection of Microbial Biodiversity, for further reading on the legal status of biodiversity conservation). With the advancements made in molecular biology and genome sequencing technologies (refer to Chapter 14: Recent Advances in Microbial Databases with Special Reference to Kinetoplastids, and Chapter 15: Advances in Sequencing Technology, Databases and Analyses Tools for the Assessment of Microbial Diversity), the science of exploring, understanding, protecting, and utilizing components of biodiversity in a sustainable manner, has emerged to be an important aspect of Conservational Biology. Various scientists from different domains such as ecology, botany, zoology, microbiology, biogeography, genetics, and other life sciences' subjects, actively engage themselves in conserving the biodiversity by chalking out strategies and implementing sustainable policies, taking into account the interests of different stakeholders involved in the conservation work. Every subject expert from the aforementioned domain explores the subject from different perspectives and with different hypotheses (Klinkenberg, 2020). They study various aspects involving population dynamics, the functioning of ecosystems, species distributions, and other factors of biodiversity, in order to understand and ascertain the scope of protection.

In recent times, a plethora of threats has been incurred upon the biological diversity of almost all the regimes over the globe resulting from unreasonable utilization and overexploitation of natural resources, with increased urbanization and unaccountable industrial development (refer to Chapter 8: Microbial Diversity at the Polluted Sites). These factors have especially undermined the viability of wild populations of flora and fauna, collectively (Wake & Vredenburg, 2008). For instance, this trend of overexploitation of resources has predominately been witnessed in the nations consisting of a rich heritage of biological diversity. In order to understand the basis of this trend, it is essential to first understand the factors forcing such countries to attain a "compulsive-state-of-mind" in which they are at vulnerable economic position and naturally, the exploitation of their natural resources is the only option left with them. However, while the abundance of natural resources in different regions of the globe quite necessarily implies a seemingly vast potential for the economic development of the region as a whole by sustainable usage of the natural components (OECD, 2011), the recent times have accounted for a rather paradoxical situation. Termed as the "Resource Curse" or the "paradox of plenty," it is a surprisingly contradictory situation, in which the countries with an abundance of nonrenewable resources experience shunted economic growth (Chandra et al., 2013). This theory is intricately applicable to the components of biodiversity as well.

The theory of Resource Curse is dependent on a couple of essential concepts, namely the Resource-Ratio Theory and the Carrying Capacity (in ecological terms). The "Resource-Ratio theory" or the "R* rule," (Wilson et al., 2007), is an essential framework in community ecology to understand and hypothesize the influence of competition for growth-limiting resources on the biological diversity of a region (Smith et al., 1998). The theory predicts in a mathematical way the coexistence of two or more species in a particular ecosystem in the presence of growth-limiting nutrients with the condition that each species is limited by such a resource, which it (the species) is least prone or able to deplete (Mazancourt & Schwartz, 2010). The second concept, called the "Carrying Capacity" (in ecological terms), is described as the limiting factor of the population size which is supported by a particular regional environment, is the availability of the natural resources in that environment (Rachlow, 2008). However, the population size can also be altered by other limiting factors such as food, predation, and diseases, which do not account for inclusion as the limiting factor in ecological carrying capacity. In simpler words, the ecological carrying capacity is the level of production (i.e., the population size of a species) that does not affect the equilibrium of the surrounding ecosystem or its environmental carrying capacity (TettPaul et al., 2015). Coming back to the implication of Resource Curse theory with respect to Microbial Ecology and Biological characteristics of a particular region, when the natural resources are abundant at a particular place, microbes tend to increase in their colony size and populations, thereby making way for strong competitive behavior and instincts for survival, a phenomenon that can be explained by the Resource-Ratio hypothesis. Regions bearing high Carrying Capacity, tend to provide survival edge on species adopting R* behavior in which there is aggression and competition among species to survive, while the low Carrying Capacity regions tend to show mutualistic behavior adopted by different species by not adopting the R* behavior, to sustain in that particular environment (Marcel, 2002).

The essence of this mentality (unaccounted exploitation of natural resources), adopted by such countries with abundant biodiversity, can lead to an increase in the environmental misfortunes and maybe threatening in the form of species elimination, that would result into unbalancing of the earth's natural diversity. Hence, a dire need can be felt for adopting a conservation strategy in order to minimize and counter the biodiversity loss. Also, we must realize that the biodiversity is not consistently spread across the globe and there are hardly any zones where we observe the exceptionally rich diversity of endemic species, which are unfortunately threatened by irresponsible human activities (Habel et al., 2019; Myers et al., 2000). So, in order to protect such regions possessing high numbers of endemic species, the concept of Biodiversity Hotspots has been upheld by the Conservationists, with the aid of participation of all the major stakeholders (governmental, nongovernmental) at various conservation levels and different strategies, culminating into a potentially fruitful and sustainable outcome in the near future (Christian, 2015).

1.3 Biodiversity hotspots: a brief overview

Put forward as a sustainable conservation strategy for biodiversity by Norman Myers and supported thereafter by Conservation International, biodiversity hotspots, simply, are miscellanies providing a unique niche with a significant reservoir of diverse flora, fauna, and landscape. Every hotspot region features unique characteristics and adaptability with significant ecological balance for organisms to thrive in. The considerably high diversity, especially with respect to endemic species present in these regions, is termed as mega diversity. Such a region necessarily constitutes an area that contains at least 0.5% of total plant species as endemic species and with the implication that the area has lost a minimum of 75% of its original vegetation (Myers, 1988, 1990; the conditions for declaring a region as a Biodiversity Hotspot are detailed in Chapter 16: Legal Protection of Microbial Biodiversity).

The idea of Biodiversity Hotspots is much more than a conservation tool as it quantifies the effect of environmental pressure due to human activities. Till now (with the last one being recently announced in February, 2016), a total of 36 hotspots have been reported and defined as places of biodiversity with vulnerability and irreplaceability (Mittermeier, 2004; Mittermeier et al., 2011). The term Hotspots' Ecosystem is a geographic niche where plants, animals, and different microorganisms interact together constituting natural energy cycles and processes, along with its weather and landscape.

1.4 Tools for systematically studying the biodiversity hotspots through various aspects

Advancements in scientific and technological progress have ushered the way biological elements and components of biodiversity were

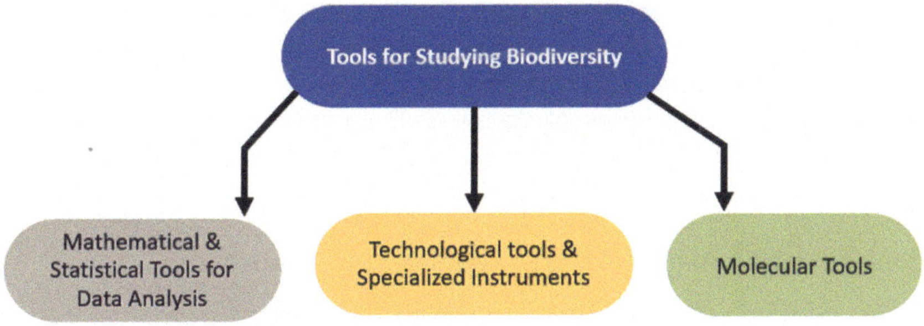

FIGURE 1.1 Tools for studying biodiversity.

previously understood. Researchers use these tools for understanding the micronature and relations among biodiversity, species, populations, and ecosystems. Fig. 1.1 explains excerpts of tools used in studying Biological diversity.

1.4.1 Mathematical and statistical tools for data analysis

Downstream analysis of ecological data can be performed by the simplest and most efficient approach by estimating the biodiversity. Large varieties of bioinformatics tools have been developed to analyze and compare the microbial diversity from the array of a microbial sample. Though ecological data have been substantially less explored in contrast to molecular research and bioinformatics, still many computational tools can be used to explore the patterns involved in studying biodiversity, as well. A classic example, that of Bayesian Networks, can be cited in this context which can be used for interpreting and studying interdependence and relationships among different organisms, especially in studying the biodiversity distribution in a particular region (Tucker & Duplisea, 2012; for more information refer to Chapter 14: Recent Advances in Microbial Databases with Special Reference to Kinetoplastids, and Chapter 15: Advances in Sequencing Technology, Databases and Analyses Tools for the Assessment of Microbial Diversity).

1.4.2 Molecular tools

In particular, molecular biology tools, especially the use of ribosomal RNA (rRNA), in particular, the 16S or small subunit (SSU) RNA and Internal Transcribed Spacer (ITS) sequences have proved to be important in the history of biodiversity exploration and its conservation (Rappé & Giovannoni, 2003; for more information refer to Chapter 13: Isolation Methods for Evaluation of Extremophilic Microbial Diversity From Antarctica Region). Apart from them, DNA markers such as mini- and microsatellite DNA sequences, Restriction Fragment Length Polymorphism (RFLPs) and genomic sequence data can also be effectively used as a reliable tool for ecological data analysis, especially in the cases of microbes, where culturing techniques are also well suited for the same purpose (Allan & Max, 2010).

1.4.3 Technological tool and specialized equipment

Other equipment such as remote sensing modules, biometric devices, etc. are proactively used in monitoring and tracking the movement of biodiversity elements, especially those like tigers, snakes, etc. Among them, the use of actively monitoring animal locations is through global positioning system (GPS) and digital images for visualization (Alexander et al., 2020).

1.5 Biodiversity hotspots and microbial ecology

Biodiversity hotspots are not only diverse in flora and fauna but may also show immense variation in microbial diversity with respect to their unique ecosystems. Microbial diversity hotspots could be deemed to necessarily include temperature, pressure, nutrients, and energy-exchange acting on the system, that is, niche, containing the microbes. For instance, Biodiversity hotspot of the Hawaiian Archipelago has substantial endemism (Donachie et al., 2004; for more information refer to Chapter 2: Habitat-Specific Microbial Community Associated With the Biodiversity Hotspot). It is well known that the relationship between species and the existing environmental conditions is embodied by the concept of ecological niche (Begon et al., 2006). The term Ecological Niche encapsulates the position of a species within an ecosystem, describing both the range of conditions its ecological role in the ecosystem (Polechová & Storch, 2008). Extending this concept for aiding the inclusion of microbes, it is imperative to note that microbial community of a particular niche has an important feature, characterized by the number of species and their composition diversity. Hence, the developments and advancements made in genome sequencing technologies with respect to their efficacy, accuracy, cost-effectiveness, and reliability (Roderic & Michiel, 2018; for more information refer to Chapter 15: Advances in Sequencing Technology, Databases and Analyses Tools for the Assessment of Microbial Diversity) and the use of environmental amplicon sequencing survey has built to a large number of biodiversity estimates for analysis of microbial studies. With the knowledge of microbial taxa and the resources available, we can march toward understanding the complexity and diversity of an organism in that particular habitat. It is important to consider the interactions between these elements (biotic and abiotic) when studying the biogeographical or spatio-temporal patterns and not just the isolated being. On a practical note instead of knowing who all are there, it will be easy to find what are they doing and later find out who they were.

1.6 Microbial hotspots: an overview

When defining microbial hotspots, the obvious difficulty will be to understand what should be the exact definition of microbial hotspots. Irrespectively the importance of measuring biodiversity lies more in adjusting the downward scale and unmapped terrain (Wilson, 1994).

The study of microbes in biodiversity hotspots for their uniqueness is somewhat discouraging as microbes survive and grow almost everywhere on earth, dramatically extending our perception of the limits of the biosphere and unclear definition of species. In line with the Baas-Becking hypothesis which states, "Everything is everywhere, but the environment selects" (Baas-Becking, 1934). Our understanding to find a specific microbe in a specific geographic location is limited. In compliance with this, whatever one has to study on Mars can be studied in the regions on earth where the spatial conditions are similar (Martiny et al., 2006; for more information refer to Chapter 2: Habitat-Specific Microbial Community Associated With the Biodiversity Hotspot). Microscale distribution is definitely a tedious job and may often get a heterogeneous result. Some studies have demonstrated a no-point connection with species richness, a threat to the habitat and endemism do not show the same geographical distribution (Orme et al., 2005). Within the currently recognized 36 biodiversity hotspot regions, few of them highlighted for the microorganisms are studied. For example, Atlantic forest is a dense ombrophilous forest that has high variations creating vegetation gradient ranging from shrubs to well-developed Montane forest (CEPF, 2001). Thermophilic Proteobacteria are unique to obsidian hot spring pool with 75°C–95°C in Southern Brazilian Atlantic forest (Faoro et al., 2010). California floristic province is a Mediterranean-type climate divided into hot-dry summers and cold-dry winters (Burge et al., 2016). *Rhodobacter capsulatus*, a purple nonsulfur bacterium is studied in the region that grows in anaerobic conditions also responsible for denitrification (Costa et al., 2017). Cape Floristic region is evergreen fire-dependent scrubland, divided in the tropical, subtropical dry broadleaf forest. Amphora diatoms are highly studied in this region. The presence of this species indicates high pH (Rea & De Stefano, 2019). The Caribbean Island consists of diverse ecosystem ranging from Montane cloud forest to cactus scrubland. *Rhodococcus rhodochrous* actinobacteria were studied in this region (CEPF, 2011). Presence of *Salinispora* and nonacid fast bacteria indicates organic contamination (Bauermeister et al., 2018). Himalaya which is rich in psychrophiles, that is, *Pseudomonas palleroniana* N26 was studied at low temperature and nitrogen-deficient ecosystem (Joshi et al., 2017). Indo-Burma hotspot is wet and dry evergreen, deciduous forest, swamps, and mangrove ecosystem. *Tepidimonas taiwanensis* and *Tepidimonas fonticoldi* thermophilic bacteria are found in the hot spring of Southern Thailand (CEPF, 2020). Coastal forest of Eastern Africa are characterized with small and fragmented forest has different diatoms species indicators like *Nitzschia asterionelloides, Nitzschia forticola,* and *Fragillaria* (Descy) (Jean-Pierre & Hugo, 2008).

1.7 Microbial ecology: microbial habitats and the distribution of microbes

As extreme and far-reaching and diverse environmental conditions can only be manifested supporting life, in reality, the microbial habitats consist of some of the wildest places on the globe including hot springs (Panosyan et al., 2018), steam-heated soils, different soil horizons (surface, subsurface, vertical, and depth), cryoconite holes (Sanyal et al., 2018), mud holes, surface waters (Bhattacharyya & Jha, 2015; Lan et al., 2019; Yoshitake et al., 2018), geothermal power plants (Hou et al., 2020), drylands (Wang et al., 2018), and coral reefs (Tout et al., 2014).

An interesting case, that of cryoconite holes in the chilling biodiversity hotspots of the Himalayan region and Antarctica provides a suitable habitat for the growth of microbial communities dominated by Proteobacteria, Bacteroidetes, Actinobacteria. Apart from these, there are many other microbial communities present in assemblages, which are specific to these regions characterized by the presence of cryoconites, which assist in shaping the microbial interactions, in part, the unique ecosystem of this region (Sanyal et al., 2018; for more information refer to Chapter 10: Bacterial Diversity From Garampani Warm Spring, Assam; Chapter 12: Biodiversity of Cold Adapted Extremophiles From Antarctica and Their Biotechnological Potential; and Chapter 13: Isolation Methods for Evaluation of Extremophilic Microbial Diversity From Antarctica Region).

Apart from these, the other commonly found ecosystems include the diverse soil types characterized with respective nutrient contents, gut and rumen of animals, aquatic environments, rhizosphere, phyllosphere, and others. A change in the habitat by natural fragmentation of the area leads to a change in the microbial interactions and networks and increases their geographical isolation (Speer et al., 2020; for more information refer to Chapter 2: Habitat-Specific Microbial Community Associated With the

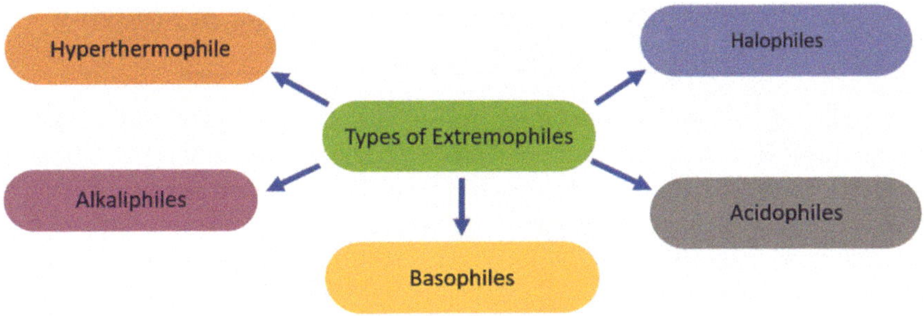

FIGURE 1.2 Types of extremophiles.

Biodiversity Hotspot). As discussed in the introductory paragraph of this part, some of the microbiota, termed as extremophiles, are located in harsh and extreme environmental conditions Fig. 1.2.

These extremophilic microbial life forms, tend to sustain in these harsh and extreme conditions, where no other life forms would ostensibly sustain and thrive for longer periods, are conditioned by virtue of their great and unique ability to survive in array of habitats like hyper extremophile environment.

At the center-stage, the microbial distribution is governed by nutrients, water, and other resources present in the habitat, habitat complexity, flora and fauna of that habitat as well as the anthropogenic activities (Young et al., 2008). One study shows microbial density is higher in the soil in the summer and spring due to increased soil moisture and nutrients which plays an important role in their distribution (Bhattacharyya & Jha, 2015). Distributions of microbes such as algae found in the red snow of Svalbard and Arctic Sweden are uncultured *Chlamydomonaceae* species, of Green snow are *Microglena* sp. and *Raphidonema sempervirens* in Svalbard and *Chloromonaspolyptera* in Sweden is diverse (Lutz et al., 2017). The distribution pattern in the environment can be beneficial for understanding the ecological theories, recognizing and predicting the changes in community structures with environmental alterations (Bhattacharyya & Jha, 2015; Valdespino-Castillo et al., 2018). Microbial distribution in psychrophilic biota and limnetic microbial mats favors a huge portion of the biomass in extreme environmental conditions with respect to inland Antarctica. While the conditions affecting microbial composition over successional and prolonged time periods with the environmental changes and imbalances may be disclosed by the distribution patterns or microbial assemblages, the microbial distribution is also altered as a result of unaccounted and overly done anthropogenic activities like agricultural processes, especially the unwarranted use of artificial pesticides which accounts for drastic pH changes in the soil of that region, thus causing a significant reduction of biodiversity and the instability of that region's ecosystem (2018).

1.8 Microbial diversity indices: application in studying community ecology

In spite of being the most commonly manifested form of explaining the Ecological Diversity in a particular region, the concept of Species Diversity subtly differs in its meaning and application, as compared to Biodiversity. For describing this sophisticated concept, ecologists tend to explain species diversity as a function of two factors—species richness and relative species abundance (Hamilton, 2005). Species richness means the number of species present in a particular region and species abundance means the number of individuals per species. Relative species abundance means the evenness of distribution among individuals among species in a community (Baillie & Upham, 2012). An important approach in measuring the species diversity is to construct the mathematical diversity indices. These indices are used for quantification of heterogeneous microbial distribution. Different microbial diversity indices are shown in Tables 1.1 and 1.2.

1.9 Microbial composition and succession

Depending upon the biogeography, the microbial composition is often determined at the community level, as microbial structures and functions/ processes are determined by their characteristic community, that is, population dependent (Martiny et al., 2006; Martiny et al., 2006; Nemergut et al., 2014). The assemblage of microbes helps in signaling between the microbial networks and also helps in the degradation of complex organic compounds to simpler ones. For instance, the degradation of xenobiotics or recalcitrant compounds by certain species of *Pseudomonas* and *Rhodococcus* is an efficient bioremediation technique being highly potentialized in contemporary times (Phale et al., 2019). Some bacterial and fungal species are however unable to convert and utilize complex molecules such as cellulose and lignin and do initiate the enzymatic process of converting these complex molecules to simpler forms by producing extracellular enzymes (Selim & Zayed, 2017; Tribedi, 2016). Apart from that, the microbial assemblages are responsible to alter microbial structures and compositions, which obstruct the stabilization of microbial communities (Martiny et al., 2006; Martiny et al., 2006; Nemergut et al., 2014).

As microbial communities play an important role in maintaining a healthy soil ecosystem, microbial succession is responsible for bringing in changes in the microbial composition. Microbial succession changes with change in the habitats, such as changes taking place when the ice melts with that the nutritional concentration changes leading to change

TABLE 1.1 Calculating microbial diversity indices using alpha diversity.

Diversity indices	Equation	Variables	Description	Reference
Alpha diversity				
Fisher's Index (S)	$S = \alpha \ln(1 + \frac{N}{\alpha})$	S = Total number of species; α = Index of diversity of population; N = Total number of individuals	Estimates logarithmic distribution of number of individuals of different species	Fisher et al. (1943)
Simpson's Index (D)	$D = \frac{1}{\sum_{i=1}^{s} Pi^2}$	D = Simpson's dominance index; S = Total number of species in the community; pi = Proportion of community represented by Operational taxonomic unit (OTU)	Quantify number of species present as well as their abundance	Simpson (1949)
Odum's Index (R_{odum})	$R_{odum} = \frac{S}{\ln(N)}$	R_{odum} = Odum's Index; S = Total number of species; N = Total number of individuals in the sample	Estimates species diversity	Odum et al. (1960)
McIntosh Index	$Mc = [N - \sqrt{(\Sigma n_i\, 2)}]/[N - \sqrt{N}]$	Mc = McIntosh Diversity Index; n_i = Number of individuals belonging to i species; N = Total number of individuals	Estimates species richness	McIntosh (1967)
McIntosh Evenness Index	$E = \frac{N - U}{N - N/\sqrt{S}}$	E = McIntosh Evenness Index; N = Total number of individuals; S = Total number of species; U = Sample size	Estimates species evenness	McIntosh (1967)
Berger–Parker Index	$D\,I_{BP} = \frac{N_{max}}{N_{tot}}$	DI_{BP} = Berger–Parker Index; N_{max} = Number of individuals of most abundant species; N_{tot} = Total number of individuals of all the species in the sample	Estimates species abundance	Berger and Parker (1970)
Species Richness Index	$E(S) = f(k, N)$	$E(S)$ = Expected value for number of species; N = Number of individuals; k = Richness index	Estimates species diversity	Peet (2003)
Pielou Evenness	$J' = \frac{H'}{H'_{max}}$	J' = Pielou evenness; H' = Number derived from Shannon Diversity Index; H'_{max} = maximum possible value of H'	Estimates species evenness	Pielou (1975)

(Continued)

TABLE 1.1 (Continued)

Diversity indices	Equation	Variables	Description	Reference
Chao's Index	$S_{(max)chao} = S_{obs} + (a^2 + b^2)$	$S_{(max)}$ = Maximum number of species; S_{obs} = Number of species observed in different samples; a = Singletons (number of species represented by one individual each); b = Doubletons (number of species represented by two individuals each)	Abundance-based estimates species richness	Chao (1984, 1988)
Shannon–Wiener Diversity Index (H)	$H' = -\sum_{i=1}^{s} pi\ln(pi)$	H' = Shannon–Wiener Diversity Index; S = Number of OTUs; pi = Proportion of the community represented by OTU	Shannon–Wiener Diversity Index measures the rarity and commonness of species in a habitat.	Lemos et al. (2011)
Menhinick's Index	$D_{mn} = \frac{S}{\sqrt{N}}$	D_{mn} = Menhinick's Index; S = Number of species; N = Total number of individuals	Estimates species richness	Kim et al. (2017)
Margalef's Index	$D_{mg} = \frac{S-1}{\ln(N)}$	D_{mg} = Margalef's Index; S = Total number of species; N = Total number of individuals in the sample	Estimates dependence between species diversity and number of organisms sampled	Kim et al. (2017)
Abundance-based coverage estimator (ACE) Index	$S_{ACE} = S_{abund} + \frac{S_{rare}}{C_{ACE}} + \frac{F_1}{C_{ACE}}\gamma^2 ACE$	S_{abund} and S_{rare} = Number of abundant and rare OTUs; C_{ACE} = Sample abundance coverage estimator; F_1 = Frequency of singletons; $\gamma^2{}_{ACE}$ = Estimated coefficient of variation for rare OTUs	Abundance-based coverage estimator of species richness	Kim et al. (2017)
Brillouin Index	$H_B = \frac{\ln\left(N - \sum_{\ln} n_i\right)}{N}$	H_B = Brillouin Index; N = Total number of individuals; n_i = The number of individuals in the ith species	Estimates species richness similar to Shannon diversity index	Kim et al. (2017)

TABLE 1.2 Beta diversity—microbial index.

Diversity indices	Equation	Variables	Description	Reference
Beta diversity				
Jaccard Index	$J = \frac{S_c}{S_a + S_b + S_c}$	J = Jaccard Index, S_a and S_b are the numbers of species unique to samples a and b, respectively, S_c = Number of species common to the two samples	It is the fraction of species shared between the samples.	Jaccard (1912)
Sorensen Index	$S_s = \frac{2_a}{2_a + 2_b + 2_c}$	S_s = Sorensen's similarity coefficient; a is the species common to both samples; b is the species in sample 1; and c is the species in sample 2	Used for comparing the similarity of two samples.	Sorenson (1948)
Bray–Curtis	$BC_{ij} = 1 - \frac{C_{ij}}{S_i + S_j}$	BC_{ij} = Bray–Curtis Index; C_j = Sum of the lesser values for only those species in common between both sites; S_i and S_j = The total number of specimens counted at both sites	Quantify the compositional dissimilarity between two different sites, based on counts at each site.	Bray and Curtis (1957)
Morista-Horn	$C_D = \frac{2\sum_{i=1}^{s} x_i y_i}{(D_x + D_y)xy}$	$C_D = 0$ if the two samples do not overlap in terms of species; $C_D = 1$ if the species occur in the same proportions in both samples; x_i = Number of times species i is represented in the total X from one sample; y_i is the number of times species i is represented in the total Y from another sample; D_x and D_y are the Simpson index values for the x and y samples, respectively. S is the number of unique species	Estimate the measure of dispersion of individuals in a population.	Morisita (1959)

(Continued)

TABLE 1.2 (Continued)

Diversity indices	Equation	Variables	Description	Reference
Wilson and Shmida Index	$\beta_T = \frac{[g(H) + l(H)]}{2\alpha}$	β_T = Beta turnover; H is habitat gradient; g = Species gained; l = Species lost; α = Average number of species found within the community samples	Estimates changes, additivity in the alpha diversity.	Wilson and Shmida (1984)
Colwell and Coddington Index	$S = S_{obs} + \frac{F_1^2}{2F_2}$	S_{obs} = Number of species in the sample; F_1 = Number of singletons (i.e., the number of species with only a single occurrence in the sample); F_2 = Number of doubletons (the number of species with exactly two occurrences in the sample)	Estimate species diversity.	Colwell and Coddington (1994)
Shannon Beta Index	$H\beta i = \sum_j \frac{C_{ij}}{C_{i++}} + \sum_k \frac{C_{jk}}{C_{ij+}} \ln\left(\frac{C_{ijk}}{C_{ij+}}\right) / \frac{C_{i+k}}{C_{+++}}$	$H(\beta i)$ = Shannon Beta Index; C_{ijk} = The sequence count for subject i, from time j and taxon k; + (plus) in the subscript denotes the summation of the counts over the specified indicator	Measures multiple alpha and beta components when community weights are unequal.	Marcon et al. (2012)

These listed microbial indices are used for analyzing the numerical composition of microbial diversity in a complex community of microbes.

in the microbial community (Valdespino-Castillo et al., 2018). Some of the bacterial species, discovered in the early stages of ecological succession with respect to the glacial soil in the mountain ranges of the Alps in Switzerland, including Proteobacteria (alpha > betaproteobacteria) has shown a gradual decline in the soil percentage with age over a prolonged time period.

The process of a generalized microbial succession is slowed down due to decreasing mineralization in the soil nutrients such as carbon and nitrogen (Bhattacharyya & Jha, 2015). The succession results in diverse microbial composition in a single habitat, for instance as observed in the deep seas. A hydrothermal vent sulfide chimney is another suitable example as the habitat showing the microbial succession of chemoautotrophic microorganisms, on and around the walls of the chimneys. Some microbes are dominant when the chimney is active as it may supply the growth conditions. The succession depends on the source of nutrients, temperature condition, and other conditions for the growth of microbes from thermophilic to mesophilic to psychrophilic microorganisms (Hou et al., 2020). The ground surface after the glacial retreat is covered by a primary succession of microbes and these stimulate physical and biological processes which lead to favorable conditions for further microbial and ecosystem succession (Yoshitake et al., 2018).

1.10 Microbial interactions

Microorganisms interact with one another and the other elements of the ecosystem in a host by ways called microbial interactions Fig. 1.3, during which the genetic material and biologic molecules are exchanged

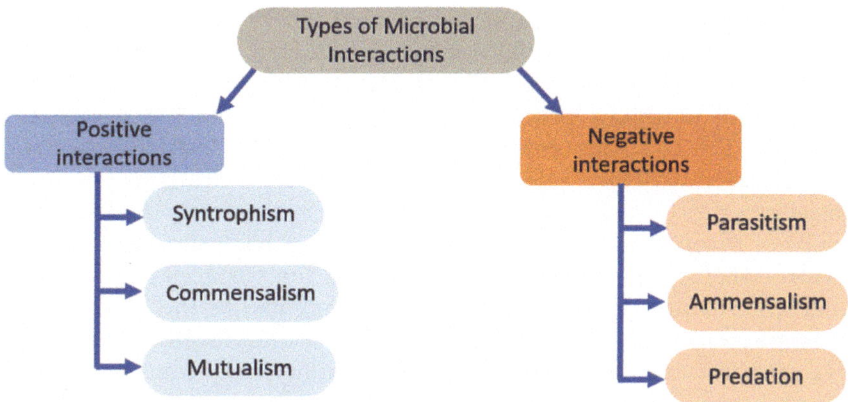

FIGURE 1.3 Types of microbial interactions.

and processed (Braga et al., 2016). These interactions can be classified as being intraspecific or interspecific (Selim & Zayed, 2017) depending upon the nature of the involvement of various interacting elements. These interactions can also be classified as being positive or negative, depending upon the benefits derived by interacting elements, or the harmful product resulting out of such involvements (Shimane et al., 2018).

The microbial communities are actively involved in creating favorable environmental conditions, that too, for other microbial communities to survive and carry out their interactions with each other and in an ecosystem as well. For instance, the snow and ice algae present in the cryoconite holes act as a source of nutrients and carbon for heterotrophic communities (Lutz et al., 2017). The ecosystem and hotspot regions characterized with cryoconite holes are governed by the microbial interactions which play an important role in influencing and vitalizing the carbon cycle in these ecosystems. The microbial interactions within the cryoconite debris form a microenvironment and matrix, promoting biochemical interactions, many of them, yet to be explored (Sanyal et al., 2018).

The microbial communities interact for certain functional and metabolic activities, the establishment and maintenance of which, determines a successful establishment of the microbial community. An outcome of these interactions is the coevolution process, which guides the microbial adaptations in different habitats (Braga et al., 2016). Microbes associated with different chemical processes may be driven by localized growth enhancement or chemotaxis into chemical microniches (Mitchell, 2002). These interactions help in improving their activities like signaling, exchange of compounds and involve the physicochemical changes, metabolites exchange and conversion, chemotaxis, and many more processes. Some microbes are known to interact with the roots and seeds through chemotaxis or chemotropism activated by plant exudates. Some processes like the root nodule factors released by *Rhizobia* are perceived by root cells leading to alteration of calcium flux and activates signaling pathway for the production of cytokinins, which also acts as symbiotic interaction of microbes (Braga et al., 2016; Moënne-Loccoz et al., 2015; Tribedi, 2016).

It is a well-established fact that the microbes, by virtue of exchanging the vital materials produced through their interactions with each other and the environment, can regulate the plant development process, indirect way, by releasing phytohormones, contributing to the ecosystem processes, producing biocontrol compounds and different signaling compounds. Diverse communities of bacteria and fungi live in close association with each other and on the surfaces of plant roots (rhizosphere) and leaves (phyllosphere) and inside the plant tissues (endosphere) (Kent & Triplett, 2002). These are highly diverse and have a fundamental role in water economy, nutrient acquisition, growth, and disease tolerance. Some examples are arbuscular mycorrhizal fungi (AMF), Basidiomycetes, *Penicillium*, *Archeospora*, Glomus, Actinomycetes, Proteobacteria.

While such microbial interactions help in stabilizing and understanding the dynamics of the biological communities, these interactions, can show both positive as well as a negative effect in their relationship with the host (Fig. 1.3).

1.11 Hotspots bioindicators/indicating unique environment of hotspots

A characteristic exhibited by every microorganism in an ecosystem to potentially indicate the environmental health of the surrounding region can be accounted on the basis that they can tolerate a limited range of chemical, physical, and biological conditions (Holt & Miller, 2010). Through the utilization of bioindicators, researchers can anticipate the common conditions for a specific region or the level and/or the degree of environmental contamination (Parmar et al., 2016). Some bioluminescent bacteria belonging to the genus Photobacterium from the North-Western Mediterranean Sea Hotspot region can effectively be used to test water for environmental toxins with the help of stress proteins. Toxins released by microbes affect the cellular metabolism of bacteria (Martini et al., 2016).

Microorganisms also encompass the oceanic biomass and are responsible for the majority of productivity and nutrient cycle in a marine ecosystem, as they have a rapid rate of growth, and react to even low levels of contaminants and other *physicochemical* and biological changes. One such example is of the bacterium *Vogesella indigofera* from the Mediterranean basin region which reacts to heavy metals quantitatively. Under conditions influenced by no metal pollution, blue pigment is a significant marker of the underlying morphological change that has occurred (Krieg, 2015). On the other hand, under the presence of hexavalent chromium ions, the production of pigment is blocked. This can be attributed to the fact that there is an essentially important relationship between the concentration of chromium ions and the generation of blue pigment by the bacterium (Aslam et al., 2012).

An interesting fact is that some microbes are adaptable to extreme environmental conditions and habitats by showing different strategies for their existence, such as the generation of pigments like violacein, carotenoids, melanin, β-carotene, etc. These pigments help the microbes to get adopted to harsh conditions like ultraviolet radiance, the genetic ability for carbon and nitrogen cycling and the capacity to degrade complex organic compounds (Sajjad et al., 2020; Valdespino-Castillo et al., 2018).

1.12 Conclusion and way forward

With the scientific advancements made in sophisticating the processes of microbial exploration and understanding their mysterious roles in the undiscovered functioning of the ecosystem, it is imperative to consider that the revelation of various interactions, networks, and intricate scope of distribution associated with microbial ecology would have a much penetrating impact in understanding their outreach as quintessential elements of biological diversity. Rather in Ed Yong's tone—It's the start of a new era when people are finally ready to embrace the microbial world.

References

Alexander, P., Alexander, K., Artem, R., & Svetlana, I. (2020). Methods for monitoring large terrestrial animals in the wild. *Forests, 11*(8), 808. Available from https://doi.org/10.3390/f11080808.

Allan, G. J., & Max, T. L. (2010). Molecular genetic techniques and markers for ecological research. *Nature Education Knowledge, 3*(10), 2.

Aslam, S., Jahan, N., & Ali, S. (2012). An innovative microwave-assisted extraction and antioxidant potential of polyphenols from different parts of *Ocimum basilicum*. *Journal of Medicinal Plants Research, 6*(11), 2150–2159.

Baas-Becking, L. G. M. (1934). *Geobiologie of inleiding tot de milieukunde*. Den Haag: W. P. Van Stockum & Zoon.

Baillie, J. E. M., & Upham, K. (2012). Species diversity within and among ecosystems. In R. A. Meyers (Ed.), *Encyclopedia of Sustainability Science and Technology*. Springer.

Bauermeister, A., Velasco-Alzate, K., Dias, T., Macedo, H., Ferreira, E. G., Jimenez, P. C., Lotufo, T. M. C., Lopes, N. P., Gaudêncio, S. P., & Costa-Lotufo, L. V. (2018). Metabolomic fingerprinting of salinispora from Atlantic Oceanic Islands. *Frontiers in Microbiology, 9*, 3021. Available from https://doi.org/10.3389/fmicb.2018.03021.

Begon, M., Townsend, C., & Harper, J. L. (2006). *Ecology: From individuals to ecosystems*. Blackwell Publishing.

Berger, W. H., & Parker, F. L. (1970). Diversity of planktonic foraminifera in deep-sea sediments. *Science, 168*(3937), 1345–1347.

Bhattacharyya, P. N., & Jha, D. K. (2015). Variations in bacterial population numbers and enzyme activities in different land-use systems of Brahmaputra valley. *International Journal of Applied Reasearch, 1*(12), 762–766.

Braga, R. M., Dourado, M. N., & Araújo, W. L. (2016). Microbial interactions: ecology in a molecular perspective. *Brazilian Journal of Microbiology, 47*, 86–98. Available from https://doi.org/10.1016/j.bjm.2016.10.005.

Bray, J., & Curtis, J. (1957). An ordination of upland forest communities of southern Wisconsin. *Ecological Monographs, 27*, 325–349.

Burge, D. O., Thorne, J. H., Harrison, S. P., O'Brien, B., & Rebman, J. (2016). Plant diversity and endemism in the California Floristic Province. *Madroño, 63*(2), 3–206.

CEPF. (2001). *Ecosystem profile. Atlantic forest biodiversity hotspot Brazil* (p. 29) CEPF Conservation International.

CEPF (2011). Protecting nature's hotpots for people and prosperity. Available from https://www.cepf.net/sites/default/files/CEPF_annualreport_2011.pdf

CEPF (2020). Indo-Burma biodiversity hotspot technical summary. Available from https://www.cepf.net/sites/default/files/indo-burma-annual-portfolio-overview-2020.pdf.

Chandra, R. B., Satyaki, S., & Ranjan, M. N. (2013). Natural resource abundance and economic performance—A literature review. *Current Urban Studies, 1*(4), 148–155. Available from https://doi.org/10.4236/cus.2013.14016.

Chao, A. (1984). Non-parametric estimation of the classes in a population. *Scandinavian Journal of Statistics, 11*, 265–270. Available from https://doi.org/10.2307/4615964.

Chao, A. (1988). Estimating the population size for capture-recapture data with unequal catchability. *Biometrics, 43*, 783–791. Available from https://doi.org/10.2307/2531532.

Christian, M. (2015). Biodiversity hotspots: A shortcut for a more complicated concept. *Global Ecology and Conservation, 3*, 297–309. Available from https://doi.org/10.1016/j.gecco.2014.12.008.

Colwell, R. K., & Coddington, J. A. (1994). Estimating terrestrial biodiversity through extrapolation. *Philosophical Transactions of the Royal Society of London. Series B, Biological Sciences, 345*, 101–118. Available from https://doi.org/10.1098/rstb.1994.0091.

Costa, S., Ganzerli, S., Rugiero, I., Pellizzari, S., Pedrini, P., & Tamburini, E. (2017). Potential of *Rhodobacter capsulatus* grown in anaerobic-light or aerobic-dark conditions as bioremediation agent for biological wastewater treatments. *Water (Switzerland), 9*(2), 108. Available from https://doi.org/10.3390/w9020108.

Donachie, S. P., Hou, S., Lee, K. S., Riley, C. W., Pikina, A., Belisle, C., Kempe, S., Gregory, T. S., Bossuyt, A., Boerema, J., Liu, J., Freitas, T. A., Malahoff, A., & Alam, M. (2004). The Hawaiian Archipelago: A microbial diversity hotspot. *Microbial Ecology, 48*(4), 509–520. Available from https://doi.org/10.1007/s00248-004-0217-1.

Faoro, H., Alves, A. C., Souza, E. M., Rigo, L. U., Cruz, L. M., Al-Janabi, S. M., Monteiro, R. A., Baura, V. A., & Pedrosa, F. O. (2010). Influence of soil characteristics on the diversity of bacteria in the southern brazilian atlantic forest. *Applied and Environmental Microbiology, 76*(14), 4744–4749. Available from https://doi.org/10.1128/AEM.03025-09.

Fisher, R., Corbet, A., & Williams, C. (1943). The relation between the number of species and the number of individuals in a random sample of an animal population. *Journal of Animal Ecology, 12*, 42–58.

Habel, J. C., Samways, M. J., & Schmitt, T. (2019). Mitigating the precipitous decline of terrestrial European insects: Requirements for a new strategy. *Biodiversity and Conservation, 28*(1799), 1–18. Available from https://doi.org/10.1007/s10531-019-01741-8.

Hamilton, A. (2005). Species diversity or biodiversity? *Journal of Environmental Management, 75*(1), 89–92.

Holt, E. A., & Miller, S. W. (2010). Bioindicators: Using organisms to measure environmental impacts. *Nature Education Knowledge, 3*(10), 8.

Hou, J., Sievert, S. M., Wang, Y., Seewald, J. S., Natarajan, V. P., Wang, F., Wang, F., Wang, F., Xiao, X., & Xiao, X. (2020). Microbial succession during the transition from active to inactive stages of deep-sea hydrothermal vent sulfide chimneys. *Microbiome, 8* (1). Available from https://doi.org/10.1186/s40168-020-00851-8.

Jaccard, P. (1912). The distribution of the flora of the Alpine zone. *New Phytologist, 11*, 37–50. Available from https://doi.org/10.1111/j.1469-8137.1912.tb05611.x.

Jean-Pierre, D., & Hugo, S. (2008). Microorganisms of the East African great lakes and their response to environmental changes. *Freshwater Reviews, 1*(1), 59–73. Available from https://doi.org/10.1608/FRJ-1.1.4.

Joshi, D., Kumar, S., Suyal, D. C., & Goel, R. (2017). *The microbiome of the Himalayan ecosystem. Mining of microbial wealth and metagenomics* (pp. 101–116). Springer Singapore. Available from https://doi.org/10.1007/978-981-10-5708-3_6.

Kent, A. D., & Triplett, E. W. (2002). Microbial communities and their interactions in soil and rhizosphere ecosystems. *Annual Review of Microbiology, 56*, 211–236. Available from https://doi.org/10.1146/annurev.micro.56.012302.161120.

Kim, B. R., Shin, J., Guevarra, R. B., Lee, J. H., Kim, D. W., Seol, K. H., Lee, J. H., Kim, H. B., & Isaacson, R. E. (2017). Deciphering diversity indices for a better understanding of microbial communities. *Journal of Microbiology and Biotechnology, 27*(12), 2089–2093. Available from https://doi.org/10.4014/jmb.1709.09027.

Klinkenberg, B. (Ed.). (2020). Biodiversity of British Columbia. http://www.biodiversity.bc.ca.

Krieg, N. R. (2015). In M. E. Trujillo, S. Dedysh, P. DeVos, B. Hedlund, P. Kämpfer, F. A. Rainey, & W. B. Whitman (Eds.), *Bergey's manual of systematics of archaea and bacteria*, 1-3. Available from https://doi.org/10.1002/9781118960608.gbm00985.

Lan, G., Wu, Z., Li, Y., & Chen, B. (2019). The drivers of soil bacterial communities in rubber plantation at local and geographic scales. *Archives of Agronomy and Soil Science*.

Lemos, L. N., Fulthorpe, R. R., Triplett, E. W., & Roesch, L. F. (2011). Rethinking microbial diversity analysis in the high throughput sequencing era. *Journal of Microbiological Methods, 86*, 42–51.

Locey, K. J., & Lennon, J. T. (2016). Scaling laws predict global microbial diversity. *Proceedings of the National Academy of Sciences of the United States of America, 113*(21), 5970–5975. Available from https://doi.org/10.1073/pnas.1521291113.

Lovejoy, T. E. (1980). The Global 2000 report to the President. *The Technical Report, 2*, 327–332.

Lutz, S., Anesio, A. M., Edwards, A., & Benning, L. G. (2017). Linking microbial diversity and functionality of arctic glacial surface habitats. *Environmental Microbiology, 19*(2), 551–565. Available from https://doi.org/10.1111/1462-2920.13494.

Marcel, C. (2002). The life-history basis of latitudinal diversity gradients: How do species traits vary from the poles to the equator? *Journal of Animal Ecology, 71*, 79–87. Available from https://doi.org/10.1046/j.0021-8790.2001.00577.x.

Marcon, E., Hérault, B., Baraloto, C., & Lang, G. (2012). The decomposition of Shannon's entropy and a confidence interval for beta diversity. *Oikos, 121*, 516–522.

Martiny J. B., Bohannan, B. J., Brown, J. H., Colwell, R. K., Fuhrman, J. A., Green, J. L., Horner-Devine, M. C., Kane, M., Krumins, J. A., Kuske, C. R., Morin, P. J., Naeem, S., Øvreås, L., Reysenbach A.-L., Smith, V. H., & Staley, J. T. (2006). Microbial biogeography: putting microorganisms on the map, *Nature Reviews Microbiology, 4*, 102–12. Available from https://doi.org/10.1038/nrmicro1341. PMID: 16415926.

Martini, S., Michotey, V., Casalot, L., Bonin, P., Guasco, S., Garel, M., & Tamburini, C. (2016). Bacteria as part of bioluminescence emission at the deep ANTARES station (North-Western Mediterranean Sea) during a one-year survey. *Deep-Sea Research Part I: Oceanographic Research Papers, 116*, 33–40. Available from https://doi.org/10.1016/j.dsr.2016.07.014.

Martiny, J. B. H., Bohannan, B. J. M., Brown, J. H., Colwell, R. K., Fuhrman, J. A., Green, J. L., Horner-Devine, M. C., Kane, M., Krumins, J. A., Kuske, C. R., Morin, P. J., Naeem, S., Øvreås, L., Reysenbach, A. L., Smith, V. H., & Staley, J. T. (2006). Microbial biogeography: Putting microorganisms on the map. *Nature Reviews Microbiology, 4*(2), 102–112. Available from https://doi.org/10.1038/nrmicro1341.

Mazancourt, C. D., & Schwartz, M. W. (2010). A resource ratio theory of cooperation. *Ecology Letters, 13*, 349–359. Available from https://doi.org/10.1111/j.1461-0248.2009.01431.x.

McIntosh, R. P. (1967). An index of diversity and the relations of certain concepts to diversity. *Ecology, 48*, 392–404. Available from https://doi.org/10.2307/1932674.

Mitchell, J. G. (2002). The energetics and scaling of search strategies in bacteria. *American Naturalist, 160*(6), 727–740. Available from https://doi.org/10.1086/343874.

Mittermeier, R. (2004). Hotspots revisited. Earth's biologically richest and most endangered terrestrial ecoregions. *Conservation International*, 392.

Mittermeier, R., Turner, W., Larsen, F., Brooks, T., & Gascon, C. (2011). Global biodiversity conservation: The critical role of hotspots. *Biodiversity Hotspots*. Available from https://doi.org/10.1007/978-3-642-20992-5_1.

Moënne-Loccoz, Y., Mavingui, P., Combes, C., Normand, P., & Steinberg, C. (2015). *Microorganisms and biotic interactions. Environmental microbiology: Fundamentals and applications* (pp. 395–444). Netherlands: Springer. Available from https://doi.org/10.1007/978-94-017-9118-2_11.

Mora, C., Tittensor, D. P., Adl, S., Simpson, A. G. B., & Worm, B. (2011). How many species are there on earth and in the ocean? *PLoS Biology, 9*(8), e100112. Available from https://doi.org/10.1371/journal.pbio.1001127.

Morisita, M. (1959). Measuring of the dispersion and analysis of distribution patterns. *Memories of the Faculty of Science, Kyushu University. Series E: Biology, 2*, 215–235.

Myers, N. (1988). Threatened biotas: Hot spots in tropical forests. *The Environmentalist, 8* (3), 187–208. Available from https://doi.org/10.1007/BF02240252.

Myers, N. (1990). The biodiversity challenge: Expanded hot-spots analysis. *The Environmentalist, 10*(4), 243–256. Available from https://doi.org/10.1007/BF02239720.

Myers, N., Mittermeler, R. A., Mittermeler, C. G., Da Fonseca, G. A. B., & Kent, J. (2000). Biodiversity hotspots for conservation priorities. *Nature, 403*(6772), 853–858. Available from https://doi.org/10.1038/35002501.

Nemergut, D., Shade, A., & Violle, C. (2014). When, where and how does microbial community composition matter? *Frontiers in Microbiology, 5*, 497. Available from https://doi.org/10.3389/fmicb.2014.00497.

Odum, H. T., Cantfon, J. E., & Kornicker, L. S. (1960). An organizational hierarchy postulate for the interpretation of species – individual distribution, species entropy, ecosystem evolution and the measuring of species variety index. *Ecology, 41*, 395–399.

OECD. (2011). *The economic significance of natural resources: Key points for reformers in Eastern Europe, Caucasus and Central Asia*. EAP Task Force.

Orme, C. D. L., Davies, R. G., Burgess, M., Eigenbrod, F., Pickup, N., Olson, V. A., Webster, A. J., Ding, T. S., Rasmussen, P. C., Ridgely, R. S., Stattersfield, A. J., Bennett, P. M., Blackburn, T. M., Gaston, K. J., & Owens, I. P. F. (2005). Global hotspots of species richness are not congruent with endemism or threat. *Nature, 436*(7053), 1016–1019. Available from https://doi.org/10.1038/nature03850.

Panosyan, H., Margaryan, A., Poghosyan, C., & Saghatelyan, A. (2018). Microbial diversity of terrestrial geothermal springs in Lesser Caucasus. In D. Egamberdieva, N. K. Birkeland, H. Panosyan, & W. J. Li (Eds.), *Extremophiles in Eurasian ecosystems: Ecology, diversity, and applications. Microorganisms for sustainability*. Springer. Available from https://doi.org/10.1007/978-981-13-0329-6_4.

Parmar, T. K., Rawtani, D., & Agrawal, Y. K. (2016). Bioindicators: The natural indicator of environmental pollution. *Frontiers in Life Science, 9*(2), 110–118.

Peet, R. (2003). The measurement of species diversity. *Annual Review of Ecology and Systematics, 5*, 285–307. Available from https://doi.org/10.1146/annurev.es.05.110174.001441.

Phale, P. S., Sharma, A., & Gautam, K. (2019). *Microbial degradation of xenobiotics like aromatic pollutants from the terrestrial environments. Pharmaceuticals and personal care products: Waste management and treatment technology emerging contaminants and micro pollutants* (pp. 259–278). Elsevier. Available from https://doi.org/10.1016/B978-0-12-816189-0.00011-1.

Pielou, E. C. (1975). *Ecological diversity* (p. 165) New York: Wiley.

Polechová, J., & Storch, D. (2008). *Ecological niche. Encyclopedia of ecology, five-volume set* (pp. 1088–1097). Elsevier Inc. Available from https://doi.org/10.1016/B978-008045405-4.00811-9.

Rachlow, J. L. (2008). *Wildlife ecology. Encyclopedia of ecology, five-volume set* (pp. 3790–3794). Elsevier Inc. Available from https://doi.org/10.1016/B978-008045405-4.00861-2.

Rappé, M. S., & Giovannoni, S. J. (2003). The uncultured microbial majority. *Annual Review of Microbiology, 57*, 369–394.

Rea, I., & De Stefano, L. (2019). Recent advances on diatom-based biosensors. *Sensors (Switzerland), 19*(23), 5208. Available from https://doi.org/10.3390/s19235208.

Roderic, G., & Michiel, de H. (2018). Recent advances in functional genome analysis. *F1000Research*, 1968. Available from https://doi.org/10.12688/f1000research.15274.1.

Sajjad, W., Din, G., Rafiq, M., Iqbal, A., Khan, S., Zada, S., Ali, B., & Kang, S. (2020). Pigment production by cold-adapted bacteria and fungi: Colorful tale of cryosphere with wide range applications. *Extremophiles: Life Under Extreme Conditions, 24*(4), 447–473. Available from https://doi.org/10.1007/s00792-020-01180-2.

Sanyal, A., Antony, R., Samui, G., & Thamban, M. (2018). Microbial communities and their potential for degradation of dissolved organic carbon in cryoconite hole environments of Himalaya and Antarctica. *Microbiological Research, 208*, 32–42. Available from https://doi.org/10.1016/j.micres.2018.01.004.

Selim, S. M., & Zayed, M. S. (2017). *Microbial interactions and plant growth, Plant-microbe interactions in agro-ecological perspectives* (Vol. 1, pp. 17–34). Singapore: Springer. Available from https://doi.org/10.1007/978-981-10-5813-4_1.

Shimane, T., Wajid, U. M., Feng, H., & Guang, Y. (2018). Current trends and potential applications of microbial interactions for human welfare. *Frontiers in Microbiology, 9*, 1156. Available from https://doi.org/10.3389/fmicb.2018.01156.

Simpson, E. H. (1949). Measurement of diversity. *Nature, 163*, 688. Available from https://doi.org/10.1038/163688a0.

Smith, V. H., Graham, D. W., & Cleland, D. D. (1998). Application of resource-ratio theory to hydrocarbon biodegradation. *Environmental Science and Technology, 32*(21), 3386–3395. Available from https://doi.org/10.1021/es9805019.

Sorenson, T. (1948). A method of establishing groups of equal amplitudes in plant sociology based on similarity of species content and its application to analyses of the vegetation on Danish commons. *Kongelige Danske Videnskabernes Selskab, Biologiske Skrifter, 5*, 1–34.

Speer, K. A., Teixeira, T., Brown, A., Perkins, Dittmar, K., Ingala, M., Wultsch, C., Krampis, K., Dick, C., Vieira, M., & Delciellos, A. (2020). Parasite species identity and local community diversity mediate effects of habitat fragmentation on bacterial microbiomes. *Authorea Preprints*.

Steven, D., Shostak, S., Fry, I., Vidal, C., et al. (2015). *The impact of discovering life beyond earth*. Cambridge University Press.

Tett, P., Black, K., Brenann, R., Cook, E., & Dawidson, K. (2015). *Coastal zones: Solutions for the 21st century*. Elsevier.

Tout, J., Jeffries, T. C., Webster, N. S., Stocker, R., Ralph, P. J., & Seymour, J. R. (2014). Variability in microbial community composition and function between different niches within a coral reef. *Microbial Ecology, 67*(3), 540–552. Available from https://doi.org/10.1007/s00248-013-0362-5.

Tribedi, P. (2016). Prediction of microbial community stability through lorenz curve: A potential tool for microbial community studies. *Journal of Applied and Fundamental Sciences, 2*(1).

Tucker, A., & Duplisea, D. (2012). Bioinformatics tools in predictive ecology: Applications to fisheries. *Philosophical Transactions of the Royal Society B: Biological Sciences, 367*(1586), 279–290. Available from https://doi.org/10.1098/rstb.2011.0184.

Valdespino-Castillo, P. M., Cerqueda-García, D., Espinosa, A. C., Batista, S., Merino-Ibarra, M., Taş, N., Alcántara-Hernández, R. J., & Falcón, L. I. (2018). Microbial distribution and turnover in Antarctic microbial mats highlight the relevance of heterotrophic bacteria in low-nutrient environments. *FEMS Microbiology Ecology, 94*(9), fiy129. Available from https://doi.org/10.1093/femsec/fiy129.

Wake, D. B., & Vredenburg, V. T. (2008). Are we in the midst of the sixth mass extinction? A view from the world of amphibians. *Proceedings of the National Academy of Sciences of the United States of America, 105*(1), 11466–11473. Available from https://doi.org/10.1073/pnas.0801921105.

Wang, S., Wang, X., Han, X., & Deng, Y. (2018). Higher precipitation strengthens the microbial interactions in semi-arid grassland soils. *Global Ecology and Biogeography, 27*(5), 570–580. Available from https://doi.org/10.1111/geb.12718.

Wilson, E. O. (1994). *Naturalist*. Island Press Shearwater Books.
Wilson, J. B., Spijkerman, E., & Huisman, J. (2007). Is there really insufficient support for Tilman's R* concept? A comment on Miller et al. *American Naturalist, 169*(5), 700–706. Available from https://doi.org/10.1086/513113.
Wilson, M., & Shmida, A. (1984). Measuring beta diversity with presence-absence data. *Journal of Ecology, 72*(3), 1055–1064. Available from https://doi.org/10.2307/2259551.
Yoshitake, S., Uchida, M., Iimura, Y., Ohtsuka, T., & Nakatsubo, T. (2018). Soil microbial succession along a chronosequence on a High Arctic glacier foreland, Ny-Ålesund, Svalbard: 10 years' change. *Polar Science (New York, N.Y.), 16*, 59–67. Available from https://doi.org/10.1016/j.polar.2018.03.003.
Young, I. M., Crawford, J. W., Nunan, N., Otten, W., & Spiers, A. (2008). Microbial distribution in soils: Physics and scaling. *Advances in Agronomy, 100*, 81–121.

CHAPTER

2

Habitat-specific microbial community associated with the biodiversity hotspot

Sarita Tiwari[1], *Sandhya Moghe*[1], *W.B. Gurnule*[2], *Devidas S. Bhagat*[3] *and Aparna Gunjal*[4]

[1]Department of Biotechnology, Kamla Nehru Mahavidyalaya, India, [2]Department of Chemistry, Kamla Nehru Mahavidyalaya, India, [3]Department of Forensic Chemistry and Toxicology, Government Institute of Forensic Science, Aurangabad, India, [4]Department of Microbiology, Dr. D. Y. Patil Arts, Commerce & Science College, Pune, India

2.1 Introduction

India is ranked as a mega-diverse country enduring four biodiversity hotspots among the 36 proclaimed sites situated worldwide (De Mandal et al., 2015; Rajkhowa et al., 2015). The four regions are the Himalayas, the Indo-Burma region, Western Ghats, and the Sundaland. The arena tagged as hotspots are known for their rich and discrete biological combo witnessing biological activity par the normal limit but having fear of extinction is marked as biodiversity hotspots. The overall survey in India reports the existence of over 91,000 species of animals, and 45,500 plant species have been documented in India. The major factors responsible for creating a diverse aura are climate, temperature, soil quality, rainfall percentage, and the presence of many rivers culpable for vegetation cover and microbes dwelling in soil (Rathour et al., 2017).

The soil in the biodiversity zone is laced with a unique natural environment in which, inhabitant special microbes with distinctive characteristics also involved in plant growth, etc. have the possibility of

Microbial Diversity and Ecology in Hotspots
DOI: https://doi.org/10.1016/B978-0-323-90148-2.00018-3

encouraging diversified species. Within the microbial community, prokaryotes are the dominant group in comparison to others and are actively involved in ecosystem functioning (Powell, 1993). In the biodiversity hotspot region, the microbial world is still to uncover which is a large pool of biodiversity on land (Vibha & Neelam, 2012). The study of microbial ecology is gaining immense interest. Members of the microbial family, that is, bacteria, fungi, and actinomycetes are at the forefront in shaping the morphology, physiology, functioning, and overall working of living systems (Braga et al., 2016; Briones & Raskin, 2003). As most of the findings are based on discovering microbes of surface soil (about 10–40 cm depth), the data on identification of the microorganisms are not sufficient (Martínez-Alonso et al., 2010).

2.2 Biodiversity hotspot in the Indian continent

India is blessed with a diverse climatic condition which is the basis for biological diversity, ultimately maintains equity relations amidst humans and organisms. The well-known biodiversity hotspots of the Indian subcontinent are the Himalayas, Indo Burma Region, Sundaland, and the Western Ghats situated in different parts of India. They are listed in the top 36 biodiversity hotspot around the world. However, these hotspots are under threat and on the verge of extinction of many rare species. Thus it is the need of time to conserve these species as it is a heritage of the country. Conserving this biodiversity amid many development projects going on in a country like India is challenging but essential for sustainable development.

2.3 Factors affecting biodiversity hotspots in India

The presence of green hills, clear water, fantastic caves, and variety had made this biodiverse area alluring. But recent activities like fraternization of trade, human, and geogenic action to a certain extent has dragged the cleaner zone of biodiversity toward containment of pollutants (Vareda et al., 2019; Yang et al., 2018). Different actions responsible for deteriorating the habitats in the hotspots are listed in Fig. 2.1. One of the activities which are done for human welfare is dam construction. The dam is a modern-day technology having double facet, that is, affirmative as well as the pessimistic outcome. Even this biodiversity zone is not left alone by today's world mechanistic engineering action. Even the emergence of toxicants like metal imparts toxicity to reachable biota belonging to nearby ranges depending on its type and chemical form. Escorted with cytotoxicity, mutagenicity, and carcinogenic character, they are a big threat to animals

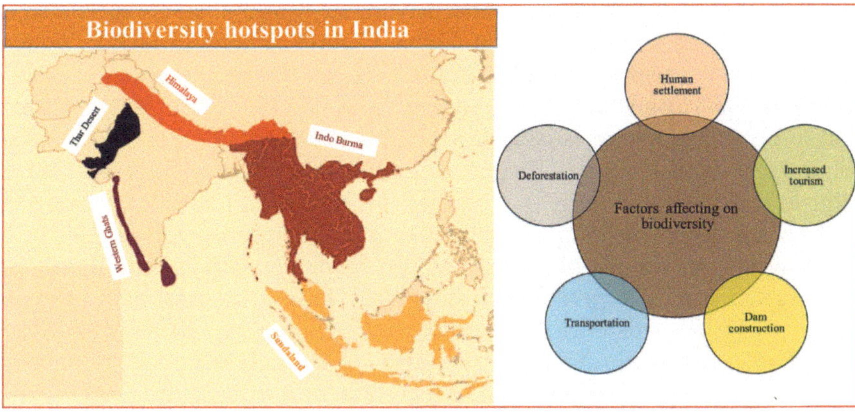

FIGURE 2.1 Factors responsible for creating disturbances in biodiversity hotspot of India.

and plants especially to human health, calling an emergency for implementation of an appropriate technique for abatement and inhibition of its entry into the natural environment (Hadjiliadis, 2012).

2.4 The Himalayas

The Himalayan region has spread in the arena of country like Bhutan, Northern India, and major parts of Nepal. They are well known for their high peaks mountain comprising about more than 100 mountains higher than 7200 m height and also marked as highest mountains in the world including Mount Everest and K2 (Rüber et al., 2020). There are about 163 species declared on the verge of extinction belonging to the group of mammals, birds, reptiles, invertebrate, and plant species. In case of plants, the Himalayan valley is densely populated with varied species of plants and about 300 varied species of animals including wild water buffalo, sloth bear, big squirrel, and requires urgent attention. The issues faced by these regions are imparting harm to local medicinal plants, black bears, etc. (Pandit et al., 2007; Tali et al., 2019). For example, some of the species are on the verge of extinction such as Namdapha's flyinal plants like Himalayan Yew, Blue-poppy, Snow Lotus plants for profiteering purposes.

2.5 The Indo-Burma region

The Indo-Burma region is the second biodiversity hotspot that belongs to the Indian subcontinent. The regions stretch over different countries including India's Northeastern states, Myanmar, Cambodia, Laos, Thailand, Vietnam, and the southern part of China, covering about 23 lakh km^2 area. This hotspot consists of 13,500 plant species, 1260 marine fishes, reptiles, mammals, etc. (Jose et al., 2016). Among these, the botanical cover is threatened to be extinct which is clear from the fact that only 1.18 lakh km^2 vegetation cover is remaining. Factors dragging the priceless rare species to extinction are farming activity, commercial plantation, fisheries, etc. which has to drop down the original habitat to 5% (Dudgeon, 2012). These areas are known for its unique biota and environ created due to its unique habitat. Also known for its superiority in biological values, provide and promote diverse niche for existence of microbes with beneficial characteristic (Jose et al., 2016). However, regrettably, because of its tough and high mountains, its extreme environment construct, researchers find it difficult to attend and explore the microbial communities dwelling in this ecological condition.

2.6 The Sundaland

The Sundaland biodiversity hotspot region exists in the Nicobar group of Islands—Borneo, Java and Sumatra, Singapore, the Philippines which covers 15 lakh km^2 area. The Nicobar Islands constitute India. These islands have been declared as the sector biosphere reserve in 2013 through the United Nations. These islands have wealthy terrestrial as nicely as marine ecosystems together with mangroves, seagrass beds, and coral reefs. Also, it is accompanied by various marine fishes, reptiles, etc. dwelling in the water bodies (Giam et al., 2012). The Sundaland comprises 25,000 species of plants, 380, 770, 450, 245, 950 mammals, birds, reptiles, amphibians, freshwater fishes, respectively. Presently only 1 lakh km^2 vegetation cover is remaining while the area under protection is 1.79 lakh km^2. Fifteen thousand plants, almost half of mammals and reptiles are endemic to the region. These islands are an abode to unique orchids and tropical plant species of high economic value, like oil palm, rubber, etc. The endangered species in this hotspot are turtles, pangolins, orangutans, tigers, rhinoceros, the Bali starling, straw-headed bulbul, and many more. The Sundaland biodiversity hotspot faces red danger and has lost most of its habitat due to deforestation, intensive commercial agriculture, man-made forest fires, extensive illegal hunting, and poaching of animals (Verma et al., 2020).

2.7 The Western Ghats

Another hotspot of India is the Western Ghats (the Sahayadri range) which is also listed as a biodiversity zone with a unique climate and environmental ecosystem. There are over 7000 plant species in the Western Ghats, of which about 5600 are endemic (Bossuyt et al., 2004). The rare and susceptible species include Nilgiri Tahr, Lion-tailed macaque, parakeets, laughing thrush birds, and more. The biotic group of Western Ghats is also under extinction and needs attention for the preservation of endangered species. Due to the proximity of this region to the ocean, they receive a higher volume of rainfall. Most of the species belonging to the group of amphibians and reptiles present in this region are rare and not found in other parts of the world. Also, another country like Srilanka, situated extreme south of India is known for rustic and wealthy species too and connected with India through an extended stretch of bridge having 140 Km. There is more than 6000 vascular flora present here belonging to 2500 genus. Within these, 3000 plants are prevalent in this area. Even the world famous spices like black pepper, cardamom have been known to be evolved from the Western Ghats. Majorly, of the total plant species, most of the plants are situated on Agasthyamalai Hills located in the acute South. The area is likewise domestic to round 450 species of birds, 140 mammals, 260 reptiles, and 175 amphibians. Such variety is pretty lovely in addition to uncommon, however, now lies at the verge of extinction. The plants on this area became firstly unfold over 190,000 rectangular kilometers, however, has decreased to 43,000 rectangular kilometers today. Presently, most effective 1.5% of the authentic woodland, nevertheless, exists in Sri Lanka. The mountains of the Western Ghats in Peninsular India are an international biodiversity hotspot (Garg et al., 2017). An exponential twofold increment withinside the counted number of species during the last decade and a half (2001–15) has evidenced that a vast majority of amphibians of this place had still to be discovered (Biju & Bossuyt, 2005), and there is probability that the original diversity is far greater in number than the estimated value (Biju & Bossuyt, 2005). In terms of species diversity, the Western Ghats positioned at the fourth position which ranked next to Indo-Burma, also grabbed the foremost spot among another hotspot region in terms of amphibian's range. The speedy disclosure of these species becomes possible due to the vast range of available advanced molecular tools and a synchronized stepwise approach.

2.8 Approach to safeguard biodiversity hotspots in India

Industralization, urbanization and other human activities are responsible for aggravating pollution, deforestration etc. which cause uncontrolled impact on their habitats, inturn causing nuisance. The conservation of biodiversity is the need of time. The negligence of the government toward precautionary investment, monitoring, and maintenance also lead to dwindling this diversity zone. Depicting the menacing state of art can overture for fastening policy and administrative action toward this crisis. In such circumstances, paramount involvement of government and policymakers for sustainable preservation model and definitive protocol for species restoration is required, also demands effective policy and financial commitments at large scale. The protection of biodiversity in India also requires the involvement of residents who indirectly get benefited from this diverse heritage. Many policies are already introduced by the Indian government such as Marine Turtle Policy and Marine Standing Policy for marine biodiversity protection from pollution. Ministry of Environment, Forest, and Climate Change is the heading agency that has been administered by the Indian Government; they also highlighted the Biological Diversity Act in 2002. International Union for Conservation of Nature has categorized the biological entity into different groups depending upon their present status as threatened, vulnerable, endangered, critically endangered, and extinct. Also, many international agencies are actively involved in funding the plan/policy for the preservation of threatened species in the hotspot region.

2.9 Habitat-specific microbes of hotspot region

The microorganisms play a vital and frequently specific position withinside the functioning of the ecosystems in preserving equality amidst biosphere and productivity. The harm in biodiversity and the approach to revive for human welfare is the thrust area seeking attention. Each habitat comprises a specific group of microbes which differ from site to site. Microbes present in this region are mostly in symbiotic association with other members of microbes pertaining in the soil (Taylor et al., 2007). For instance, the coexistence of algae and bacterial diversity determines the number of the carbon source utilized by bacteria, which directly symbolizes direct relation of nutrient cycling with microbial exploitation suggesting linkage between bacterial diversity and algal production. The major reason for inhabitant of special community is competition between microbial species, resultant into decrement in the microbial diversity of specific habitats. Specific habitat is a battle field where microbes are engaged in a fight for dominance defeating existence of other species. In case of some microbes, they are backed with genetic setup which makes them behave in ecologically aggressive manner. The characteristics laced with these microbes are high resistance against abiotic stress, tolerance against viral infection, varied substrate intake, out-grow competitors. However, there are some concepts which are unexplained concerned to adherence of particular microbial species within communities (Yadav et al., 2015).

A genetically diverse group of microbes depending upon their metabolic need, environmental conditions, etc. persist in a particular niche (Kühl et al., 2012). Invertebrate-microbe symbioses additionally play essential roles in host ecological achievement via the supply of supplemental vitamins and the manufacturing of protective secondary metabolites. The total pull of DNA in soil signifies cumulative genomes of microbes, small eukaryotes, and many plants and animals, which is used for tracking by the individual. In the Indo-Burman region, the presence of strains Trichoderma has been reported to synthesize three important enzymes, namely, chitinase, protease, and β-1,3-glucanase. Tayang and Jha (2010) detected the presence of an endophytic fungus *Fusarium* sp. having antimicrobial effect against most of the pathogens. In another study by Jose et al. (2016), they identified three cellulolytic fungal isolates displaying excessive FPase pastime and their information had been submitted in NCBI, GenBank as *Talaromyces verruculosus* SGMNPf3 (KC937053), *Trichoderma gamsii* SGSPf7 (KC937055), and *Trichoderma atroviride* SGBMf4 (KC937054). Bhattacharyya et al. (2014) used a metagenomics approach to assess the different soil bacterial communities of five varied zones in terms of geology and hydrology, targeting mainly floor and subsurface soil habitats of Brahmaputra valley, tagged as mega-biodiversity hotspot. Identifying the total microbial species persistence in this unique zone can definitely add on to the diverse beneficial role of microbes performed with the aid of using extraordinary microbial companies inside an ecosystem.

Microbial diversity also had a pivotal involvement the diversification of a particular ecosystem. The Northeastern part of the Indian continent is covered by natural forest and many parts of the zone are untouched and uncovered by the discoverer/researcher and from industrial interference. This region provides varied niches that trigger the propagation of novel microorganisms in the soil which are still unexplored. Many microbes are unidentified, undiscovered, and many are unculturable which are putting a big constrain in exploring. According to the study of De Mandal et al. (2015), Indo-Burman hotspot consists of a total of 29 phyla, within which the dominant phyla have been Acidobacteria (39.45%) accompanied with the aid of using Proteobacteria (26.95%), Planctomycetes (7.81%), Actinobacteria (7.18%), Bacteroidetes (6.65%), Chloroflexi (4.11%), and Nitrospirae (3.33%) (Bhattacharyya et al., 2014) while the study on a microbial pool of Indo-Burman hotspot used a culture-independent metagenomic approach for the identification of microbial population. By following 16s rRNA housekeeping gene for identification of microbes, the authors found dominancy of microbes belonging to the genera *Proteobacteria, Acidobacterium,* and

Comamonadaceae. These findings could open the prospect of exploring undiscovered species having a significant role in maintaining homeostatic in biodiversity ecosystems. Also, Jose et al. (2016) isolated culturable cellulolytic bacteria such as *T. verrucous* SGMNPf3, *T. gamsii* SGSPf7, and *T. atroviride* which was found dominant in the Indo-Burman region laced with unique capability which can be utilized for industrial application. In the mountains of the Himalayas, most prominently the group of psychrophilic and psychrotolerant microorganism is living withinside the variety of 8.35×10^5 to 1.0×10^3 CFU/g sediment or M/L water. The bacterial population prevails more on a snow-covered mountain in comparison to uncovered mountains of the Himalayas. In a study by Yadav et al. (2015), identified a total of 232 microorganisms from four different sites within north-west Himalaya region. ARDRA consequences provided a grouping of all diagnosed isolates into 45, 54, 39, and 31 clusters from Chandratal Lake, Rohtang Pass, Dashair Lake, and Beas River, respectively. Identification by molecular methods through 16s rRNA categorized the isolates based on genera into four phyla, *Actinobacteria, Bacteroidetes, Firmicutes,* and *Proteobacteria*. The persistence of such unique organisms in cold regions and high altitudes with plant growth-promoting attributes can increase the propagation of plants in a particular region and enhance the soil quality.

2.10 Factors responsible for the diversity of habitat-specific microbial community in the Western Ghats

The physicochemical parameter, nutritional values of soil, variations in monsoon, and three different seasons (rainy, winter, and summer) are responsible for promoting the diversity of the microorganisms, plants, and animals in the Western Ghats (Velmourougane, 2017). Different species of microbes are present in different regions of Western Ghats due to their complex functional interactions with nutrients in the ecosystem. The various regions of the Western Ghats have different physicochemical parameters responsible for the diversity of the habitat-specific microbial community in this region (Selvan et al., 2015). The favorable physicochemical parameters like pH of the soil, temperature, chemical composition of the soil, and water availability of water resources effects growth and diversity of the microbial network withinside the Western Ghats (Singh et al., 2011). Microbial diversity also depends on the geographical location and soil nutritional values (Nannipieri et al., 2003).

Climate conditions in the Western Ghats vicissitudes with altitudinal gradation and distance from the equator. The average annual temperature range in th Western Ghats is 15°C–20°C (Prasad et al., 2008). The geographical location of Western Ghats, the hill ranges from Peninsular India lie in between 8°20′ North–21°40′ North and 73°–77° East. The length of Western Ghats is 1600 km which runs parallel to the west coastline of India from Tapti valley Gujarat to Kanyakumari southern endpoint (Deshpande et al., 2001). The monsoon season in the Western Ghats lies from June to October, the annual average rainfall in this region is within the range of 240–250 cm (K. et al., 2012). The suitable climatic and environmental conditions and adequacy of physicochemical parameters favored the diversity of the habitat-specific microbial communities in the Western Ghats.

2.11 Habitat-specific microbes

Since long duration members of microbial communities like bacteria, fungi, and actinomycetes play are known for their important contribution in shaping the structure, characteristic, and usual sustainability of dwelling systems. This biodiversity zone has a high concentration of varied microbial types which also have a significant role in maintaining homeostatic of this diversified arena. One of the aspect answerable for the nearby richness and network composition of microbes, is essentially pushed through their host desire or specificity. As seen previously, these hotspot regions are blessed with varied climatic conditions which are an important parameter in molding the microbial community of a particular zone. To date, several microorganisms belonging to these four diverse zones have been reported as a dominant group or laced with a special character. The details of such microbes are listed in Table 2.1. Fungus and bacteria are the dominant groups unraveled by the researchers, within which fungal endophytes are very rarely studied species of microbes that is prevalent in inner regions of plant's part without inflicting any harm or damage to the host plant. Varied works carried out so far regarding the role of endophytes in host plants vividly explained its role in plant growth promotion, resistance against pathogen, empower host plants against various abiotic stressor, and recycle nutrients (Tiwari et al., 2016). Endophytic fungi dwelling without any symptom in plant tissues have also been recognized as a repository of novel secondary metabolites having medicinal properties. In the Indo-Burman region, Goyari et al. (2014) observed that ratios of culturable microorganism to culturable fungi had been extra withinside the woodland soil, final results became especially primarily based totally at the conventional microbiological techniques. In his study, his key finding was the high

TABLE 2.1 List of Indian biodiversity hotspot habitat-specific microbes.

Biodiversity hotspots name	Source	Microbes type	Genera/species	Characteristics	References
Indo-Burma	Bark of *Taxus baccata*	Fungal endophytes	*Fusarium, Penicillium, Aspergillus*	Antimicrobial activity	
Indo-Burma	Soil samples	Fungus	*Talaromyces verruculosus, Trichoderma gamsii, Trichoderma atroviride*	Cellulose production	Goyari et al. (2014)
Indo-Burma	Soil samples	Bacteria	*Proteobacteria, Acidobacterium*, and *Comamonadaceae*	Dominant group	Bhattacharyya et al. (2014)
Western Ghats	Water samples	Bacteria	*Proteobacteria, Actinobacteria, Bacteroidetes, Firmicutes*	Dominant group	
Western Ghats	Soil samples	Bacteria	*Pontibacter niistensis, Paracoccus niistensis*	Dominant group	
Western Ghats	Soil samples	Bacteria	300 isolates of *Bacillus thuringenesis*	Production of parasporal crystals	
Himalayan region	Cold desert	Bacteria	*Arthrobactor nicotianae, Brevundimonas terrae, Paenibacillus tylopili*, and *Pseudomonas cedrina*	Dominant group	Yadav et al. (2015)
Himalayan region	Brackish water	Bacteria	*Methyllophaga*	Dominant group	Rathour et al. (2017)

population density of culturable cellulolytic bacteria in comparison to the fungi in the forest soil. The cellulolytic capability of *T. verruculosus* SGMNPf3, *T. gamsii* SGSPf7, and *T. atroviride* SGBMf4 reported from the in place of remoted from the Indo-Burma biodiversity hotspot was explored and authors gave futuristic target to further explore the activity of their purified cellulases and its possible implication at the industrial level.

2.12 The utility of microbial diversity

Soil is a sink of nutrition which is essential for the growth of wide variety of microorganisms that inculcate unique capability to degrade a vast variety of complex organic compounds due to their metabolic activity. In soil, the recycling of inorganic nutrients such as phosphorus, oxygen, carbon, nitrogen, and sulfur elements are carried out by microbes by the process of degradation of the residual substrate like dead and decaying plant and animal remains which in turn replenish the environment. The replenishment is crucial attributable to the reuse of important factors and therefore microbes acting as green decomposers and scavengers to easy up the biosphere. Microorganisms actually have a sizeable function in preserving soil fertility consisting of its preservation with the aid of using adopting unique processes. They can act as signs for the evaluation of sustainable land use in area and time. It acts pointers for the valuation of ecological terrestrial use in planetary time. Optimization and upgradation of a different approach for sustainability assists in cumulative action in functional eco-physiology of microbiota. The natural habitat of the microbial community is disturbed by human encroachments through the utilization of pesticides with a positive approach, but it exerts negative action by affecting the microbiota enduring in the soil which ultimately affects the action of biotransformation reaction initiated by them in soil.

2.13 Importance of habitat-specific microorganisms in agriculture

In the present scenario, modern agriculture is mainly dependent on chemical fertilizers which are synthesized by nongreen chemistry approach which are harmful, unfriendly, deteriorating field quality, and costly approach. Utilization of biological inoculant over chemical fertilizer that involves consortia of wide microbes specially belonging to mycorrhizae, rhizobium a, cyanobacteria, some symbiotic and nonsymbiotic microbes, microbes with plant growth promoting activity are more preferred. Many researchers are harassed on selective usage of the prevailing variety of microbial species and their function, which gift possibilities to enhance agricultural production. Rhizobacteria from genera *Streptomyces, Achromobacter, Arthrobacter, Azotobacter, Azospirillum, Bacillus, Enterobacter, Pseudomonas,* and *Serratia* have been reported having plant progression peculiarity even on metal abundant soil (Asad et al., 2019). The presence of mycorrhizal fungi in symbiosis or nonsymbiotic association in the soil are responsible for the uptake of P and other nutrients. Further, mycorrhizae offer amino acids and hormones, which can be secreted in excessive quantities accordingly growing the durability of plants. The range of soil microbial groups can be key to the capability of soils to suppress soil-borne plant diseases. The fitness of soil may be described in phrases of its microbiological potential to suppress the interest of plant pathogenic microbes. Some microbes like fluorescent pseudomonads are stated to provide antifungal antibiotic 2,4-diacetyl phloroglucinol (DAPG) withinside the wheat monocropping gadget and accordingly specific, suppressiveness of take-all ailment in wheat that multiplied the yield. Recently in India, DAPG manufacturing through *P. fluorescence* turned into stated to suppress rice bacterial blight. Rhizosphere microorganism will be used as biocontrol sellers for plant diseases. The rhizosphere is the region where an alliance between plant roots, soil, and microorganism exists; furthermore, this environment supports the root's accessibility to nutrients (Mishra et al., 2017).

Rhizosphere confronts corroboration of vital chemical reactions in soil occurring with time and arena, conjointly endure more variation in the region through root's activity in terms of size, type, location, and ages (Zhang et al., 2018).

2.14 Techniques used for assessing microbial diversity

The traditional microbiological method is unable to culture about 99% of bacteria that can be visualized under a microscope but is not culturable in laboratory conditions. To get over the troubles related to noncultural microorganisms, several techniques were advanced to become aware of and observe those microorganisms. Phospholipid fatty acid profiles provide touchy reproducible measurements for characterizing the numerically predominant soil microbial groups without cultivating the organisms. The method offers estimates of each microbial network composition and biomass size. The cheapest method for identification is the phenotypic methods buts it is laborious and time taking. The modern time researcher gives preference to molecular biology technique for characterization of microbial ecology because of its quick, specific, and nonlaborious approach. It also removes the constraints faced in the microbiological technique of initial culturing of microbes through the advances of metagenomics approach (Handelsman, 2004). It was evident from the literature that the microbial count uncovered by phylogenetic analysis are more to that of traditional culture-dependent approach (Webster et al., 2001). Molecular methods have definitely add-on information in databases of microbial culture. The most common approach for the identification of microbial communities is by molecular biology technique, especially metagenomics. The metagenomics method is a complicated approach of the present day that is primarily based totally on subculture impartial for the characterization of bacterial network from a selected sample. Bhattacharyya et al. (2014) followed subculture impartial metagenomic method to get to the bottom of the soil bacterial network composition of five physiologically and hydrologically numerous floor and subsurface soil habitats of Brahmaputra valley, Assam, North-East India, an Indo-Burma mega-biodiversity hotspot.

A modern method like molecular sequencing strategy has assisted in coming across a huge variety of microorganisms and fungi living in any such biodiversity zone (Teo & Wang, 2014). In an observation by Nampoothiri et al. (2013), the authors applied high-throughput techniques for unraveling the bacterial groups of the Western Ghats. Also, this enhanced device will help in figuring out the genera and species of residing bacterial variety of unique zone. The trending molecular method like next-generation sequencing era have correctly applied to resolve bacterial groups in exclusive environments which includes bat guano, warm spring, lake, cave sediments, etc. (De Mandal et al., 2015; Panda et al., 2015). An impartial metagenomics technique become hired withinside the gift research to symbolize the variety of soil bacterial network composition in five geochemically and hydrologically exclusive floor and subsurface soil habitats of Brahmaputra valley, Assam, North-East India, an Indo-Burma mega-biodiversity hotspot.

2.14.1 Habitat biogeography and habitats in different hotspots

The deterministic and stochastic (includes various ecological processes, including selection, dispersal, diversification, and drift) process have played a key role in microbial ecosystems (Monard et al., 2016). The habitat-specific microorganisms show various ecological responses to change with climatic conditions (Kneitel, 2018). Thus the deterministic and stochastic mechanism of microorganisms that show distribution patterns can help to know the habitat-specific microbial community associated with the biodiversity hotspot. Sustainable environmental and climatic factors are responsible for the conservation of biodiversity hotspots various distribution patterns, and their driving factors around the globe (Ren et al., 2018). For example, Mount Wutai Natural Reserve, which is located in northeastern Shanxi Province, is one of the preserved ecosystems and also the highest peak in North China. The change in climate and topography accounts for the major vertical zonation of forest types in Mount Wutai (Li et al., 2020). These features make Mount Wutai the best area to study microbial biogeography and its assembly. The elevational patterns of microorganisms present in the soil have attracted many scientists for research in this area (Zhang et al., 2018). The elevational distribution patterns of microorganisms in the soil with various niche breadth characters across elevational gradients have yet to be studied. The unmeasured variables which influence microbial population include plant structures, dispersion, environmental factors which are spatially structured, habitat-heterogeneity which influence microbial interactions, and relationships. Poor habitat quality is responsible for reducing diversity of microorganisms. The contrasting mechanisms, that is, poor habitat and reduced dispersal on smaller islands reduce diversity of fungi (Li et al., 2020). The biodiversity conservation of hotspot marked area needs intense information on biogeographical patterns and the ecological relationships underpinning them (Crain & Fernández, 2020). Presently, the major scientific group aims to unravel the biogeography especially the description and explanation of microbial distribution patterns, as well as in search of centers of high species diversity of the hotspot zones (Martiny et al., 2006). This habitat biogeography will help to understand the taxa of microorganisms. The macroorganism biogeography is influenced by factors such as colonization, diversification, extinction, etc. (Martiny et al., 2006). Also, the current situation demands connecting the patterns of diversity biogeography includes the study of geographical space through geological timescale. Conservation biologists are mainly focusing on biogeographical research due to the rapid increase in human-induced pressures on biodiversity, also facing constraints related to conservation resources such as financial and human capital. Many organizations, agencies suggest integrating diversity hotspots into conservation planning strategies with a motto to preserve biodiversity all over the world (Zheng et al., 2019). The Hawaiian Archipelago is a rich biodiversity hotspot where eukaryotes evolved due to different locations and topography (Donachie et al., 2004). There is a report on soil microbial communities in habitats of differing soil—water balance on the Tibetan plateau (Li et al., 2017). The pH and moisture are also known to influence the microbial population.

2.15 Conclusion

The biodiversity zone of India is the heritage of the country and safeguarding its biological entity is the responsibility of all. Nowadays, inclination toward anthropogenic activities has put an alarm on the extinction of many species of this zone. It is high time to preserve this legacy of national importance and impart urgent policy and approach for its preservation. In this region, different biological species such as plants, animals, reptiles, birds are discovered and accountable but microbial diversity, which is an inseparable entity, the study is very scarce. Microbes also assist in promoting biological species diversity and plant growth. Revealing the microbial communities of the hotspot area will add to the knowledge base of the researcher and can be further utilized for maintaining the uniqueness of the highlighted arena.

2.16 Future prospects

The biodiversity marked hotspot is a heritage for the country, and initiatives for better preservation and management is the need of time and also measures should be taken involving local people and government policy to safeguard these biospheres. Microbes resemble large integration of DNA that needs to be restored in unchangeable form for the future generation. As these biodiversity hotspots are under threat for the extinction of their unique biological entities' application of impactful and innovative protocols should be launched for effective involvement of research groups as well as local citizens with the moto to safeguard ecological biodiversity zone. The novel approach referred to above may be changed and tailored withinside the destiny for isolation and characterization of bacterial groups of the extreme surroundings, a good way to upload directly to the brilliant expertise approximately diverse roles performed via way of means of distinct microbial companies inside an ecosystem. Also, a literature survey highlighting the prevailing nation on bacterial network shape in herbal ecosystems. Psychrotrophic microbes may also be implied withinside the destiny with more price in farming as biofertilizer and biocontrol retailers in bloodless conditions. The use of psychrophiles as biofertilizers, biocontrol retailers, and bioremediation could be of exceptional use in agriculture below bloodless climatic conditions. The restoration approach can provide us the information about the range of species that are still to be discovered.

References

Asad, S. A., Farooq, M., Afzal, A., & West, H. (2019). Integrated Photobiol heavy metal remediation strategies for a sustainable clean environment-a review. *Chemosphere, 217*, 925–941.

Bhattacharyya, P. N., Tanti, B., Barman, P., & Jha, D. K. (2014). Culture-independent metagenomic approach to characterize the surface and subsurface soil bacterial community in the Brahmaputra valley, Assam, North-East India, an Indo-Burma mega-biodiversity hotspot. *World Journal of Microbiology and Biotechnology*, *30*(2), 519–528. Available from https://doi.org/10.1007/s11274-013-1467-1.

Biju, S. D., & Bossuyt, F. (2005). New species of Philautus (Anura: Ranidae, Rhacophorinae) from Ponmudi Hill in the Western Ghats of India. *Journal of Herpetology*, *39*(3), 349–353. Available from https://doi.org/10.1670/133-04A.1.

Bossuyt, F., Meegaskumbura, M., Beenaerts, N., Gower, D. J., Pethiyagoda, R., Roelants, K., Mannaert, A., Wilkinson, M., Bahir, M. M., Manamendra-Arachchi, K., Ng, P. K. L., Schneider, C. J., Oommen, O. V., & Milinkovitch, M. C. (2004). Local endemism within the Western Ghats-Sri Lanka biodiversity hotspot. *Science (New York, N.Y.)*, *306*(5695), 479–481. Available from https://doi.org/10.1126/science.1100167.

Braga, R. M., Dourado, M. N., & Araújo, W. L. (2016). Microbial interactions: Ecology in a molecular perspective. *Brazilian Journal of Microbiology*, *47*, 86–98. Available from https://doi.org/10.1016/j.bjm.2016.10.005.

Briones, A., & Raskin, L. (2003). Diversity and dynamics of microbial communities in engineered environments and their implications for process stability. *Current Opinion in Biotechnology*, *14* (3), 270–276. Available from https://doi.org/10.1016/S0958-1669(03)00065-X.

Crain, B. J., & Fernández, M. (2020). Biogeographical analyses to facilitate targeted conservation of orchid diversity hotspots in Costa Rica. *Diversity and Distributions*, *26*(7), 853–866. Available from https://doi.org/10.1111/ddi.13062.

De Mandal, S., Zothansanga., Lalremsanga, H. T., & Kumar, N. S. (2015). Bacterial diversity of Murlen National Park located in Indo-Burman Biodiversity hotspot region: A metagenomic approach. *Genomics Data*, *5*, 25–26. Available from https://doi.org/10.1016/j.gdata.2015.04.025.

Deshpande, A. U., Apte, G. S., Bahulikar, R. A., Lagu, M. D., Kulkarni, B. G., Suresh, H. S., Singh, N. P., Rao, M. K. V., Gupta, V. S., Pant, A., & Ranjekar, P. K. (2001). Genetic diversity across natural populations of three montane plant species from the Western Ghats, India revealed by intersimple sequence repeats. *Molecular Ecology*, *10*(10), 2397–2408. Available from https://doi.org/10.1046/j.0962-1083.2001.01379.x.

Donachie, S. P., Hou, S., Lee, K. S., Riley, C. W., Pikina, A., Belisle, C., Kempe, S., Gregory, T. S., Bossuyt, A., Boerema, J., Liu, J., Freitas, T. A., Malahoff, A., & Alam, M. (2004). The Hawaiian Archipelago: A microbial diversity hotspot. *Microbial Ecology*, *48*(4), 509–520. Available from https://doi.org/10.1007/s00248-004-0217-1.

Garg, S., Suyesh, R., Sukesan, S., & Biju, S. D. (2017). Seven new species of Night Frogs (Anura, Nyctibatrachidae) from the Western Ghats Biodiversity Hotspot of India, with remarkably high diversity of diminutive forms. *PeerJ*, *2017*(2). Available from https://doi.org/10.7717/peerj.3007.

Giam, X., Koh, L. P., Tan, H. H., Miettinen, J., Tan, H. T. W., & Ng, P. K. L. (2012). Global extinctions of freshwater fishes follow peatland conversion in Sundaland. *Frontiers in Ecology and the Environment*, *10*(9), 465–470. Available from https://doi.org/10.1890/110182.

Goyari, S., Devi, S. S., Kalita, M. C., & Talukdar, N. C. (2014). Population, diversity, and characteristics of cellulolytic microorganisms from the Indo-Burma Biodiversity hotspot. *Springer Plus*, *3*(1), 1–9.

Hadjiliadis, N. (2012). Cytotoxic mutagenic and carcinogenic potential of heavy metals related to human environment. Springer Science & Business Media, *26*.

Handelsman, J. (2004). Metagenomics: Application of genomics to uncultured microorganisms. *Microbiology and Molecular Biology Reviews*, *68*(4), 669–685. Available from https://doi.org/10.1128/MMBR.68.4.669-685.2004.

Jose, R. C., Goyari, S., Louis, B., Waikhom, S. D., Handique, P. J., & Talukdar, N. C. (2016). Investigation on the biotrophic interaction of *Ustilago esculenta* on *Zizania latifolia* found in the Indo-Burma biodiversity hotspot. *Microbial Pathogenesis*, *98*, 6–15. Available from https://doi.org/10.1016/j.micpath.2016.06.021.

K., R., A., K., J., S., R., M., & R., K. (2012). Monsoon circulation interaction with Western Ghats orography under changing climate. *Theoretical and Applied Climatology*, 555–571. Available from https://doi.org/10.1007/s00704-012-0690-2.

Kneitel, J. M. (2018). Occupancy and environmental responses of habitat specialists and generalists depend on dispersal traits. *Ecosphere*, *9*(3). Available from https://doi.org/10.1002/ecs2.2143.

Kühl, M., Behrendt, L., Trampe, E., Qvortrup, K., Schreiber, U., Borisov, S. M., Klimant, I., & Larkum, A. W. D. (2012). Microenvironmental ecology of the chlorophyll b-containing symbiotic cyanobacterium Prochloron in the didemnid ascidian *Lissoclinum patella*. *Frontiers in Microbiology*, *3*. Available from https://doi.org/10.3389/fmicb.2012.00402.

Li, S. P., Wang, P., Chen, Y., Wilson, M. C., Yang, X., Ma, C., Lu, J., Chen, X. Y., Wu, J., Shu, W. S., & Jiang, L. (2020). Island biogeography of soil bacteria and fungi: Similar patterns, but different mechanisms. *ISME Journal*, *14*(7), 1886–1896. Available from https://doi.org/10.1038/s41396-020-0657-8.

Li, Y., Adams, J., Shi, Y., Wang, H., He, J. S., & Chu, H. (2017). Distinct soil microbial communities in habitats of differing soil water balance on the Tibetan Plateau. *Scientific Reports*, *7*. Available from https://doi.org/10.1038/srep46407.

Martínez-Alonso, M., Escolano, J., Montesinos, E., & Gaju, N. (2010). Diversity of the bacterial community in the surface soil of a pear orchard based on 16s rRNA gene analysis. *International Microbiology*, *13*(3), 123–134. Available from https://doi.org/10.2436/20.1501.01.117.

Martiny, J. B. H., Bohannan, B. J. M., Brown, J. H., Colwell, R. K., Fuhrman, J. A., Green, J. L., Horner-Devine, M. C., Kane, M., Krumins, J. A., Kuske, C. R., Morin, P. J., Naeem, S., Øvreås, L., Reysenbach, A. L., Smith, V. H., & Staley, J. T. (2006).). Microbial biogeography: Putting microorganisms on the map. *Nature Reviews Microbiology*, *4*(2), 102–112. Available from https://doi.org/10.1038/nrmicro1341.

Mishra, J., Singh, R., & Arora, N. K. (2017). Alleviation of heavy metal stress in plants and remediation of soil by rhizosphere microorganisms. *Frontiers in Microbiology*, *8*. Available from https://doi.org/10.3389/fmicb.2017.01706.

Monard, C., Gantner, S., Bertilsson, S., Hallin, S., & Stenlid, J. (2016). Habitat generalists and specialists in microbial communities across a terrestrial-freshwater gradient. *Scientific Reports*, *6*. Available from https://doi.org/10.1038/srep37719.

Nampoothiri, K. M., Ramkumar, B., & Pandey, A. (2013) Western Ghats of India: a rich source of microbial biodiversity.

Nannipieri, P., Ascher, J., Ceccherini, M. T., Landi, L., Pietramellara, G., & Renella, G. (2003). Microbial diversity and soil functions. *European Journal of Soil Science*, *54*(4), 655–670. Available from https://doi.org/10.1046/j.1351-0754.2003.0556.x.

Panda, A. K., Bisht, S. S., Kumar, N. S., & De Mandal, S. (2015). Investigations on microbial diversity of Jakrem hot spring, Meghalaya, India using cultivation-independent approach. *Genomics Data*, *4*, 156–157. Available from https://doi.org/10.1016/j.gdata.2015.04.016.

Pandit, M. K., Sodhi, N. S., Koh, L. P., Bhaskar, A., & Brook, B. W. (2007). Unreported yet massive deforestation driving loss of endemic biodiversity in Indian Himalaya. *Biodiversity and Conservation*, *16*(1), 153–163. Available from https://doi.org/10.1007/s10531-006-9038-5.

Powell, M. J. (1993). Looking at mycology with a Janus face: A glimpse at Chytridiomycetes active in the environment. *Mycologia*, *85*(1), 1–20. Available from https://doi.org/10.2307/3760471.

Prasad, V. K., Badarinath, K. V. S., & Eaturu, A. (2008). Effects of precipitation, temperature and topographic parameters on evergreen vegetation greenery in the Western Ghats, India. *International Journal of Climatology*, *28*(13), 1807–1819. Available from https://doi.org/10.1002/joc.1662.

Rajkhowa, D. J., Bhattacharyya, P. N., Sarma, A. K., & Mahanta, K. (2015). Diversity and distribution of earthworms in different soil habitats of Assam, North-East India, an Indo-Burma Biodiversity Hotspot. *Proceedings of the National Academy of Sciences India Section B - Biological Sciences, 85*(2), 389–396. Available from https://doi.org/10.1007/s40011-014-0380-1.

Rathour, R., Gupta, J., Kumar, M., Hiloidhari, M., Mehrotra, A. K., & Thakur, I. S. (2017). Metagenomic sequencing of microbial communities from brackish water of Pangong lake of the Northwest Indian Himalayas. *Genome Announcements, 5*(40). Available from https://doi.org/10.1128/genomeA.01029-17.

Ren, C., Zhang, W., Zhong, Z. K., Han, X., Yang, G., Feng, Y., & Ren, G. (2018). Differential responses of soil microbial biomass, diversity, and compositions to altitudinal gradients depend on plant and soil characteristics. *Science of the Total Environment, 610–611*, 750–758. Available from https://doi.org/10.1016/j.scitotenv.2017.08.110.

Rüber, L., Tan, H. H., & Britz, R. (2020). Snakehead (Teleostei: Channidae) diversity and the Eastern Himalaya biodiversity hotspot. *Journal of Zoological Systematics and Evolutionary Research, 58*(1), 356–386. Available from https://doi.org/10.1111/jzs.12324.

Selvan, P. S., Jebanesan, A., Divya, G., & Ramesh, V. (2015). Diversity of mosquitoes and larval breeding preference based on physico-chemical parameters in Western Ghats, Tamilnadu, India. *Asian Pacific Journal of Tropical Disease, 5*(1), S59–S66. Available from https://doi.org/10.1016/S2222-1808(15)60858-1.

Singh, J. S., Pandey, V. C., & Singh, D. P. (2011). Efficient soil microorganisms: A new dimension for sustainable agriculture and environmental development. *Agriculture. Ecosyst. Environ., 140*(3), 339–353.

Tali, B. A., Khuroo, A. A., Nawchoo, I. A., & Ganie, A. H. (2019). Prioritizing conservation of medicinal flora in the Himalayan biodiversity hotspot: An integrated ecological and socioeconomic approach. *Environmental Conservation, 46*(2), 147–154. Available from https://doi.org/10.1017/S0376892918000425.

Taylor, M. W., Radax, R., Steger, D., & Wagner, M. (2007). Sponge-associated microorganisms: evolution, ecology, and biotechnological potential. *Microbiol Mol Biol Rev, 71*, 295–347.

Teo, J. K. C., & Wong, C. M. V. L. (2014). Analyses of soil bacterial diversity of the Schirmacher Oasis, Antarctica. *Polar Biol, 37*(5), 631–640.

Tiwari, S., Sarangi, B. K., & Thul, S. (2016). Identification of arsenic resistant endophytic bacteria from Pteris vittata roots and characterization for arsenic remediation application. *J Environ Manag, 180*, 359–365.

Vareda, J. P., Valente, A. J. M., & Durães, L. (2019). Assessment of heavy metal pollution from anthropogenic activities and remediation strategies: A review. *Journal of Environmental Management, 246*, 101–118. Available from https://doi.org/10.1016/j.jenvman.2019.05.126.

Velmourougane, K. (2017). Shade trees improve soil biological and microbial diversity in coffee based system in Western Ghats of India. *Proceedings of the National Academy of Sciences India Section B - Biological Sciences, 87*(2), 489–497. Available from https://doi.org/10.1007/s40011-015-0598-6.

Verma, M., William, S. S., James, E. M. W., Kendall, R. J., James, R. A., Oscar, V., Frank, E. R., David, P. E., & Luis, R. C. (2020). Severe human pressures in the Sundaland biodiversity hotspot. *Conservation Science and Practice, 2*(3), e169.

Vibha, B., & Neelam, G. (2012). Importance of exploration of microbial biodiversity. *International Research Journal of Biological Sciences, 1*(3), 78–83.

Webster, N. S., Wilson, K. J., Blackall, L. L., & Hill, R. T. (2001). Phylogenetic diversity of bacteria associated with the marine sponge Rhopaloeides adorable. *Applied Environmental Microbiology, 67*, 434–444.

Yadav, A. N., Sachan, S. G., Verma, P., & Saxena, A. K. (2015). Prospecting cold deserts of north western Himalayas for microbial diversity and plant growth promoting attributes. *Journal of Bioscience and Bioengineering, 119*(6), 683–693. Available from https://doi.org/10.1016/j.jbiosc.2014.11.006.

Yang, Q., Li, Z., Lu, X., Duan, Q., Huang, L., & Bi, J. (2018). A review of soil heavy metal pollution from industrial and agricultural regions in China: Pollution and risk assessment. *Science of the Total Environment, 642*, 690–700. Available from https://doi.org/10.1016/j.scitotenv.2018.06.068.

Zhang, Q., Goberna, M., Liu, Y., Cui, M., Yang, H., Sun, Q., Insam, H., & Zhou, J. (2018). Competition and habitat filtering jointly explain phylogenetic structure of soil bacterial communities across elevational gradients. *Environmental Microbiology, 20*(7), 2386–2396. Available from https://doi.org/10.1111/1462-2920.14247.

Zheng, Y., & Gong, X. (2019). Niche differentiation rather than biogeography shapes the diversity and composition of microbiome of Cycas panzhihuaensis. *7*(1), 1–19.

CHAPTER

3

Microbial diversity and ecology of saline environments from India

Rebecca S. Thombre[1,2] *and Amitsinh V. Mangrola*[3]

[1]Blue Marble Space Institute of Science, Seattle, WA, United States, [2]Department of Biotechnology, Modern College of Arts, Science and Commerce, India, [3]Department of Biochemistry, Shri Alpesh. N. Patel P. G. Institute of Science & Research, Charotar Education Society, India

3.1 Introduction

The Earth's surface has the occurrence of different types of saltwater bodies. In general, freshwater contains less than 0.05% NaCl (salt) and saline habitats contain more than 0.05% NaCl. Brackish water contains 0.05%–3% NaCl and saline water contains 3%–5% NaCl. Water bodies that contain more than 5% NaCl and contain up to 28% NaCl are hypersaline and are called brines. Examples of such hypersaline environments in the world include the Dead Sea (Israel), Lake Magadi (Kenya), Mono Lake (California), Wadi Natrun Lake (Egypt), and Great Salt Lake (United States) (DasSarma, 2006). These hypersaline environments vary in their salt composition, mineral composition, the chemical composition which affects the microbial diversity of that system. Some of these saline water bodies are marine in nature and others fall in the "Inland water" category. Inland waters consist of lakes, rivers, reservoirs, and wetlands. Deserts are extreme terrestrial habitats on the Earth with high levels of salinity, low levels of nutrients, and low water retention capacity (Rewald et al., 2012). Examples of such deserts are the Atacama Desert, Thar Desert (India), and the Rann of Kutch (India).

Microbial Diversity and Ecology in Hotspots
DOI: https://doi.org/10.1016/B978-0-323-90148-2.00010-9

The hypersaline environments that originate from marine environments are called thalassohaline, for example, Solar salterns or salt ponds. Others that differ in their salt composition from the marine environments are called thalassohaline environments which are terminal lakes or inland water bodies whose salt composition is affected by geographical location and geology.

Saline ecosystems encompass salt mines, saline lakes, rock salt deposits, and solar salterns in caves and deserts (Thombre, Shinde, Oke, et al., 2016). Hypersaline lakes offer a diverse opportunity to the ecologists for the comprehensive study of halophiles from this harsh environment (Gunde-Cimerman et al., 2018). The superior metabolic adaptability of halophiles, their low nutritious needs, and their genetic tools of adaptation to severe conditions, such as high solar radiation, dehydration, high ionic strength, and nutrition deficiency make them a favorable member of extremophiles for drug discovery in the medical field (Charlesworth & Burns, 2015). Conventional microbiology uses culture-dependent techniques that can isolate only a minor proportion of the microbial population in the hypersaline habitats. The practice of culture-independent methods (metagenomics) compiled with high-throughput sequencing can help us gain a better insight of the microbial universe of salt lakes, salterns, and crystallizer lakes (Oren, 2015).

The salt-loving organisms that inhabit these saline systems are known as halophiles. These halophiles occur in all three domains, namely, Archaea, (*Haloarcula*), bacteria (*Halomonas*), and Eucaryotes (*Dunaliella salina*). Halophiles are classified on the basis of their optimal requirement of salt (NaCl) concentration. Extreme halophiles require more than 15% (15%–30% NaCl), moderate halophiles grow between 3% and 10% NaCl, slight halophiles grow between 1% and 3% NaCl. Most of the extreme halophiles are from domain archaea and are generally called "haloarchaea." Cultured haloarchaea have around 47 genera with 165 species in the family *Halobacteriaceae* (Oren, 2014). These extremophiles inhabit extreme inhospitable habitats on earth. To survive in such a harsh environment, halophiles produce many unique compounds, for example, glycine betaine, polyols, stress proteins, enzymes, vesicles, pigments (bacterioruberin and bacteriorhodopsin), and exopolysaccharides. These unique microbial products have enormous biotechnological potential and industrial applications. Halophiles also contribute to environmental management by potential application in bioremediation of saline or alkaline toxic wastes and wastewaters. Hence, it is imperative to study the microbial diversity of halophiles in saline habitats. In this chapter, we have presented an overview of the microbial diversity studies of saline lakes and saline habitats present in India.

3.2 Microbial diversity of saline lakes of India

3.2.1 Lonar Lake, Maharashtra

Lonar Lake is an extreme alkaline and saline lake in Western Maharashtra, India, formed by the accumulation of water in a crater created by the impact of a meteorite in basalt (Fig. 3.1A). The impact crater (latitude 19° 58′, longitude 76° 36′) was formed ~50 million years is bowl shaped with a diameter of 1.83 km and depth of 130 m (Thombre, 2012). The microbial diversity of Lonar Lake has been studied and it consists of extremophiles including alkaliphiles, alkali tolerant, halophiles, and haloalkaliphiles. The lake recently turned red overnight and the change of color is perused to be attributed to increase in salinity that may have favored growth of saline algae like *Dunaliella* salina or haloarchaea as observed in other saline red lakes such as Hillier (Australia) and Laguna de Pena Hueca (Spain). The microbial diversity of Lonar Lake has been studied (Joshi et al., 2008; Paul et al., 2016) and extremophiles such as *Rhodobaca bogoriensis, Halomonas campisalis, Marinobacter alkaliphilus, Roseinatronobacter monicus, Alkalimonas delamerensis,* and other organisms such as *Planococcus, Dietzia, Vagococcus, Exigobacterium, Alcaligenes, Stenotrophomonas,* and *Paracoccus* have been isolated from the lake.

The alkaliphiles from Lonar Lake have also been sourced for extremozymes and bioactives such as glycosyl transferase, proteases, and poly hydroxyl alkanoates that have industrial applications (Joshi et al., 2008; Thombre, 2012). This highly alkaline lake located in the Deccan Volcanic Province is the only meteorite impact structure in basalt on earth.

3.2.2 Chilika Lake, Odisha

Chilika Lake (19° 28′–19° 54′ N; 85° 06′–85° 35′ E) is a brackish water lagoon present in Odisha, India. The lake has a depth of 2 m, is 906 km^2 in summer and is 1065 km^2 in the monsoon season. This lagoon on the

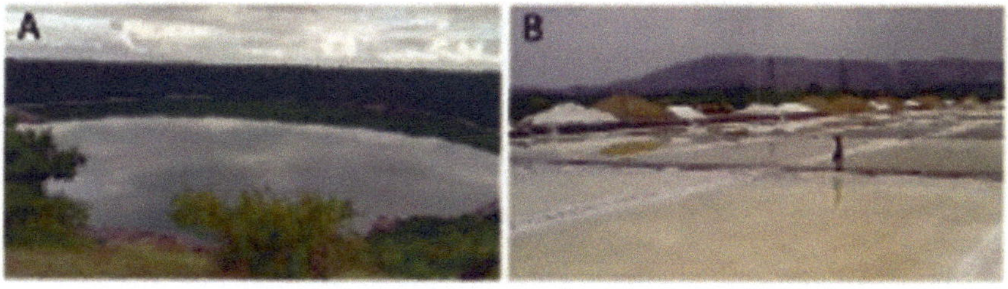

FIGURE 3.1 (A) Lonar Lake, India; (B) salt pans from Mumbai (Thombre et al., 2020).

Eastern coast of India is considered the biggest saltwater lake in Asia. Chilika Lake is a biodiversity hotspot and is considered a Wetlands of International importance. This water body is a combination of river, marine, and estuarine habitats hence, the microbial diversity of this lake is unique. Culture-independent investigation of the microbial diversity of this lake has revealed the presence of microbial phyla including *Firmicutes*, Proteobacteria, *Chlamydiae*, *Spirochetes*, *Planctomycetes*, and *Tenericutes Shewanella chilikensis* sp. nov., a novel alkaliphilic Proteobacterium, *Streptomyces chilikensis* sp. nov., a halophilic actinomycete and *Streptomyces barkulensis* sp. nov. have been isolated from this lake (Pramanik et al., 2015).

3.2.3 Pulicat Lake

The brackish saline lake Pulicat is present at the Coromandel coast bordering Tamil Nadu and Andhra Pradesh with outflow in Bay of Bengal. The lake harbors diverse flora and fauna and a salt manufacturing unit. Sahay et al. (2011) have studied the microbial diversity of Pulicat Lake using culture-dependent approach and 16S rDNA-based analyses. The study yielded 198 morphotypes and 29 strains that were allocated into *Firmicutes* and Proteobacteria. The bacteria were related to family *Bacillaceae*, such as *Virgibacillus*, *Rummelibacillus*, *Bacillus*, *Alkalibacillus*, *Halobacillus*, and *Salimicrobium halophilum*. The Proteobacteria group showed the presence of *Halomonas*, for example, *Halomonas salina*, *H. shengliensis*, *H. salifodinae*, *H. pacifica*, *H. aquamarina*, and *H. halophila*. Many strains produced enzymes such as amylase, protease, xylanase, and cellulase.

3.2.4 Pangong Tso Lake, Ladakh

The Pangong Tso (33°43′04.59″N:78°53′48.48″E) is a saline water lake situated at the height of 4350 m, in Ladakh (Indian Tibet). The salinity of this lake is ~11.02 g/L and the lake is frozen in winters. Rathour et al. (2017, 2020) have studied the taxonomic diversity of the Pangong Lake sediments using Shotgun metagenomic sequencing. Their study revealed the presence of psychrophilic and psychrotolerant organisms. Most of the organisms belonged to the phyla Proteobacteria. Besides, the other phyla were *Halomonas*, *Methylophaga*, and *Marinobacter*. Rathour et al. (2017) have isolated Proteobacteria and *Bacteroides* from the microbial diversity of Pangong Lake using metagenomic sequencing of lake water. The functional studies revealed the presence of enzymatic pathways like nitrogen and methane metabolism, benzoate, and xylene degradation and sulfur reduction. Besides, many other genes for stress response to pH, salt, and cold temperature were also observed.

3.2.5 Lunsu and Gumma, Himachal Pradesh

In Himachal Pradesh, there are two prominent saltwater bodies, one in Lunsu and the other in Gumma (Gupta et al., 2015). Lunsu is a natural saltwater body in the Kangra district, whereas Gumma is located in the Mandi district in Himachal (Gupta et al., 2015). The Lunsu water body has its origin in the rocky mountains and is located in the foothills of the Himalayan range. Gupta et al. (2015) have isolated halophiles from Lunsu Lake and studied them using random amplification of polymorphic DNA (RAPD) and sequencing. Sequence analysis confirmed the presence of halophiles such as *Halobacillus trueperi, Halomonas venusta, Shewanella algae,* and *Marinomonas* sp. There are very few studies that have been conducted to characterize microbial diversity of the saltwater bodies of Himachal and more studies are warranted in this direction.

3.2.6 Sambhar Lake, Rajasthan

Sambhar Lake is the largest haloalkaline lake in Rajasthan, India, and few reports are available on microbial diversity from the same area (Sahay et al., 2012) (Fig. 3.2). Prior research studies revealed the isolation of gram-negative halophilic strains (SL-1 to SL-8), from Sambhar Salt Lake, capable of growing at salt supplies between 3% and 15%. These strains were screened for the production of extracellular enzymes such as amylases, proteases, and lipases (Gaur & Mohan, 2014). The research has then been carried out to isolate culturable, aerobic, and heterotrophic *Bacillus* and *Bacillus*-derived genera from the Sambhar Lake (Yadav et al., 2014). Previously, the culture-dependent studies were attempted to isolate *Piscibacillus* sp. from Sambhar Lake and was used to evaluate anticancer activity against breast cancer cell lines (Neelam et al., 2019). Singh et al. (2019) have documented the antimicrobial activity of the novel halophiles, *Paenibacilllus* sp. SMB1 from Sambhar Lake (Singh et al., 2019). In most recent times, 12 bacterial genera and 13 archaea have been isolated from Sambhar Lake with *Alkalibacillus* and *Natronococcus* as major microbiota (Kajale et al., 2020).

3.2.7 Pachpadra Salt Lake, Rajasthan

Pachpadra Salt Lake is a unique untapped hypersaline habitat in Rajasthan, India (Fig. 3.2). An exhaustive study on isolation and identification of halocin-producing bacteria from this lake has been carried out using a culture-dependent method (Kumar et al., 2016). The identification and characterization results revealed the presence of the *Haloferax larsenii* strain HA1 with their capability to survive at 15% salt (w/v), 42°C temperature, and pH 7.2. These properties of halocin HA1 isolates may be useful for the long-term preservation of salted foods (Kumar et al., 2016). Recently, halocin-producing archeal strain *Haloferax larsenii* HA4 was isolated from the

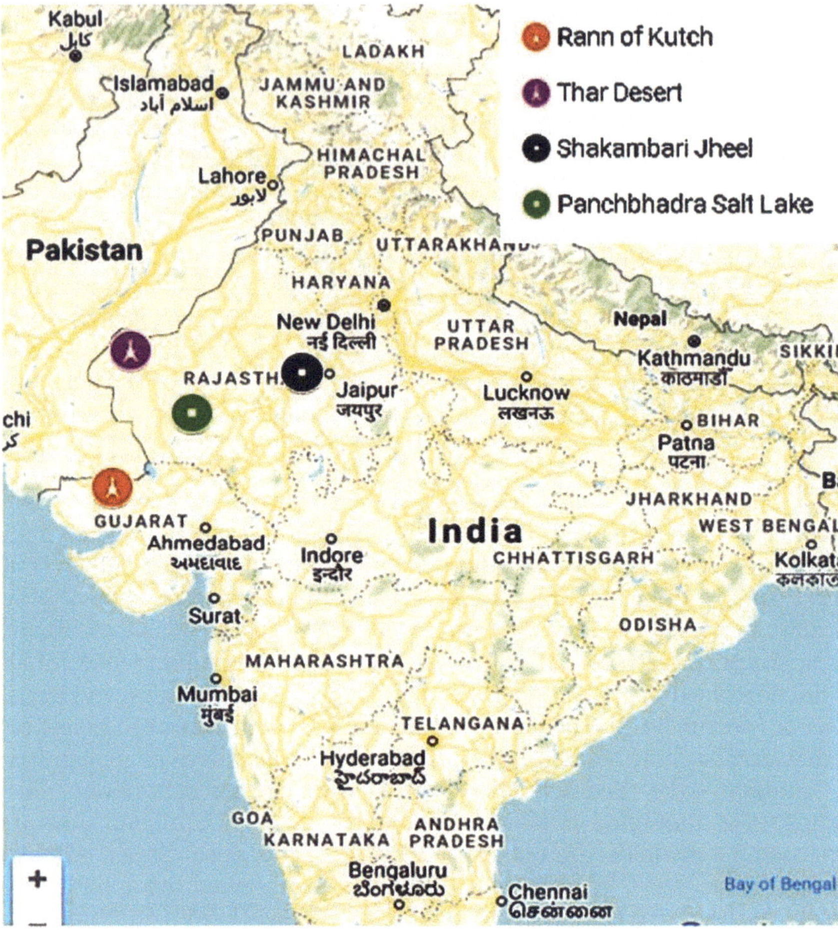

FIGURE 3.2 Geological distribution of hypersaline environments in India. *Note:* The map was constructed using the My Maps feature of Google Maps.

Pachpadra Salt Lake using a culture-dependent method and characterized for their stability, sensitivity, and antimicrobial activity (Kaur & Tiwari, 2020). The geological distribution of hypersaline environments and distinctive hypersaline spots in India are depicted in Fig. 3.2.

3.3 Microbial diversity of Indian deserts

3.3.1 Thar Desert

Thar Desert is a unique hypersaline territory, stretching over an area of about 320,000 km^2 of which about 60% lies in the Rajasthan state of

India (Gehlot et al., 2014) (Fig. 3.2). Despite the possibility of searching the novel microbes with huge commercial and industrial demands, limited reports are available on the microbial diversity of this terrain. The researchers have investigated the unculturable microbial diversity in the Thar Desert using denaturing gradient gel electrophoresis (DGGE) (Kumar et al., 2007). Sharma et al. (2013) have reported the isolation of osmotolerant bacteria from sand dunes of the Thar Desert. The bacterial species isolated were *Bacillus licheniformis* and *Bacillus subtilis* based on the 16S rRNA gene sequencing method (Sharma et al., 2013). This research presented the insight that sand dunes may be a rich source of osmotolerant bacteria and can be explored for plant growth promotion activity in the agriculture sector (Sharma et al., 2013). Rao et al. (2016) have performed microbial diversity analysis from the monsoon desert of Thar. The microbial community analysis of this hotspot showed that Actinobacteria and alpha-Proteobacteria can survive in semiarid soils, whereas *Chloroflexi* and alpha-Proteobacteria thrived in the arid soils of Thar Desert. Some fungal ascomycete and Cyanobacteria have also been observed in dunes of the Thar Desert (Rao et al., 2016). Sharma (2016) has isolated the actinomycetes such as *Streptomyces variabilis* and *Streptomyces rubiginosus* from the Thar Desert, and these isolates were further screened for the production of biosurfactants and enzymes (Sharma, 2016). The culture-dependent studies from diverse arid areas of the Thar Desert demonstrated the prevalence of thermophilic bacteria with the potential of extracellular enzymes such as amylases and lipases (Gaur et al., 2017). Later studies also brought forth the presence of thermophilic species such as *Geobacillus thermodenitificans*, *Bacillus paralicheniformis*, and *Aeribacillus pallidus* from the dunes of the Thar Desert. These bacterial species have been further explored for several biotechnological applications such as the production of enzymes like protease, lipase, asparaginase, and amylase (Neelam et al., 2018). Likewise, the thermophilic actinomycetes have been isolated from the Thar Desert and were exploited for the production of various enzymes such as amylases, proteases, catalases, and lipases (Agarwal et al., 2019). Presently, halophilic bacteria were isolated from the Thar Desert and were used for the qualitative screenings for the lipolytic enzymes (Verma et al., 2020).

3.3.2 Rann of Kutch

Gujarat state is situated in the Western region of India and it has the largest shoreline in the country. The Kutch Peninsula is located in Gujarat and it builds the westernmost part of India comprising Kutch deserts. The Rann of Kutch is partitioned into the Little Rann and Great Rann (Fig. 3.2). The Indian Wild Ass Sanctuary is situated in Little Rann of Kutch, Gujarat and is a unique ecosystem that includes abundant faunal diversity (Patel et al., 2015). The Little Rann of Kutch extends northeast from the Gulf of Kutch occupying an area of around 5100 km^2 (Gupta & Ansari., 2014). Moreover, the Kutch desert has a typical and myriad demography from other deserts of the world due to its vicinity to the ocean and low-lying areas, through which ocean water shows up into the enormous region of this habitat. In this way, the Kutch desert is a mixture of saline, marshy, and coastal habitats with saline water and soil systems. This property makes the desert a special territory from the biodiversity perspective since it harbors its own and unique environment (Bhatt & Singh, 2017; Gupta & Ansari, 2014). Numerous expeditions have been conducted to study the microbial diversity of the saline desert of Kutch, Gujarat, India using culture-dependent and culture-independent (metagenomic) approaches. In one study, researchers unveiled the microbial diversity in the Great Rann of Kutch using a metagenomic method (Pandit et al., 2014). This investigation revealed the presence of halophilic phyla such as Proteobacteria (18.6%–47.7%), Euryarchaeota (7.9%–40.1%), *Bacteroidetes* (9.1%–18.7%), *Firmicutes* (5.8%–8.4%), Actinobacteria (2.1%–5.8%), and Cyanobacteria (2%–4.2%) (Pandit et al., 2014). The microbial diversity of the Great Rann of Kutch has also been investigated using a metagenomic approach and a study showed more than 60% of archaeal genera were found in this environment (Pandit et al., 2015). The microbial diversity study of the Little Rann of Kutch has been analyzed using a metagenomic method that leads to the detection of 43 diverse halophilic phyla with leading *Marinobacter* (48.7%) and *Halobacterium* (4.6%) genus (Patel et al., 2015). Recently, metagenomic studies have been performed to divulge lichen species in the Greater Rann of Kutch using an ion proton sequencing method (Puvar et al., 2020). In recent studies, the microbial diversity of the Kutch desert has also been explored using a culture-dependent approach. This investigation showed the prevalence of 52 bacterial, 17 archaea, and 2 fungal strains with their multienzyme production capacity and antimicrobial activity (Josh & Trivedi, 2020).

3.3.3 Cold deserts of Leh Ladakh

Leh Ladakh is a cold desert and microbial diversity has been studied in various regions of Pangong Lake, Chumathang Lake, Confluence of Indus and Zanskar River, and Khardungla Pass (Yadav et al., 2015). The microbial diversity studies in these systems showed genera of *Pseudomonas, Aurantimonas, Citricoccus, Cellulosimicrobium, Brevundimonas, Desemzia, Psychrobacter, Sporosarcina, Sinobaca, Stenotrophomonas,* and *Sanguibacter* (Gupta et al., 2016; Singh et al., 2015; Yadav et al., 2015). Numerous psychrophiles have been isolated from the microbial communities in the Himalayan glaciers. Microorganisms such as *Exiguobacterium indicum* sp. nov., have been reported from the Hamta glacier, Himalayas and *Paenibacillus glacialis* sp. nov., *Bacillus cecembensis* sp. nov., and *Cryobacterium pindariense* sp. nov., have been reported from the Kafni and Pindari glaciers of the Indian Himalayas, respectively (Chaturvedi & Shivaji, 2006; Pradhan et al., 2010; Kishore et al., 2010; Reddy et al., 2008, 2010).

3.4 Microbial diversity of Indian solar salterns and halite deposits

Solar salterns are thalassohaline marine environments that have a discontinuous salinity gradient and are chiefly used for commercially salt (NaCl) production. Because of the high concentration of salt at various levels of crystallization (between 3% and 30%), different groups of microorganisms thrive in salterns at varying levels of salinity. Most organisms are halophilic archaea and other groups such as halotolerant bacteria, halophilic fungi, and halophilic algae also comprise the microbial diversity of salterns. There are many saline habitats, deposits, and lakes that are used for the production of salt in India. These sites are located in Rajasthan, coastal of Gujarat, Maharashtra, Goa, Orissa, West Bengal, Tamil Nadu, and Andhra Pradesh (Thombre, Joshi, & Oke, 2016; Thombre et al., 2019, 2020). It is reported that there are approximately 10,000 salt-production sites and units present all over India. An example of the largest marine chemical units of salt production in Southeast Asia is at the plant at Tata Chemicals Limited (TCL) in Mithapur, West coast of Gujarat. Other salterns in Gujarat are located in Kutch and Saurashtra at Kandla, Jamnagar, Maliya, Mithapur, Porbandar, and Bhavnagar. Halophiles such as *Geomicrobium halophilum*, *Halobacillus kuroshimensis*, *Virgibacillus halodenitrificans*, and *Chromohalobacter israelensis* have been reported from salterns in Gujarat (Table 3.1).

In Maharashtra, a number of salterns are located along the coastal areas of Mumbai and Konkan (Fig. 3.1B). The microbial diversity of salterns in Mumbai has been studied by Thombre, Joshi, and Oke (2016). Thombre, Joshi, and Oke (2016) have reported extremophilic halophilic archaea such as *Haloarcula argentinensis*, *Haloarcula marismortui*, *Haloferax alexandrines*, and *Haloferax prahovense* from salterns of Mumbai, India. Other salterns have been explored in Rajasthan, Tuticorin, Marakkanam, South India, Goa, and West Bengal. Thombre, Bhalerao, Shinde, et al. (2017) have studied microbial diversity of rocks trapped in salt deposits and Arabian sea. *Natronococcus jeotgali* RR17 and *Halovivax asiaticus* RT-5 have been isolated from a laterite rock and is an example of isolation of organisms from rock microbial community. These rocks submerged in the Arabian sea have been collected from Vengurla port, in India (Thombre, Bhalerao, Shinde, et al., 2017; Thombre, Joshi, and Oke, 2016). Similarly, Manikandan et al. (2009) have reported members of genera *Haloferax*, *Halorubrum*, *Haloarcula*, *Halobacterium*, and

TABLE 3.1 Representative extremophiles isolated from hypersaline environments in India.

No.	Hypersaline environment	Location	Representative genera	Reference(s)
1	Solar salterns	Mumbai	*Haloarcula argentinensis, Haloarcula marismortui, Haloferax alexandrines,* and *Haloferax prahovense*	Thombre, Joshi, and Oke (2016); Thombre et al. (2014, 2019)
2	Solar salterns	Goa	*Halococcus, Halorubrum* *Haloarcula* and *Haloferax*	Mani, Salgaonkar, and Braganca (2012); Mani, Salgaonkar, Das, et al. (2012)
3	Solar salterns	Tamil Nadu	*Haloferax, Halorubrum,* *Haloarcula, Halobacterium,* and *Halogeometricum*	Vijayanand et al. (2012); Manikandan et al. (2009)
4	Seawater, solar salterns, and saline soil	Gujarat	*Geomicrobium halophilum* *Halobacillus kuroshimensis* *Virgibacillus halodenitrificans* *Chromohalobacter israelensis*	Dave and Desai (2006); Kumar et al. (2012)
5	Sambar Salt Lake	Rajasthan	*Natronobacterium* sp. *Geomicrobium halophilum*	Upasani and Desai (1990)
6	Pulicat Lake	Andhra Pradesh	*Bacillus, Virgibacillus, Rummelibacillus, Alkalibacillus,* and *Halobacillus*	Sahay et al. (2011)
7	Chilika Lake	Orissa	*Shewanella chilikensis* sp. nov. *Streptomyces chilikensis* sp. nov.	Pramanik et al. (2015)
8	Lunsu Lake	Himachal Pradesh	*Halobacillus trueperi, Shewanella algae, Halomonas venusta, Marinomonas* sp. SS8	Gupta et al. (2015)
9	Lonar Lake	Maharashtra	*Alkalimonas delamerensis, Rhodobaca bogoriensis, Halomonas campisalis, Marinobacter excellens, Marinobacter alkaliphilus, Roseinatronobacter monicus*	Joshi et al. (2008)
10	Pangong Lake	Ladakh	*Halomonas, Methylophaga,* and *Marinobacter*	Rathour et al. (2017, 2020)

Halogeometricum from salterns of Tamil Nadu using cultural methods, culture-independent methods, polymerase chain reaction, and 16S rRNA molecular phylogenetic approaches.

The representative halophiles isolated from these salterns have been presented in Table 3.1.

3.5 Conclusion and future outlook

It is known that microbial life can survive and thrives in extremely harsh conditions. This chapter summarizes the various saline lakes and ecosystems in India and presents an overview of microbial diversity studies that have been conducted on such saline systems. Some of the saline systems in India have been studied well, however there are many other saline lakes and deposits that still have not been explored for its microbial diversity. More studies are warranted in this direction of microbial diversity of saline to ecosystems in India to understand the interactions and survival mechanisms of these microbial communities, to characterize the enzymes and bioactives for potential industrial application and to explore the genetic factors responsible for adaption in these environments.

Acknowledgment

RT thanks Prof. Nigel Mason, University of Kent, United Kingdom, and Principal, Modern College of Arts, Science and Commerce (Autonomous), Pune, India, for their support.

References

Agarwal, A., Mathur, N., Singh, A., Sharma, J., & Yadav, D. (2019). Study of actinomycetes isolated from the Thar desert, India for their potential to produce extracellular enzymes. *Journal of Pharmaceutical, Chemical and Biological Sciences, 7*(1), 28–35.

Bhatt, H. B., & Singh, S. P. (2017). *Phylogenetic and phenogram based diversity of haloalkaliphilic bacteria from the saline desert. Microbial biotechnology—Technological challenges and developmental trends* (pp. 373–386). New York: Apple Academic Press Chapter 19. Available from https://doi.org/10.1201/b19978.

Charlesworth, J. C., & Burns, B. P. (2015). Untapped resources: Biotechnological potential of peptides and secondary metabolites in archaea. *Archaea (Vancouver, BC), 2015*, 1–7. Available from https://doi.org/10.1155/2015/282035.

Chaturvedi, P., & Shivaji, S. (2006). *Exiguobacterium indicum* sp. nov., a psychrophilic bacterium from the Hamta glacier of the Himalayan mountain ranges of India. *International Journal of Systematic and Evolutionary Microbiology, 56*, 2765–2770.

DasSarma, S. (2006). Extreme halophiles are models for astrobiology. *Microbe, 1*(3), 120–126.

Dave, S. R., & Desai, H. B. (2006). Microbial diversity at marine saltern near Bhavnagar, Gujarat, India. *Current Science, 90*, 497–500.

Gaur, A., & Mohan, D. (2014). Diversity of moderate halophilic bacteria from Sambhar Salt Lake. *International Journal of Current Microbiology and Applied Sciences, 3*(7), 518–525.

Gaur, D., Jain, P. K., & Bajpai, V. (2017). Production of extracellular α-amylase by thermophilic *Bacillus* sp. isolated from arid and semi-arid region of Rajasthan, India. *Journal of Microbiology and Biotechnology, 2*(5), 675–684.

Gehlot, P., Sharma, R., & Sharma, K. (2014). Diversity of wild mushroom flora from Indian Thar desert. In *Proceedings of the 8th international conference on mushroom biology and mushroom products (ICMBMP8)* (pp. 92–97).

Gunde-Cimerman, N., Plemenitaš, A., & Oren, A. (2018). Strategies of adaptation of microorganisms of the three domains of life to high salt concentrations. *FEMS Microbiology Reviews, 42*(3), 353–375. Available from https://doi.org/10.1093/femsre/fuy009.

Gupta, P., Manjula, A., Rajendhran, J., Gunasekaran, P., & Vakhlu, J. (2016). Comparison of metagenomic DNA extraction methods for soil sediments of high elevation Puga hot spring in Ladakh, India to explore bacterial diversity. *Geomicrobiology Journal, 34*(4), 289–299. Available from https://doi.org/10.1080/01490451.2015.1128995.

Gupta, S., Sharma, P., Dev, K., Srivastava, M., & Sourirajan, A. (2015). A diverse group of halophilic bacteria exist in Lunsu, a natural salt water body of Himachal Pradesh, India. *Springer Plus, 4*, 274. Available from https://doi.org/10.1186/s40064-015-1028-1.

Gupta, V., & Ansari, A. A. (2014). Geomorphic portrait of the Little Rann of Kutch. *Arabian Journal of Geosciences, 7*(2), 527–536. Available from https://doi.org/10.1007/s12517-012-0743-y.

Josh, C. M., & Trivedi, N. S. (2020). Isolation, identification and characterization of microorganisms isolated from unexplored saline regions of Kutch, Gujarat, India. *Acta Scientific Microbiology, 3*(7), 3–9.

Joshi, A. A., Kanekar, P. P., Kelkar, A. S., Shouche, Y. S., Vani, A. A., Borgave, S. B., et al. (2008). Cultivable bacterial diversity of alkaline Lonar Lake, India. *Microbial Ecology, 55*, 163–172. Available from https://doi.org/10.1007/s00248-007-9264-8.

Kajale, S., Deshpande, N., Shouche, Y., & Sharma, A. (2020). Cultivation of diverse microorganisms from hypersaline lake and impact of delay in sample processing on cell viability. *Current Microbiology, 77*, 716–721. Available from https://doi.org/10.1007/s00284-019-01857-8.

Kaur, R., & Tiwari, S. K. (2020). Identification and characterization of a halocin-producing haloarchaeon isolated from Pachpadra Salt Lake. *Letters in Applied Microbiology, 71*, 620–626. Available from https://doi.org/10.1111/lam.13377, Epub ahead of print.

Kishore, K. H., Begum, Z., Pathan, A. A. K., & Shivaji, S. (2010). Paenibacillus glacialis sp. nov., isolated from the Kafni glacier of the Himalayas, India. *International Journal of Systematic and Evolutionary Microbiology, 60*, 1909–1913.

Kumar, G. R., Kumar, N. V., Mohan, M. K., Sasmal, D., & Ghosh, P. (2007). Extraction of bulk DNA from Thar desert soils for optimization of PCR-DGGE based microbial community analysis. *Electronic Journal of Biotechnology, 10*(3), 400–408.

Kumar, S., Karan, R., Kapoor, S., Singh, S. P., & Khare, S. K. (2012). Screening and isolation of halophilic bacteria producing industrially important enzymes. *Brazilian Journal of Microbiology, 43*(4), 1595–1603. Available from https://doi.org/10.1590/S1517-83822012000400044.

Kumar, V., Saxena, J., & Tiwari, S. K. (2016). Description of a halocin - producing Haloferax larsenii HA1 isolated from Pachpadra Salt Lake in Rajasthan. *Archives of Microbiology, 198*(2), 181–192. Available from https://doi.org/10.1007/s00203-015-1175-3.

Mani, K., Salgaonkar, B. B., & Braganca, J. M. (2012). Culturable halophilic archaea at the initial and crystallization stages of salt production in a natural solar saltern of Goa, India. *Aquatic Biosystems, 8*(1), 15. Available from https://doi.org/10.1186/2046-9063-8-15.

Mani, K., Salgaonkar, B. B., Das, D., & Bragança, J. M. (2012). Community solar salt production in Goa, India. *Aquatic Biosystems, 8*(1), 30. Available from https://doi.org/10.1186/2046-9063-8-30.

Manikandan, M., Kannan, V., & Pašić, L. (2009). Diversity of microorganisms in solar salterns of Tamil Nadu, India. *World Journal of Microbiology and Biotechnology, 25*, 1007–1017. Available from https://doi.org/10.1007/s11274-009-9980-y.

Neelam, D. K., Agrawal, A., Tomer, A. K., Bandyopadhayaya, S., Sharma, A., Jagannadham, M. V., Mandal, C. C., & Dadheech, P. K. (2019). A *Piscibacillus* sp. isolated from a soda lake exhibits anticancer activity against breast cancer. *Microorganisms, 7*(34), 1–16. Available from https://doi.org/10.3390/microorganisms7020034.

Neelam, D. K., Agrawal, A., Tomer, A. K., & Dadheech, P. K. (2018). Characterization, phylogenetic analysis and potential applications of heterotrophic bacteria inhabit sand dunes of Thar desert, India. *Journal of Pure and Applied Microbiology, 12*(4), 1887–1898. Available from https://doi.org/10.22207/JPAM.12.4.24.

Oren, A. (2014). Halophilic archaea on Earth and in space: Growth and survival under extreme conditions. *Philosophical Transactions of the Royal Society A: Mathematical, Physical and Engineering Sciences, 372*, 1–12. Available from https://doi.org/10.1098/rsta.2014.0194.

Oren, A. (2015). Halophilic microbial communities and their environments. *Current Opinion in Biotechnology, 33*, 119–124. Available from https://doi.org/10.1016/j.copbio.2015.02.005.

Pandit, A. S., Joshi, M. N., Bhargava, P., Ayachit, G. N., Shaikh, I. M., Saiyed, Z. M., Saxena, A. K., & Bagatharia, S. B. (2014). Metagenomes from the saline desert of Kutch. *Genome Announcements, 2*(3), 1. Available from https://doi.org/10.1128/genomeA.00439-14.

Pandit, A. S., Joshi, M. N., Bhargava, P., Shaikh, I., Ayachit, G. N., Raj, S. R., Saxena, A. K., & Bagatharia, S. B. (2015). A snapshot of microbial communities from the Kutch: One of the largest salt deserts in the world. *Extremophiles: Life Under Extreme Conditions, 19*, 973–987. Available from https://doi.org/10.1007/s00792-015-0772-z.

Patel, R., Mevada, V., Prajapati, D., Dudhagara, P., Koringa, P., & Joshi, C. G. (2015). Metagenomic sequence of saline desert microbiota from wild ass sanctuary, Little Rann of Kutch, Gujarat, India. *Genomics Data, 3*, 137–139. Available from https://doi.org/10.1016/j.gdata.2015.01.003.

Paul, D., Kumbhare, S. V., Mhatre, S. S., Chowdhury, S. P., Shetty, S. A., Marathe, N. P., Bhute, S., & Shouche, Y. S. (2016). Exploration of microbial diversity and community structure of Lonar Lake: The only hypersaline meteorite crater lake within basalt rock. *Frontiers in Microbiology, 6*, 1553. Available from https://doi.org/10.3389/fmicb.2015.01553.

Pradhan, S., Srinivas, T. N. R., Pindi, P. K., Kishore, K. H., Begum, Z., Singh, P. K., Singh, A. K., Pratibha, M. S., Yasala, A. K., Reddy, G. S., & Shivaji, S. (2010). Bacterial biodiversity from Roopkund glacier, Himalayan mountain ranges, India. *Extremophiles: Life Under Extreme Conditions, 14*, 377–395.

Pramanik, A., Basak, P., Banerjee, S., Sengupta, S., Chattopadhyay, D., & Bhattacharyya, M. (2015). Pyrosequencing based profiling of the bacterial community in the Chilika Lake, the largest lagoon of India. *Genomics Data, 4*, 112–114. Available from https://doi.org/10.1016/j.gdata.2015.04.003.

Puvar, A., Mootapally, C., Joshi, C., & Joshi, M. (2020). Metagenomic dataset on lichen *Dirinaria* sp. from the Great Rann of Kutch and tropical moist deciduous Dang forest of Gujarat. *Data in Brief, 30*, 1–8. Available from https://doi.org/10.1016/j.dib.2020.105504.

Rao, S., Chan, Y., Bugler-Lacap, D. C., Bhatnagar, A., Bhatnagar, M., & Pointing, S. B. (2016). Microbial diversity in soil, sand dune and rock substrates of the Thar monsoon desert, India. *Indian Journal of Microbiology, 56*(1), 35–45. Available from https://doi.org/10.1007/s12088-015-0549-1.

Rathour, R., Gupta, J., Kumar, M., Hiloidhari, M., Mehrotra, A. K., & Thakur, I. S. (2017). Metagenomic sequencing of microbial communities from brackish water of Pangong Lake of the Northwest Indian Himalayas. *Genome Announcements, 5*(40), e01029-17. Available from https://doi.org/10.1128/genomeA.01029-17.

Rathour, R., Gupta, J., Mishra, A., Rajeev, A. C., Dupont, C. L., & Thakur, I. S. (2020). A comparative metagenomic study reveals microbial diversity and their role in the biogeochemical cycling of Pangong lake. *Science of the Total Environment, 731*, 1–14. Available from https://doi.org/10.1016/j.scitotenv.2020.139074.

Reddy, G. S. N., Pradhan, S., Manorama, R., & Shivaji, S. (2010). Cryobacterium pindariense sp. nov., a psychrophilic bacterium from a Himalayan glacier. *International Journal of Systematic and Evolutionary Microbiology, 60*, 866–870.

Reddy, G. S. N., Uttam, A., & Shivaji, S. (2008). Bacillus cecembensis sp. nov., isolated from the Pindari glacier of the Indian Himalayas. *International Journal of Systematic and Evolutionary Microbiology, 58*, 2330–2335.

Rewald, B., Eppel, A., Shelef, O., Hill, A., Degu, A., Friedjung, A., & Rachmilevitch, S. (2012). Hot desert environments. In E. Bell (Ed.), Life at extremes—Environments, organisms, and strategies for survival (pp. 196–218). Wallingford, UK: CABI Publishing.

Sahay, H., Mahfooz, S., Singh, A. K., Singh, S., Kaushik, R., Saxena, A. K., & Arora, D. K. (2012). Exploration and characterization of agriculturally and industrially important haloalkaliphilic bacteria from environmental samples of hypersaline Sambhar Lake, India. *World Journal of Microbiology & Biotechnology, 28*, 3207–3217. Available from https://doi.org/10.1007/s11274-012-1131-1.

Sahay, H., Singh, S., Kaushik, R., Saxena, A. K., & Arora, D. K. (2011). Characterization of halophilic bacteria from environmental samples from the brackish water of Pulicat Lake, India. *Biologia (Lahore, Pakistan), 66*(5), 741–747.

Sharma, R. (2016). Isolation and characterization of thermophilic actinomycetes with extracellular enzyme and bio-surfactant production potential from Thar desert, India. *International Journal of Microbiology Research, 8*(4), 743–746.

Sharma, R., Manda, R., Gupta, S., Kumar, S., & Kumar, V. (2013). Isolation and characterization of osmotolerant bacteria from Thar desert of Western Rajasthan (India). *International Journal of Tropical Biology, 61*(4), 1551–1562.

Singh, H., Kaur, M., Jangra, M., Mishra, S., Nandanwar, H., & Pinnaka, A.K. (2019). Antimicrobial properties of the novel bacterial isolate Paenibacilllus sp. SMB1from a halo-alkaline lake in India. Scientific Reports. USA: Springer. https://doi.org/10.1038/s41598-019-47879-x.

Singh, R., Chopra, C., & Kumar, V. (2015). Purification and characterization of CHpro1, a thermotolerant, alkali-stable and oxidation-resisting protease of Chumathang hot spring. *Science Bulletin, 60*, 1252–1260. Available from https://doi.org/10.1007/s11434-015-0834-8.

Thombre, R. (2012). *Studies on production of cyclodextrin glycosyl transferase (CGTase) from alkaliphilic bacteria* (pp. 23–25). Ph.D. Thesis. Savitribai Phule Pune University, India. http://hdl.handle.net/10603/77945.

Thombre, R., Bhalerao, A., Shinde, V., Dhar, S., & Shouche, Y. (2017). Response of haloalkaliphilic archaeon natronococcus Jeotgali RR17 to hypergravity. *Microgravity Science and Technology, 29*(3), 191–200. Available from https://doi.org/10.1007/s12217-017-9538-9.

Thombre, R., Joshi, V., & Oke, R. (2016). Halophiles: Pharmaceutical potential and biotechnological applications. In D. Thangadurai, & S. Jeyabalan (Eds.), *Industrial biotechnology: Sustainable production and bioresource utilization* (pp. 111–139). Waretown, NJ: Apple Academic Press Inc.

Thombre, R., Shivakarthik, E., Sivaraman, B., Vaishampayan, P., Seuylemezian, A., Meka, J. K., Vijayan, S., Kulkarni, P., Pataskar, T., & Patil, B. (2019). Survival of extremotolerant bacteria from the Mukundpura meteorite impact crater. *Astrobiology, 19*, 785–796. Available from https://doi.org/10.1089/ast.2018.1928.

Thombre, R., Vaishampayan, P., & Gomez, F. (2020). Applications of extremophiles in astrobiology. In R. Salwan, & V. Sharma (Eds.), *Physiological and biotechnological aspects of extremophiles*. USA: Elsevier.

Thombre, R. S., Shinde, V. D., Oke, R. S., Dhar, S. K., & Shouche, Y. S. (2016). Biology and survival of extremely halophilic archaeon Haloarcula marismortui RR12 isolated from Mumbai salterns, India in response to salinity stress. *Scientific Reports, 6*, 25642. Available from https://doi.org/10.1038/srep25642.

Upasani, V., & Desai, S. (1990). Sambhar Salt Lake. Chemical composition of the brines and studies on haloalkaliphilic archaebacteria. *Archives of Microbiology, 154*, 589–593. Available from https://doi.org/10.1007/BF00248841.

Verma, S., Kumar, R., Kumar, P., & Sharma, D. (2020). Cloning, characterization, and structural modeling of an extremophilic bacterial lipase isolated from saline habitats of the Thar desert. *Applied Biochemistry and Biotechnology, 192*, 557–572. Available from https://doi.org/10.1007/s12010-020-03329-3.

Vijayanand, S., Hemapriya, J., Selvin, J., & Kiran, S. (2012). Biodiversity of extremely halophilic bacterial strains isolated from solar salterns of Tuticorin, Tamilnadu, India. *International Journal of Water Resources and Environmental Sciences, 1*, 01–07.

Yadav, A. N., Sachan, S. G., Verma, P., Tyagi, S. P., Kaushik, R., & Saxena, A. K. (2015). Culturable diversity and functional annotation of psychrotrophic bacteria from cold desert of Leh Ladakh (India). *World Journal of Microbiology and Biotechnology, 31*, 95–108. Available from https://doi.org/10.1007/s11274-014-1768-z.

Yadav, A. N., Verma, P., Kumar, M., Pal, K. K., Dey, R., Gupta, A., Padaria, J. C., Gujar, G. T., Kumar, S., Suman, A., Prasanna, R., & Saxena, A. K. (2014). Diversity and phylogenetic profiling of niche-specific Bacilli from extreme environments of India. *Annals of Microbiology, 65*, 611–629. Available from https://doi.org/10.1007/s13213-014-0897-9.

CHAPTER

4

Marine microbial hotspots—especially related to corals

Shruti Gupta[1,*], *Julius Eyiuche Nweze*[1,2,*] *and Sharad Dnyandev Subugade*[3]

[1]Soil and Water Research Infrastructure, Biology Centre CAS, Ceske Budejovice, Czech Republic, [2]Department of Ecosystem Biology, Faculty of Science, University of South Bohemia, Ceske Budejovice, Czech Republic, [3]Department of Biotechnology, Vidya Pratishthan's Art, Science & Commerce College, Pune, India

4.1 Introduction

Corals belong to the invertebrates group, exclusively marine with diverse morphology. They are categorized into two main subtypes, stony and soft corals. Interestingly, stony corals are worldwide referred to contribute to the largest existing calcareous skeletal structure created by the scleractinian corals (Rosenberg et al., 2007). Predominantly, they live in the tropical waters, that receive optimum sun rays exposure. Besides, some of them can be found in cold water. According to functional behavior, corals have been providing shelter and food for fishes and other organisms. And, also provide a place to reproduce and raise the offsprings. The functioning symbiosis between the microorganisms and the partner (plant or animals) is important in maintaining the individual's integrity and health. Similarly, tropical corals have an intimate relationship with their symbiotic microorganisms and become an integral part of the marine ecosystem (Oppen & Blackall, 2019). This association between the corals and their microorganism is termed as "coral

* These authors contributed equally.

Microbial Diversity and Ecology in Hotspots
DOI: https://doi.org/10.1016/B978-0-323-90148-2.00009-2

holobiont." Notably, coral harbors highly dynamic and diverse microbial consortia including archaea, bacteria, fungi, viruses, and protists including photosynthetic dinoflagellates throughout the mucus, coral tissue, skeleton, and gastrovascular cavity (Oppen & Blackall, 2019). Thus it collectively considered the "microbiome." Particularly, this symbiotic interaction is bidirectional between coral and their microbiome. Consequently, coral receives metabolic byproducts from their symbionts and residing organisms through photosynthesis and several metabolic pathways including nitrogen fixation (Ricci et al., 2019).

The recent studies that have been carried out in past decades prove the dependency of coral reef-building on the associated microbiome (Oppen & Blackall, 2019). Although, the coral microbiome benefits their partner in mechanisms such as, ease the digestion of complex matter and preventing infection by developing resistance against pathogens (Rosenberg et al., 2007). Also, they mutually involve in stringent nutrient recycling and mineralization. In addition, the coral's microbiome also prevents the destruction of coral reefs (Rosenberg et al., 2007). Eventually, they help in holobiont homeostasis, coral's building (shaping), and modulating the coral's health and development. Furthermore, coral–microbe interaction has also facilitated the origin and evolution of the host and associated microbial consortia (Rosenberg et al., 2007). Predominantly, photosynthetic dinoflagellate, *Symbiodinium* contributed to the coral microbiome that is known to mediate the carbon fixation and provide the organic matter to their host. Likewise, coral also harbors rich bacterial diversity. Although, the availability of data on fungal diversity in corals is limited (Góes-Neto et al., 2020), according to the multiomics studies, ascomycetes and basidiomycetes are dominantly present and reported to derive the ammonification (Oppen & Blackall, 2019). However, viruses are highly diverse in corals as compared to marine diversity and contribute to coral holobiont (Sweet & Bulling, 2017). They are diversely present and expected to infect the available microbial consortia including archaea, bacteria, and other eukaryotes. Thus with regard to viruses, a healthy coral microbiome is dominated by bacteriophages and archaeal phages revealed by the metagenomic approach (Sweet & Bythell, 2017).

The ecosystem of coral reefs is significantly sensitive to anthropogenic activities and abiotic factors induced by climate change. It has been drastically declined in the last two decades with the effect of heat waves due to global warming (Osman et al., 2020). Although, global warming which highly depends on greenhouse gases has been decimated by the coral ecosystem by disturbing their regular mechanisms and interaction between symbiotic partners (algae or bacteria). Consequently, the diversity of corals has declined worldwide due to bleaching. Nevertheless, thermal resistance in corals is acquired by its holobiont population.

Similar to the plant's rhizosphere, corals are capable of influencing the microbial consortia by releasing organic molecules and mucus into the surroundings (Sun et al., 2016). The host partner potentially exerts control on their microbiota (McLoughlin et al., 2016). Similarly, the alteration in the coral microbiome is greatly affected by the status of host health, which has been observed in both culture-dependent and independent analyses (Sun et al., 2016). However, 16S rRNA gene sequencing of culture-independent (metagenomics) methods reported the significantly high novel microbial diversity in coral holobionts under stressed conditions (Oppen & Blackall, 2019). According to the studies, the bacterial microbiome of corals has significantly promoted the acclimatization of the host against environmental stress. Further, they participate in maintaining host physiology and integrity, assist in nutrient cycling, and keep up the host coral immunity. Moreover, studies also revealed the diverse bacterial population in host coral including volcanic vents, deep-sea, and warmer pools (Osman et al., 2020).

4.2 Symbiosis within the coral holobionts

The coral holobionts symbiosis plays a crucial role to maintain homeostasis in the marine ecosystem. The coral-associated microbial population potentially exhibits chemotaxis behavior toward the released chemicals. These signaling pathways contribute to coral–microbe interaction and establish a symbiosis within coral holobionts. In this regard, several specific receptors adapt to work and recognize the chemotaxis. Furthermore, in order to establish symbiosis in coral holobiont, it can be acquired by inheritance through vertical transmission or significant selection from the surrounding environment through horizontal transmission. Symbiosis is not strict to one of them, it can be acquired from both the mechanisms (Bourne et al., 2016). In horizontal transmission, coral offsprings are probably found to maintain the acquisition in the early phase of life that provides the protection and nutrition from their symbiotic bacterial partner (Bourne et al., 2016). As the mechanism of symbiosis in coral holobionts is not yet clear, therefore this is a thrust of the research area.

4.3 Microhabitats of the coral microbiome

Coral microbiome covers coral-associated microbial assemblage in the system. The diverse microbial communities confined to distinct microhabitat of corals, that drive the multitude of services for corals' health, metabolism, nutrient cycling, and antimicrobial defense (Rädecker et al., 2015). In general, the site-specific microbial assemblages have been reported in vertebrates (including humans) as well as in invertebrates. Consequently, distinguished microhabitat considered for the selection of study and their relative analysis. Similarly, the coral surface mucopolysaccharide layer, tissue, gastrovascular cavity, and skeleton comprise diverse and distinct microbial assembly composition and richness (Engelen et al., 2018). Further, these coral-associated bacterial communities involved in coral acclimatization against environmental stress (Osman et al., 2020) and providing several life-saving benefits to the host partner (Engelen et al., 2018).

4.3.1 Surface mucopolysaccharide layer and epidermal surface

In coral, surface mucopolysaccharide layer (SML) is produced by endosymbiotic dinoflagellates including the genus *Symbiodinium*. This is rich in nitrogen and organic matter that support the growth of the microbial communities (10^6–10^8 cells/mL) (Garren & Azam, 2011). Particularly, it is composed of glycoproteins along with polypeptides (enzymes) and sulfated oligosaccharides (Bourne et al., 2016). Commonly reported microbial communities of SML are Photobacterium, Endozoicimonaceae, and Firmicutes. Notably, SML-associated microbial assemblage providing nutritional benefits to the partner (host), acts as a defensive first-line barrier against the invasion of a pathogen by releasing antibiotics (Engelen et al., 2018), and also protecting from ultraviolet (UV) radiation (sunlight) by producing UV-absorbing pigments (Bourne et al., 2016; Engelen et al., 2018). Further, SML attributed the chemical signaling that induces to attract the bacteria chemotactically. Thus the SML surface densely colonizes the microbial assembly which is crucial for coral health and resilience.

4.3.2 Coral tissues

According to Pollock et al. (2018), the richness, composition, and diversity of microbial assemblies vary anatomically within the coral. This variation is achieved by environmental and host conditions. For instance, unlike SML, coral tissue (epithelium and gastrodermis) inhabiting host-specific symbionts (Bourne et al., 2016). Several extrinsic and intrinsic factors contribute to shaping coral-associated microbial assemblies. Similarly, commonly reported microbial assemblies of coral tissues are *Ralstonia*, *Actinobacteria*, and *Endozoicomonas* (Engelen et al., 2018). Notably, *Actinobacteria* also detected within the endosymbiotic *Symbiodinium* and hence, contributing as a repository of photosymbionts (Bourne et al., 2016). In *Stylophora pistillata*, aggregates of *Endozoicomonas* have been identified at the interphase tissue of the epidermis and gastrodermis (Neave et al., 2016).

4.3.3 Gastrovascular cavity

It is a primitive type of gastrovascular system involved in the digestion of plankton. Thus this mechanism provided the nutritional supplement (photosynthates) to the partner (Bourne et al., 2016). With correspond to the human gut, the gastrovascular system of coral's is involved in catabolizing the ingested material including organic matter, recycle nutrients, and also fulfilling the nutritional requirement by providing vitamins to the partner (Engelen et al., 2018). Similarly, in *Galaxea fascicularis'* gastric cavity provides the crucial inorganic and vitamins to the host and endosymbionts, although details of the digestion mechanism and the participation of the gut microbiome are insufficient. In addition, it has been noted in oxygen-deficient and with elevated concentration of inorganic nutrients and vitamin B12. However, the presence of vitamin B12 in the surrounding environment is very limiting, suggesting it is produced by the bacteria of the gastric cavity (Bourne et al., 2016).

4.3.4 Coral skeleton

Coral holobionts are mainly studied for microbial communities in mucus and tissues. Thus microbial communities related to the skeleton are poorly studied. Nevertheless, it is a unique structural habitat for microbial assemblage. The ecological, physiological, and recently metagenomic studies of the coral skeleton gain attention and become a crucial hotspot for microbial diversity (Ricci et al., 2019). Diverse microbial assemblies including eukaryotes and prokaryotes are abundantly reported in the skeleton as compared to tissues and mucus. Collectively these assemblies are called endoliths (Bourne et al., 2016; Ricci et al., 2019). The skeleton of corals significantly harbors the microbial communities and is a reservoir that promotes dysbiosis and coral reef resilience (Ricci et al., 2019). Furthermore, skeleton microbiota involved in nutrient recycling including nitrogen and sulfur, shaping the corals and maybe substitute as energy is also significant (Ricci et al., 2019). In addition, they contribute to coral holobiont health and fitness. For instance, commonly known endolithic assemblies are *Ostreobium* (chlorophytes), sulfur-reducing bacteria, and fungi that are present in an equilibrium manner (Bourne et al., 2016).

4.4 Profiling of coral-associated microbial diversity

Microorganisms exist in corals as diverse microbial communities, where they maintain the symbiotic relationship in order to participate in ecological activities. They can be found in various niches in coral such as the mucus layer (Bourne et al., 2016), calcium carbonate skeleton gaps, and coral tissues (Fig. 4.1) (Naohisa et al., 2019). All these microecological niches certainly have specific physicochemical environments that affect the structure of the composition and diversity of the microorganisms in corals. Studies have shown that the metabolism of the coral mucus constituents (sugar contents) facilitates the rapid growth of microorganisms (Bourne et al., 2016). Interestingly, the physiological cycles exist within the coral ecosystem that changes from autotrophy (day) to heterotrophy (night). Such physiological shift provides different physicochemical conditions to the coral-associated microbial aggregates.

Numerous culture-based and molecular techniques have been used to understand the nature of microbial communities within coral ecological niches (Sun et al., 2016). Moreover, the dynamics in the genomic

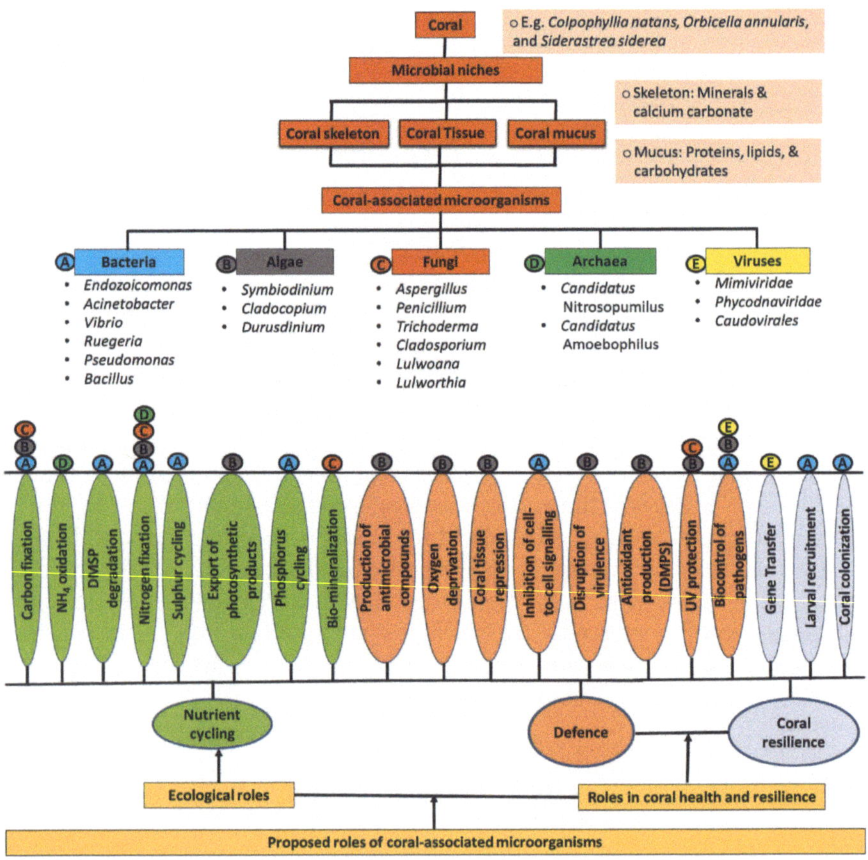

FIGURE 4.1 A schematic representation of microbial diversity and functions in corals.

diversity of these coral microorganisms have been discovered through metagenomic analyses (Sun et al., 2016). Integrating different sequences have revealed the existence of highly diverse relatedness among coral-associated microbes (Huggett & Apprill, 2019) (Fig. 4.2). For example, Alvarez-Yela et al. (2019) have reported vast coral-associated microbial communities in *Colpophyllia natans*, *Orbicella annularis*, and *Siderastrea siderea*. In Australia, a survey of the bacterial and archaeal groups in different coral species has demonstrated anatomic compartmentalization. This shows that the coral microbial diversity, richness, and responses to both the internal and external physiological changes vary according to the anatomical location such as skeleton, tissue, and mucus of holobionts (Pollock et al., 2018). The density of the coral microbial community in the tissues of *Acropora hyacinthus* obtained from the various sites in Japan and Australia was found to be correlated negatively with the distance from the coastline (Naohisa et al., 2019).

4.4.1 Bacteria

Diverse bacterial communities abundantly reported in corals (Yang et al., 2017), where about total of 39 of them belong to bacterial phyla

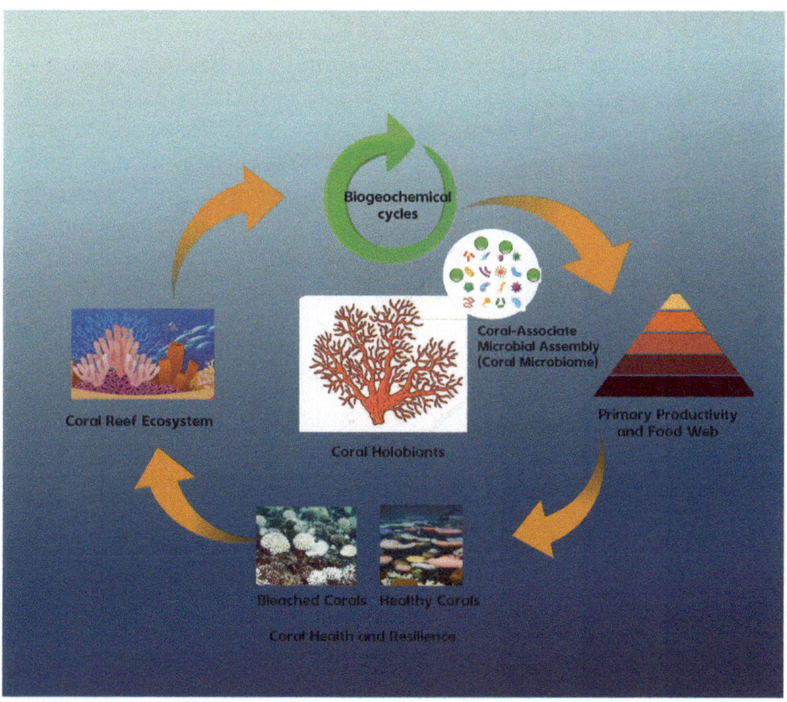

FIGURE 4.2 The functional and diverse role of coral-associated microorganisms.

(Huggett & Apprill, 2019). They are the key coral symbionts acquired across the life history of corals (Bernasconi et al., 2019). From all the studies so far, sequences from the bacterial community generally match with Proteobacteria (especially alpha-proteobacteria and gamma-proteobacteria), Firmicutes, Bacteroidetes, and Cyanobacteria (Table 4.1) (Huggett & Apprill, 2019). In the Gulf of Aqaba, the coral *Stylophora pistillata*'s mucus has revealed bacterial diversity abundantly present and especially dominated by the phylum Proteobacteria (Hussien et al., 2019). Across six typical South China Sea, samples from *Galaxea fascicularis, Acropora* sp., *Porites lutea, Favosites speciosa, Platygyra lamellina*, and *Sarcophyton glaucum* showed that the coral microorganisms were dominated by the α- (19%) and γ-proteobacteria (32%), Firmicutes (14%), Bacteroidetes (14%), Actinobacteria (6%), and Cyanobacteria (2%) (Gong et al., 2020). The comparison of microbial sequences from three popular deep-water corals viz., *L. pertusa, P. arborea*, and *M. oculata*, has, respectively, shown the predominance of alpha-proteobacteria (Jensen et al., 2019). Though the abundance and uniformity of these bacterial communities differ between different coral reefs, no significant differences have been observed between distinct coral lineages or corals harboring different subclades of *Symbiodinium* (Pantos et al., 2015).

In mid-Atlantic Bight, Baltimore Canyon, United States, cold-water corals studied using the 16S rRNA gene pyrosequencing showed that Alteromonadales and Pirellulales dominated the bacterial communities in *Alcyonium grandiflorum*, whereas Oceanospirillales and Spirochaetes dominated the bacterial communities in *Anthothela grandiflora* and *Anthothela* sp. (Lawler et al., 2016). In the tissues of four different coral species, *Acropora formosa, Acropora aspera, Isopora* sp., and *Cyphastrea* sp., Proteobacteria and Firmicutes are the most abundant followed by Bacteroidetes together with Actinobacteria, Fusobacteria, and Lentisphaerae as minor taxa (Mhuantong et al., 2019). Seasonal variations have been reported to alter the bacterial diversity patterns in corals depending on the sampling months (Epstein et al., 2019). For instance, seasonal variations influenced the diversity of microbiome in *Astrangia poculata*. In the Gulf of Thailand, the coral species viz., *Acropora millepora, Porites lutea, Acropora humilis*, and *Platygyra sinensis* had more abundance of bacteria (*Psychrobacter, Pseudoalteromonas*, and *Cobetia*) (Bulan et al., 2018). Regardless of the bacterial symbiotic state in *Astrangia poculata*, four seasonal time points have been reported to be dominated by the Proteobacteria, Flavobacteria, Cytophagia, and Thaumarchaea in the Fort Wetherill State Park, Jamestown, United States (Sharp et al., 2017). Interestingly, *Endozoicomonas* are the dominant and commonly detected bacterial species across the globe (Neave et al., 2016), but are exceedingly rare or probably not present in

TABLE 4.1 Coral-associated microorganisms of coral hotspots.

No.	Location	Coral name	Part of coral holobiont's	Dominant bacterial species	Reference
1	Northern Great Barrier Reef, Australia	*Seriatopora hystrix*	Tissue	Alpha-proteobacteria, gamma-proteobacteria, and Bacteroidetes	Pantos et al. (2015)
2	South China Sea	*Galaxea fasciculari*, *Acropora* sp., *Porites lutea*, *Favosites speciosa*, *Platygyra lamellina*, and *Sarcophyton glaucum*	Whole colonies	Alpha-proteobacteria (19%) and gamma-proteobacteria (32%), Firmicutes (14%), Bacteroidetes (14%), Actinobacteria (6%), and Cyanobacteria (2%)	Gong et al. (2020)
3	Mid-Atlantic Bight included Baltimore Canyon, United States	*Alcyonium grandiflorum*	Whole colonies	Alteromonadales and Pirellulales	Lawler et al. (2016)
		Anthothela grandiflora and *Anthothela* sp.	Whole colonies	Oceanospirillales and Spirochetes	Lawler et al. (2016)
4	Indonesian Sea in Central Java and Karimunjawa	*Acropora formosa*, *Acropora aspera*, *Isopora* sp., and *Cyphastrea* sp.	Whole colonies	Proteobacteria and Firmicutes	Mhuantong et al. (2019)
5	Kenting, Yehliu, and Lyudao, in Taiwan	*Stylophora pistillata*	Whole colonies	*Endozoicomonas*	Yang et al. (2017)
6	Gulf of Thailand	*Acropora millepora*, *Porites lutea*, *Acropora humilis*, and *Platygyra sinensis*	Whole colonies	*Psychrobacter*, *Pseudoalteromonas*, and *Cobetia*	Bulan et al. (2018)
7	Fort Wetherill State Park, Jamestown, United States	*Astrangia poculata*	Whole colonies	Proteobacteria, Flavobacteria, Cytophagia, and Thaumarchaea	Sharp et al. (2017)
8	Gulf of Aqaba, Northern-Eastern extension of the Red Sea	*Stylophora pistillata*	Coral mucus	Proteobacteria	Hussien et al. (2019)
9	Mid-Norwegian shelf	*Lophelia pertusa*	Whole colonies	Alpha-proteobacteria affiliated to members of the Rhizobiales	(Jensen et al., 2019)
		Paragorgia arborea, *Madrepora oculata*	Whole colonies	Mollicutes affiliated with the *Mycoplasma*, *Flavobacteria* affiliated with members of the *Elisabethkingia*	Jensen et al. (2019)
10	Western Atlantic, United States	*Lophelia pertusa*	Whole colonies	Proteobacteria (75%–96%)	Kellogg et al. (2017)
11	Parque Natural Municipal do Recife de Fora" of Porto Seguro, Bahia, Brazil	*Mussismilia hispida*	Mucus, tissue, and skeleton	*Acinetobacter* sp.	Leite et al. (2018)

1. Geographical profiling of dominant coral-associated microorganisms. 2. Profiling of the coral hotspot and their associated microorganisms.

deep-sea corals. Besides, the population of *Endozoicomonas* depends on the property of the corals such as soft or a hard coral (Kellogg, 2019).

The importance of separating core-symbiotic bacterial groups from other diverse coral-associated communities has been advocated especially when studying the functional contributions of the symbiotic bacteria. The two most dominant phylotypes specifically coexist with the coral endosymbiotic algae and symbiont-containing host cells. These symbionts may be facilitating the success of endosymbiosis in the coral Dinoflagellates (Ainsworth et al., 2015).

Although some strong patterns of diversity exist among different studies, there is currently no unified or decisive arrangement for coral microbiome species diversity. Some studies have demonstrated that certain coral-associated microorganisms exhibit site specificity, with microbial diversity varying between collection sites rather than the coral host species (Dunphy et al., 2019). This simply indicates that some ecological factors have some roles in shaping the microbial communities in corals.

4.4.2 Archaea

Though archaea-associated corals are rarely investigated, they are greatly outnumbered by the coral-associated bacteria (Oppen & Blackall, 2019). Probably due to the inefficiency of available primers to amplify archaeal rRNA genes compared to bacterial rRNA genes (Bahram et al., 2019). The most common coral-associated archaeal phyla belong to Thaumarchaeota and Euryarchaeota (Huggett & Apprill, 2019). Pollock et al. (2018) reported that archaea represented about 2% of the total observed sequences, and *Candidatus nitrosopumilus* and *Candidatus amoebophilus* are the two most important coral archaeal-associated species (Huggett & Apprill, 2019). A recent metagenomic study of the coral species, *Porites* sp. recovered a high-quality genome of *Thermarchaeota* (Robbins et al., 2019). This kind of study is often ignored because a large percentage of coral-associated microbial community surveys are solely based on bacteria (Irus, n.d.). Like bacteria, the archaeal populations associated with Caribbean coral surface mucus varies according to the specificity of host and reefs community turnover (Frade et al., 2016).

4.4.3 Dinoflagellates

Apart from the enduring partnerships with bacteria and archaea, corals live in symbiotic association with phototrophic Dinoflagellates of the family Symbiodiniaceae (Parkinson et al., 2016). Dinoflagellates can live freely in the surrounding coral reefs including waters and sediments. Significantly, helps in maintaining the environmental pool for colonization of new corals (Frommlet et al., 2015). Nineteen distinct endosymbiotic Dinoflagellates from the Northern Red Sea, namely, *Cladocopium*, *Symbiodinium*, and *Durusdinium* (family: Symbiodiniaceae) have been identified from six species of corals, *Sarcophyton trocheliophorum*, *Favia favus*, *Porites nodifera*, *Pocillopora damicornis*, *Xenia umbellata*, and *Seriatopora hystrix* (Osman et al., 2020). The genus *Symbiodinium* has been subjected to extensive studies in recent decades because its relationship with corals is central to the survival of the tropical and subtropical coral reef ecosystems (LaJeunesse et al., 2018). The genus contains nine large genetic groups or clades that are highly distinct genetically (clades A–I) with each clade containing genetically and ecologically distinct subclades or strains (Rouzé et al., 2016). The highest diversity has been reported in a South Western Atlantic oceanic island and they are commonly referred to as Zooxanthellate corals (Teschima et al., 2019). In Pacific Ocean, corals from the genus *Pocillopora*, the diversity of *Symbiodinium* based on the 28S and ITS1 rDNA found that most of the Zooxanthellae belong to clade C, subclade C1, with little differentiation between the populations (Magalon et al., 2007).

4.4.4 Fungi

Studies on coral-associated microbes have been focused primarily on the interaction between corals and their endosymbiotic Zooxanthellae (Baker et al., 2016), as well as their prokaryotic communities (Marcelino et al., 2017), while neglecting the distinctiveness, diversity, and ecological functions of the fungal communities (Góes-Neto et al., 2020). In coral holobionts, fungi counted as an important member, although the knowledge about their taxonomic diversity is still limited (Góes-Neto et al., 2020). Fungi are known to form relationships with various corals either as potential symbionts or pathogens (Paulino et al., 2020). For instance, in the Northeast Brazilian reefs, 43 species of fungi (16 genera) belonging to phylum Ascomycota were found on the basis of the sequencing of ITS rDNA regions, where *Aspergillus*, *Penicillium*, *Trichoderma*, and *Cladosporium* were observed to be the most dominant (Paulino et al., 2020). From Papua New Guinea and Australia, 24 most distinct fungal genera in the phylum Ascomycota and Basidiomycota have been identified. Besides *Lulwoana* and *Lulworthia* (Ascomycota) which were reported to be the most abundant and prevalent genera, others that were thought to have no form of association with corals or reef ecosystems, such as *Flammulina* (Basidiomycota), *Ophiosphaerella* (Ascomycota), and *Geranomyces* (Chytriomycota) have also been reported (Góes-Neto et al., 2020). As pathogens, anamorphic fungi (especially ascomycetes) have been implicated as the etiological agents of diseases in skeletons and tissues of hard corals and tissues of soft corals (Gleason et al.,

2017). *Aspergillus sydowii* (ascomycete) is a typical disease-causing agent and is the most extensively researched coral fungus in the Caribbean and Pacific coasts of Colombia. It has been recognized as the causative agent of Aspergillosis, although it can be found in healthy corals (Kim & Rypien, 2015). They can also exist in coral reef-building as either epibionts or endobionts (Kershaw et al., 2018), with some (endolithic fungi) being capable of drilling and living in the coral calcium carbonate skeleton (Gleason et al., 2017; Góes-Neto et al., 2020; Paulino et al., 2020).

In the South China Sea, two soft corals, *Sarcophyton tortuosum* and *Cladiella krempfi* harbor a highly diverse and abundant fungal community with five of the isolates sharing 83%–95% similarity with the genera *Mucor* and *Phaeoshaeria* (Zhang et al., 2019). Similarly, culture-dependent characterization has been used to identify 18 fungi in the phyla Ascomycota, Deuteromycota, and Mucoromycota. Five genera, viz., *Penicillium*, *Aspergillus*, *Byssochlamys* (Eutrotiales), *Rhizopus* (Mucorales), and *Curvularia* (Moniliales) were later identified (Abd El-Rahman et al., 2020).

4.4.5 Viruses

Metagenomic and electron microscopic studies have shown that the viral populations in coals are at a high level of diversity and abundance (Sweet & Bythell, 2017). For example, in all the tissue layers from the coral genus *Acropora,* viruses related to Retro, Herpes, and nucleocytoplasmic large DNA viruses have been identified using transmission electron microscopy imaging and virome analysis. Megavirus-like particles have been identified within the cells of coral-associated dinoflagellate algae (Correa et al., 2016). A diverse coral viral community often establish symbiotic relationships with the algal endosymbionts in the genus *Symbiodinium* (Brüwer et al., 2017). According to Sweet and Bythell (2017), no coral species-associated viral pathogens have been specifically identified, but the presence of different known invertebrate viruses such as Oyster Herpesvirus (OsHV-1) indicates that there is a high possibility of viral pathogens with various coral diseases. Coral *Symbiodinium* type C1 cultures have been shown to have a nucleocytoplasmic dsDNA virus affiliated to *Mimiviridae* and *Phycodnaviridae,* a new filamentous virus with unidentified phylogenetic affiliation, and a ssRNA virus affiliated to Retroviruses (Weynberg et al., 2017). In the central Great Barrier Reef, viral communities from seven coral species order *Caudovirales* (tailed bacteriophages) found to be dominant, and also the prevalence of the *Microviridae* (ssDNA viruses) had been reported. Most sequences matched other viruses reported earlier except *Herpesvirales,* which were rare (Weynberg et al., 2017). The mucus of *Fungia repanda* (disk coral) and *Acropora formosa* (staghorn coral) **can** serve as a viral proliferation hotspot, supporting the development of both kind of phages virulent as well as temperate (Nguyen-Kim et al., 2015; Sweet & Bythell, 2017).

4.5 Importance of coral microbiome for marine ecosystem

The coral reefs are highly productive, diverse, and endangered ecosystems. In addition, they create structural habitats and become reservoirs for other marine life including microorganisms, algae, mollusks, and fishes. Similarly, they host a wide range of phylogenetically different microbiome (Lima et al., 2020). They maintain the mutualistic symbiosis with the microorganisms in order to satisfy their nutritional and physiological requirement as well as participate in health and reef buildup (Wilkins et al., 2019; Rosenberg et al., 2007). Likewise to ecosystem engineers, these microbiomes are crucial to maintaining the coral reef's health. The microbiome of holobiont is extremely sensitive to the environment (Oppen & Blackall, 2019). Thus several intrinsic and extrinsic factors of holobiont organisms shape the coral-associated microbiome.

4.5.1 Contribution to primary productivity

The tropical coral reef system contributed to the most productive ecosystem, with the primary production and supporting the complex food web in the marine ecosystem (Fig. 4.2).

4.5.2 Functional role of coral-associated microbiome

Corals reefs are known to present in a nutrient-limited environment. Nevertheless, they have high productivity. This contradiction has come under the "Darwin Paradox" explanation. This can be achieved by the internalization and nutrients recycling in coral holobionts (Rädecker et al., 2015). Although the microhabitat of coral broadly categorizes into the SML, coral tissue, gastrovascular cavity, and coral skeleton, diverse microbial consortium lie in associated microhabitat. This support allows the functioning of coral's holobiont metabolism. Similarly, coral's photosynthetic symbionts, *Symbiodinium* helps to fix the carbon. Significantly, they translocate the photosynthates and other essential nutrients to the corals and their associated microbiome. Effectively, it structures the tripartite relationship between coral, symbiotic Symbiodinium, and coral-associated microbiome (Bourne et al., 2016). Consequently, coral holobionts establish the various functional pathways that drive the nutrients cycle including carbon, nitrogen, sulfur, and phosphate (Fig. 4.2). In addition, they also efficiently maintain the crucial passage (such as trace metals, vitamins, and other cofactors) (Bourne et al., 2016). According to the studies, vertical transfer facilitates host-specific evolution between corals and associated microbial communities (Rädecker et al., 2015; Rosenberg et al., 2007).

4.5.3 Maintenance of the coral health and resilience

Coral-associated microorganisms have been demonstrated to play an increasingly recognized role in coral health and survival (McDevitt-Irwin et al., 2017). These microorganisms can have a useful or harmful effect on the coral hosts depending on their taxonomic composition, functionality, and potential characteristics of the emerging microbial populations (Rosenberg et al., 2007). Corals and their intricate dependency on endosymbiotic Dinoflagellates and other associated microorganisms (i.e., bacteria, archaea, fungi, etc.) are important to maintain homeostasis. So far, the dependency of the coral host on their microbial consortium is not fully understood (Peixoto et al., 2017). Corals are highly responsive to environmental and surroundings fluctuation. This leads to the dysbiosis of the dominant microbiome population (Peixoto et al., 2017; Rosenberg et al., 2007). Typically, elevated seawater temperatures and coral bleaching are linked to changes to opportunistic and/or pathogenic bacteria with the subsequent decline in coral health (Tout et al., 2015) (Fig. 4.2). For instance, a high abundance of *Vibrionales*, *Flavobacteriales*, *Rhodobacterales*, *Alteromonadales*, *Rhizobiales*, *Rhodospirillales*, and *Desulfovibrionales* during stress shows that the coral host has reduced ability to regulate its microbiome (McDevitt-Irwin et al., 2017). Eventually, it develops diseases and leads to mortality (Egan & Gardiner, 2016). Corals possess the innate or natural immunodefense mechanism which helps in maintaining their health and integrity. Their defense mechanisms start from the physical barrier (epidermis and mucus), immunocompetent cells (phagocytic cells), and soluble factors including organic acids and antimicrobial components (Rosenberg et al., 2007). In terms of immune defenses, microorganisms make a vital contribution by inhabiting and secreting antimicrobial peptides, especially at the entries to different coral niches (Bourne et al., 2016). Coral algae can influence other associated microorganisms by repressing coral tissues or continuously depriving them of sufficient oxygen (Barott et al., 2012). Some algae can achieve that by poisoning them with secondary metabolites (Morrow et al., 2012) or sheltering pathogens (Tout et al., 2015), whereas others produce secondary metabolites or soluble organic compounds to promote or restrict microbial propagation (Barott et al., 2012) (Fig. 4.1). The defensive microbial composition depends on the coral health state, life stage, water acidity, and temperature, pollution, nutrient levels (Bettina et al., 2017), macroalgal community, the intensity of light, depth, or seasonal changes (Bourne et al., 2016). Maintaining such a stable microbiota is considered important for the well-being of corals, as it has been shown that destabilization in the structure and operation of their associated microbial diversity occurs in diseased states (Sweet & Bythell, 2017) and under adverse environmental conditions (Bourne et al., 2016; Rosenberg et al., 2007).

4.6 Conclusion and future perspectives

Coral reefs are a highly diverse and productive ecosystem and contribute to the largest existing structure in the world. However, anthropogenic activities disturbed the coral reefs and the entire ecosystem. For the survivability of coral reefs and marine ecosystems, the coral-associated microbiome (such as dinoflagellates, archaea, bacteria, and fungi) is crucial. Interestingly, coral holobionts accommodate diverse microbial assembly through symbiosis and maintaining health and resilience. Regardless, the establishment and maintenance of symbiosis interaction is yet not studied well or data are insufficient. So far, 39 bacterial phyla, 2 archaeal, and 3 fungal were reported in coral. Although the coral-associated bacteria are studied since more than 60 years, the concept of coral microbiome came later. Thus coral microbiome studies started in the last 15 years. Despite, coral associate microbial data are insufficient. Recently, multiomics metagenomics approaches are introduced to study coral holobiont. Using these approaches, significantly provides more insight into microbial profiling and the function of coral holobionts. A high-resolution study of distinct microhabitat is recommended in order to understand the function and interactions within and between these microhabitats. In addition, it will probably fulfill the current knowledge gap in the upcoming years. Such a framework for understanding the specificity of coral microbes has been established in the form of a database.

References

Abd El-Rahman, T. M. A., Tharwat, N. A., Abo El-Souad, S. M. S., El-Beih, A. A., & El-Diwany, A. I. (2020). Biological activities and variation of symbiotic fungi isolated from coral reefs collected from Red Sea in Egypt. *Mycology, 11*(3), 243–255. Available from https://doi.org/10.1080/21501203.2020.1741470.

Ainsworth, T. D., Krause, L., Bridge, T., Torda, G., Raina, J. B., Zakrzewski, M., Gates, R. D., Padilla-Gamiño, J. L., Spalding, H. L., Smith, C., Woolsey, E. S., Bourne, D. G., Bongaerts, P., Hoegh-Guldberg, O., & Leggat, W. (2015). The coral core microbiome identifies rare bacterial taxa as ubiquitous endosymbionts. *ISME Journal, 9*(10), 2261–2274. Available from https://doi.org/10.1038/ismej.2015.39.

Alvarez-Yela, C., Mosquera-Rendón, J., Noreña, P. A., Cristancho, M., & López-Alvarez, D. (2019). Microbial diversity exploration of marine hosts at Serrana Bank a Coral Atoll of the seaflower biosphere reserve. *Frontiers in Marine Science 6*, Frontiers Media SA.

Andrew C. Baker, Adrienne M. S. Correa, Ross Cunning. Diversity Distribution and Stability of Symbiodinium in Reef Corals of the Eastern Tropical Pacific. 405–420 In *Coral Reefs of the Eastern Tropical Pacific*. Springer Netherlands, 2016.

Bahram, M., Anslan, S., Hildebrand, F., Bork, P., & Tedersoo, L. (2019). Newly designed 16S rRNA metabarcoding primers amplify diverse and novel archaeal taxa from the environment. *Environmental Microbiology Reports, 11*(4), 487–494. Available from https://doi.org/10.1111/1758-2229.12684.

Baker A.C, Correa A.M.S, Cunning R., Diversity Distribution and Stability of Symbiodinium in Reef Corals of the Eastern Tropical Pacific. 405–420 In *Coral Reefs of the Eastern Tropical Pacific*. Springer Netherlands, 2016.

Barott, K. L., Rodriguez-Mueller, B., Youle, M., Marhaver, K. L., Vermeij, M. J. A., Smith, J. E., & Rohwer, F. L. (2012). Microbial to reef scale interactions between the reef-building coral Montastraea annularis and benthic algae. *Proceedings of the Royal Society B: Biological Sciences, 279*(1733), 1655–1664. Available from https://doi.org/10.1098/rspb.2011.2155.

Bernasconi, R., Stat, M., Koenders, A., Paparini, A., Bunce, M., & Huggett, M. J. (2019). Establishment of coral-bacteria symbioses reveal changes in the core bacterial community with host ontogeny. *Frontiers in Microbiology, 10*, 1529. Available from https://doi.org/10.3389/fmicb.2019.01529.

Bettina, G., S., W. N., & G., B. D. (2017). Microbial indicators as a diagnostic tool for assessing water quality and climate stress in coral reef ecosystems. *Marine Biology, 164*, 91. Available from https://doi.org/10.1007/s00227-017-3097-x.

Bourne, D. G., Morrow, K. M., & Webster, N. S. (2016). Insights into the coral microbiome: Underpinning the health and resilience of reef ecosystems. *Annual Review of Microbiology, 70*, 317–340. Available from https://doi.org/10.1146/annurev-micro-102215-095440.

Brüwer, J. D., Agrawal, S., Liew, Y. J., Aranda, M., & Voolstra, C. R. (2017). Association of coral algal symbionts with a diverse viral community responsive to heat shock. *BMC Microbiology, 17*(1), 174. Available from https://doi.org/10.1186/s12866-017-1084-5.

Bulan, D. E., Wilantho, A., Krainara, P., Viyakarn, V., Chavanich, S., & Somboonna, N. (2018). Spatial and seasonal variability of reef bacterial communities in the Upper Gulf of Thailand. *Frontiers in Marine Science, 5*, 1–15. Available from https://doi.org/10.3389/fmars.2018.00441.

Correa, A. M. S., Ainsworth, T. D., Rosales, S. M., Thurber, A. R., Butler, C. R., & Vega Thurber, R. L. (2016). Viral outbreak in corals associated with an in situ bleaching event: Atypical herpes-like viruses and a new megavirus infecting symbiodinium. *Frontiers in Microbiology, 7*, 127. Available from https://doi.org/10.3389/fmicb.2016.00127.

Dunphy, C. M., Gouhier, T. C., Chu, N. D., & Vollmer, S. V. (2019). Structure and stability of the coral microbiome in space and time. *Scientific Reports, 9*(1), 6785. Available from https://doi.org/10.1038/s41598-019-43268-6.

Egan, S., & Gardiner, M. (2016). Microbial dysbiosis: Rethinking disease in marine ecosystems. *Frontiers in Microbiology, 7*, 991. Available from https://doi.org/10.3389/fmicb.2016.00991.

Engelen, A. H., Aires, T., Vermeij, M. J. A., Herndl, G. J., Serrão, E. A., & Frade, P. R. (2018). Host differentiation and compartmentalization of microbial communities in the azooxanthellate cupcorals Tubastrea coccinea and Rhizopsammia goesi in the Caribbean. *Frontiers in Marine Science*, 9. Available from https://doi.org/10.3389/fmars.2018.00391.

Epstein, H. E., Smith, H. A., Cantin, N. E., Mocellin, V. J. L., Torda, G., & van Oppen, M. J. H. (2019). Temporal variation in the microbiome of acropora coral species does not reflect seasonality. *Frontiers in Microbiology, 10*, 1775. Available from https://doi.org/10.3389/fmicb.2019.01775.

Frade, P. R., Roll, K., Bergauer, K., & Herndl, G. J. (2016). Archaeal and bacterial communities associated with the surface mucus of caribbean corals differ in their degree of host specificity and community turnover over reefs. *PLoS One*, *11*(1), e0144702. Available from https://doi.org/10.1371/journal.pone.0144702.

Frommlet, J. C., Sousa, M. L., Alves, A., Vieira, S. I., Suggett, D. J., & Serôdio, J. (2015). Coral symbiotic algae calcify ex hospite in partnership with bacteria. *Proceedings of the National Academy of Sciences of the United States of America*, *112*(19), 6158–6163. Available from https://doi.org/10.1073/pnas.1420991112.

Garren, M., & Azam, F. (2011). Corals shed bacteria as a potential mechanism of resilience to organic matter enrichment. *The ISME Journal 6*, 1159–1165. Springer Science and Business Media LLC.

Gleason, F. H., Gadd, G. M., Pitt, J. I., & Larkum, A. W. D. (2017). The roles of endolithic fungi in bioerosion and disease in marine ecosystems. II. Potential facultatively parasitic anamorphic ascomycetes can cause disease in corals and molluscs. *Mycology*, *8*(3), 216–227. Available from https://doi.org/10.1080/21501203.2017.1371802.

Góes-Neto, A., Marcelino, V. R., Verbruggen, H., da Silva, F. F., & Badotti, F. (2020). Biodiversity of endolithic fungi in coral skeletons and other reef substrates revealed with 18S rDNA metabarcoding. *Coral Reefs*, *39*(1), 229–238. Available from https://doi.org/10.1007/s00338-019-01880-y.

Gong, S., Jin, X., Ren, L., Tan, Y., & Xia, X. (2020). Unraveling heterogeneity of coral microbiome assemblages in tropical and subtropical corals in the South China Sea. *Microorganisms*, *8*(4), 604. Available from https://doi.org/10.3390/microorganisms8040604.

Huggett, M. J., & Apprill, A. (2019). Coral microbiome database: Integration of sequences reveals high diversity and relatedness of coral-associated microbes. *Environmental Microbiology Reports*, *11*(3), 372–385. Available from https://doi.org/10.1111/1758-2229.12686.

Hussien, E., Juhmani, A. S., AlMasri, R., Al-Horani, F., & Al-Saghir, M. (2019). Metagenomic analysis of microbial community associated with coral mucus from the Gulf of Aqaba. *Heliyon*, *5*(11), e02876. Available from https://doi.org/10.1016/j.heliyon.2019.e02876.

Irus B., The Coral Holobiont. 201–230 In *Coral Whisperers*. University of California Press, 2018. https://doi.org/10.2307/j.ctvq4c11g.

Jensen, S., Hovland, M., Lynch, M. D. J., & Bourne, D. G. (2019). Diversity of deep-water coral-associated bacteria and comparison across depth gradients. *FEMS Microbiology Ecology*, *95*(7), fiz091. Available from https://doi.org/10.1093/femsec/fiz091.

Kellogg, C. A., Goldsmith, D. B., & Gray, M. A. (2017). Biogeographic comparison of lophelia-associated bacterial communities in the Western Atlantic reveals conserved core microbiome. *Frontiers in Microbiology 8*, Frontiers Media SA.

Kellogg, C. A. (2019). Microbiomes of stony and soft deep-sea corals share rare core bacteria. *Microbiome*, *7*(1). Available from https://doi.org/10.1186/s40168-019-0697-3.

Kershaw, S., Munnecke, A., & Jarochowska, E. (2018). Understanding palaeozoic stromatoporoid growth. *Earth-Science Reviews*, *187*, 53–76. Available from https://doi.org/10.1016/j.earscirev.2018.08.003.

Kim, K., & Rypien, K. (2015). *Aspergillosis of Caribbean Sea fan corals*, Gorgonia *spp*. *Diseases of coral* (pp. 236–241). John Wiley and Sons Inc. Available from https://doi.org/10.1002/9781118828502.ch16.

LaJeunesse, T. C., Parkinson, J. E., Gabrielson, P. W., Jeong, H. J., Reimer, J. D., Voolstra, C. R., & Santos, S. R. (2018). Systematic revision of symbiodiniaceae highlights the antiquity and diversity of coral endosymbionts. *Current Biology*, *28*(16), 2570–2580. Available from https://doi.org/10.1016/j.cub.2018.07.008, e6.

Lawler, S. N., Kellogg, C. A., France, S. C., Clostio, R. W., Brooke, S. D., & Ross, S. W. (2016). Coral-associated bacterial diversity is conserved across two deep-sea Anthothela species. *Frontiers in Microbiology*, 7, 458. Available from https://doi.org/10.3389/fmicb.2016.00458.

Leite, D. C. A., Salles, J. F., Calderon, E. N., Castro, C. B., Bianchini, A., Marques, J. A., van Elsas, D. D., & Peixoto, R. S. (2018). Coral bacterial-core abundance and network complexity as proxies for anthropogenic pollution. *Frontiers in Microbiology 9*, Frontiers Media SA.

Lima, L. F. O., Weissman, M., Reed, M., Papudeshi, B., Alker, A. T., Morris, M. M., Edwards, R. A., De Putron, S. J., Vaidya, N. K., & Dinsdalea, E. A. (2020). Modeling of the coral microbiome: The influence of temperature and microbial network. *mBio, 11*(2). Available from https://doi.org/10.1128/mBio.02691-19, e02691-19.

Magalon, H., Flot, J. F., & Baudry, E. (2007). Molecular identification of symbiotic dinoflagellates in Pacific corals in the genus Pocillopora. *Coral Reefs, 26*(3), 551–558. Available from https://doi.org/10.1007/s00338-007-0215-0.

Marcelino, V. R., Oppen, M. J. Hv, & Verbruggen, H. (2017). Highly structured prokaryote communities exist within the skeleton of coral colonies. *The ISME Journal, 12*(1), 300–303.

McDevitt-Irwin, J. M., Baum, J. K., Garren, M., & Vega Thurber, R. L. (2017). Responses of coral-associated bacterial communities to local and global stressors. *Frontiers in Marine Science, 4*, 262. Available from https://doi.org/10.3389/fmars.2017.00262.

McLoughlin, K., Schluter, J., Rakoff-Nahoum, S., Smith, A. L., & Foster, K. R. (2016). Host selection of microbiota via differential adhesion. *Cell Host and Microbe, 19*(4), 550–559. Available from https://doi.org/10.1016/j.chom.2016.02.021.

Mhuantong, W., Nuryadi, H., Trianto, A., Sabdono, A., Tangphatsornruang, S., Eurwilaichitr, L., Kanokratana, P., & Champreda, V. (2019). Comparative analysis of bacterial communities associated with healthy and diseased corals in the Indonesian sea. *PeerJ, 7*, e813. Available from https://doi.org/10.7717/peerj.8137.

Morrow, K. M., Ritson-Williams, R., Ross, C., Liles, M. R., & Paul, V. J. (2012). Macroalgal extracts induce bacterial assemblage shifts and sublethal tissue stress in Caribbean corals. *PLoS One, 7*(9), e44859. Available from https://doi.org/10.1371/journal.pone.0044859.

Naohisa, W., Mizuki, I., Taeko, M., Joseph, P. F., Sen-Lin, T., D., A. T., L., W. B., Nobuhiro, M., & G., B. D. (2019). Characterization of coral-associated microbial aggregates (CAMAs) within tissues of the coral Acropora hyacinthus. *Scientific Reports, 9*(1), 14662. Available from https://doi.org/10.1038/s41598-019-49651-7.

Neave, M. J., Apprill, A., Ferrier-Pagès, C., & Voolstra, C. R. (2016). Diversity and function of prevalent symbiotic marine bacteria in the genus Endozoicomonas. *Applied Microbiology and Biotechnology, 100*(19), 8315–8324. Available from https://doi.org/10.1007/s00253-016-7777-0.

Nguyen-Kim, H., Bettarel, Y., Bouvier, T., Bouvier, C., Doan-Nhu, H., Nguyen-Ngoc, L., Nguyen-Thanh, T., Tran-Quang, H., & Brune, J. (2015). Coral mucus is a hot spot for viral infections. *Applied and Environmental Microbiology, 81*(17), 5773–5783. Available from https://doi.org/10.1128/AEM.00542-15.

Oppen, M. J. Hv, & Blackall, L. L. (2019). Coral microbiome dynamics functions and design in a changing world. *Nature Reviews Microbiology, 17*(9), 557–567.

Osman, E. O., Suggett, D. J., Voolstra, C. R., Pettay, D. T., Clark, D. R., Pogoreutz, C., Sampayo, E. M., Warner, M. E., & Smith, D. J. (2020). Coral microbiome composition along the northern Red Sea suggests high plasticity of bacterial and specificity of endosymbiotic dinoflagellate communities. *Microbiome, 8*(1). Available from https://doi.org/10.1186/s40168-019-0776-5.

Pantos, O., Bongaerts, P., Dennis, P. G., Tyson, G. W., & Hoegh-Guldberg, O. (2015). Habitat-specific environmental conditions primarily control the microbiomes of the coral Seriatopora hystrix. *ISME Journal, 9*(9), 1916–1927. Available from https://doi.org/10.1038/ismej.2015.3.

Parkinson, J. E., Baumgarten, S., Michell, C. T., Baums, I. B., LaJeunesse, T. C., & Voolstra, C. R. (2016). Gene expression variation resolves species and individual strains among coral-associated dinoflagellates within the genus symbiodinium. *Genome Biology and Evolution, 8*(3), 665–680. Available from https://doi.org/10.1093/gbe/evw019.

Paulino, G. V. B., Félix, C. R., & Landell, M. F. (2020). Diversity of filamentous fungi associated with coral and sponges in coastal reefs of northeast Brazil. *Journal of Basic Microbiology*, *60*(2), 103–111. Available from https://doi.org/10.1002/jobm.201900394.

Peixoto, R. S., Rosado, P. M., Leite, D. Cd. A., Rosado, A. S., & Bourne, D. G. (2017). Beneficial microorganisms for corals (BMC): Proposed mechanisms for coral health and resilience. *Frontiers in Microbiology*, *8*, 341. Available from https://doi.org/10.3389/fmicb.2017.00341.

Pollock, F. J., McMinds, R., Smith, S., Bourne, D. G., Willis, B. L., Medina, M., Thurber, R. V., & Zaneveld, J. R. (2018). Coral-associated bacteria demonstrate phylosymbiosis and cophylogeny. *Nature Communications*, *9*(1), 4921. Available from https://doi.org/10.1038/s41467-018-07275-x, 9(1).

Rädecker, N., Pogoreutz, C., Voolstra, C. R., Wiedenmann, J., & Wild, C. (2015). Nitrogen cycling in corals: The key to understanding holobiont functioning? *Trends in Microbiology*, *23*(8), 490–497. Available from https://doi.org/10.1016/j.tim.2015.03.008.

Ricci, F., Rossetto Marcelino, V., Blackall, L. L., Kühl, M., Medina, M., & Verbruggen, H. (2019). Beneath the surface: Community assembly and functions of the coral skeleton microbiome. *Microbiome*, *7*(1), 159. Available from https://doi.org/10.1186/s40168-019-0762-y.

Robbins, S. J., Singleton, C. M., Chan, C. X., Messer, L. F., Geers, A. U., Ying, H., Baker, A., Bell, S. C., Morrow, K. M., Ragan, M. A., Miller, D. J., Forêt, S., Ball, E., Beeden, R., Berumen, M., Aranda, M., Ravasi, T., Bongaerts, P., ... Hoegh-Guldberg, O. (2019). A genomic view of the reef-building coral Porites lutea and its microbial symbionts. *Nature Microbiology*, *4*(12), 2090–2100. Available from https://doi.org/10.1038/s41564-019-0532-4.

Rosenberg, E., Koren, O., Reshef, L., Efrony, R., & Zilber-Rosenberg, I. (2007). The role of microorganisms in coral health, disease and evolution. *Nature Reviews Microbiology*, *5* (5), 355–362. Available from https://doi.org/10.1038/nrmicro1635.

Rouzé, H., Lecellier, G., Saulnier, D., & Berteaux-Lecellier, V. (2016). Symbiodinium clades A and D differentially predispose Acropora cytherea to disease and Vibrio spp. colonization. *Ecology and Evolution*, *6*(2), 560–572. Available from https://doi.org/10.1002/ece3.1895.

Sharp, K. H., Pratte, Z. A., Kerwin, A. H., Rotjan, R. D., & Stewart, F. J. (2017). Season but not symbiont state, drives microbiome structure in the temperate coral Astrangia poculata. *Microbiome*, *5*(1), 120.

Sun, W., Anbuchezhian, R., & Li, Z. (2016). *Association of coral-microbes, and the ecological roles of microbial symbionts in corals. The Cnidaria, past, present and future: The world of Medusa and her sisters* (pp. 347–357). Springer International Publishing. Available from https://doi.org/10.1007/978-3-319-31305-4_22.

Sweet, M., & Bulling, M. T. (2017). The role of viruses in coral health and disease. *Journal of Invertebrate Pathology*, *147*, 136–144. Available from https://doi.org/10.1016/j.jip.2016.12.005.

Teschima, M. M., Garrido, A., Paris, A., Nunes, F. L. D, & Zilberberg, C. (2019). Correction: Biogeography of the endosymbiotic dinoflagellates (Symbiodiniaceae) community associated with the brooding coral *Favia gravida* in the Atlantic ocean. *PLoS One*, *14*(4), e0215167. Available from https://doi.org/10.1371/journal.pone.0215167.

Tout, J., Siboni, N., Messer, L. F., Garren, M., Stocker, R., Webster, N. S., Ralph, P. J., & Seymour, J. R. (2015). Increased seawater temperature increases the abundance and alters the structure of natural vibrio populations associated with the coral Pocillopora damicornis. *Frontiers in Microbiology*, *6*, 432. Available from https://doi.org/10.3389/fmicb.2015.00432.

Weynberg, K. D., Laffy, P. W., Wood-Charlson, E. M., Turaev, D., Rattei, T., Webster, N. S., & van Oppen, M. J. H. (2017). Coral-associated viral communities show high levels of diversity and host auxiliary functions. *PeerJ*, *5*, e4054. Available from https://doi.org/10.7717/peerj.4054.

Wilkins, L. G. E., Leray, M., Dea, A. O., Yuen, B., Peixoto, R., Pereira, T. J., Bik, H. M., Coil, D. A., Duffy, J. E., Herre, E. A., Lessios, H., Lucey, N. M., Mejia, L. C., Rasher, B.,

Sharp, K., Sogin, E. M., Thacker, R. W., Vega Thurber, R., Wcislo, W. T., Wilbanks, E. G., & Eisen, J. A. (2019). Host-associated microbiomes drive structure and function of marine ecosystems. *PLoS Biol. 12*, 1–15. Available from https://doi.org/10.1371/journal.pbio.3000533.

Yang, S. H., Tseng, C. H., Huang, C. R., Chen, C. P., Tandon, K., Lee, S. T. M., Chiang, P. W., Shiu, J. H., Chen, C. A., & Tang, S. L. (2017). Long-term survey is necessary to reveal various shifts of microbial composition in corals. *Frontiers in Microbiology, 8*, 1094. Available from https://doi.org/10.3389/fmicb.2017.01094.

Zhang, X. Y., Hao, H. L., Lau, S. C. K., Wang, H. Y., Han, Y., Dong, L. M., & Huang, R. M. (2019). Biodiversity and antifouling activity of fungi associated with two soft corals from the South China sea. *Archives of Microbiology, 201*(6), 757–767. Available from https://doi.org/10.1007/s00203-019-01639-7.

CHAPTER

5

Phyllosphere microbiomes: implications and ecofunctional diversity

Mohammad Yaseen Mir[1], *Saima Hamid*[1,2] *and Javid A. Parray*[3]

[1]Centre of Research for Development, University of Kashmir, Hazratbal, India, [2]Department of Environmental Sciences, University of Kashmir, Hazratbal, India, [3]Department of Environmental Sciences, Government Degree College Eidgah, India

5.1 Introduction

The current state of knowledge about phyllosphere microbiomes is insufficient to fully comprehend the relationship between phyllospheric microorganisms and plants. Phyllospheric microorganisms, including rhizospheric microorganisms, play a role in the contribution of many beneficial factors such as defense from physiological and biological factors, nitrogen fixation, and phytohormone biosynthesis. The microbial load on the phyllospheric layer's surface is thought to be one of the world's largest microbial habitats. From a 1 cm^2 plant tissue area, it was estimated that 10^5–10^7 microorganisms could be isolated. The current state of knowledge about phyllosphere microbiomes is insufficient to fully comprehend the relationship between phyllospheric microorganisms and plants. Phyllospheric microorganisms, including rhizospheric microorganisms, play a role in the contribution of many beneficial factors such as defense from physiological and biological factors, nitrogen fixation, and phytohormone biosynthesis. The microbial load on the phyllospheric layer's surface is thought to be one of the world's largest

DOI: https://doi.org/10.1016/B978-0-323-90148-2.00005-5

microbial habitats. From a 1 cm^2 plant tissue area, it was estimated that 10^5–10^7 microorganisms could be isolated. The phyllosphere's microbiome includes viruses, bacteria, filamentous fungi, yeasts, algae, and, on rare occasions, protozoa and nematodes. Bacteria are the most numerous microorganisms in the phyllosphere, while fungi and archaea are far less numerous. Due to the difficulties encountered in culture-based methods for identifying different microbial communities on the phyllosphere, knowledge of phyllosphere microbiomes is restricted. The problem with culture-based methods of identification is that microbial communities do not evolve in vitro. As a result, this approach was unable to investigate the full extent of microbiome identification on the phyllosphere. Several alternative methods for estimating and identifying the various microorganisms on a plant surface, including the phyllosphere, have been established in recent years as a result of advances in science and technology. Advances in sequencing technologies and metagenomic approaches are reducing the complexities associated with culture-based identification methods. The genes associated with phyllospheric fitness of *Pantoea agglomerans* 299R were recently discovered using high-throughput Illumina sequencing. This strain has been widely used as a model bacterium to better understand epiphyte adaptation to the phyllosphere climate. *P. agglomerans* 299R genome sequencing revealed several adaptations that are common in plant leaf-associated bacteria, such as sugar utilization, DNA repair systems, and osmoprotectant processing. A gene for bacterial rhodopsin (S. Belkin, personal communication) is also found in the genome of 299R, which allows this bacterium and other phyllosphere colonizers to derive energy from sunlight (Atamna-Ismaeel et al., 2012). Jumpponen and Jones (2009) looked at the fungal populations in the phyllosphere of oak trees (*Quercus macrocarpa*) under different management strategies, including rural and urban. The foliar fungal group composition was revealed by high-throughput pyrosequencing of the fungal internal transcribed spacer 1 (ITS1), and unique taxa that were immune to these two management practices were established. *Devriesia, Mycosphaerella, Ramularia, Stenella, Dioszegia, Paraphaeosphaeria, Phaeosphaeria,* and *Sphaceloma* were more common in the oak trees in the rural environments, while *Aureobasidium, Davidiella, Didymella,* and *Microsphaeropsis* occurred more frequently in urban environments.

5.2 Leaf surface and microbial growth

Environmental conditions at the leaf surface have a huge impact on microbial communities in the phyllosphere, which in turn influences the plant–microbiome interactions that may occur. Water and nutrients, as well as exposure to high levels of ultraviolet (UV) radiation, often restrict growth on the leaf surface (Knief et al., 2012; Vorholt, 2012). External constraints for colonizing microorganisms are imposed by variations in the physical and chemical landscape of the leaf itself, as well as its packing on the leaf surface (Remus-Emsermann et al., 2012). As the topography of the leaf changes with age, plant–microbial interactions are influenced by both time and the host plant's unique characteristics (Kinkel et al., 2000). Moisture availability is a major limiting factor for microbial growth on the leaf surface. The cuticle protects the leaf surface from water loss. To overcome this barrier, bacteria can form aggregates and biofilms, resulting in extracellular polymeric substances that can withstand desiccation (Morris & Kinkel, 2002). Surfactants are generated by other bacteria to increase the leaf's wettability and reduce the cuticle's ability to prevent water absorption (Schreiber, 2005). In addition to these techniques, the leaf surface has a boundary layer with a microclimate that is usually higher in humidity than the wider phyllosphere, which reduces the desiccation pressure for microbiota in that layer. Microbial populations usually increase after precipitation, resulting in significant changes in the phyllosphere microbiome's diversity and composition (Copeland et al., 2015). While phyllosphere microorganisms have evolved strategies to overcome potential moisture limitations, the leaf surface's nutrient-poor nature means that their growth is still constrained by available nutrients primarily carbon and nitrogen (Mercier & Lindow, 2000). Microorganisms on the leaf surface are usually oligotrophs, meaning they can survive in low-nutrient environments, or they can associate with the host plant to gain more nutrients (Beattie & Lindow, 1999). Although cuticular waxes are normally impervious to chemical movement, some plant metabolites can migrate to the leaf surface to aid microbial development (Mercier & Lindow, 2000). These compounds can reach the leaf surface through excretion from leaf cells or osmotic pressure when the leaf is moist. Plants often release volatile organic compounds that help particular microorganism populations; for example, *Methylobacterium extorquens*, which metabolizes single-carbon compounds derived from plants, is a common constituent of the phyllosphere of many plant species (Delmotte et al., 2009). Phyllosphere microorganisms may acquire nitrogen by leaching plant-based amino acids to the leaf surface or by inorganic forms of nitrogen leaching from apoplasts. Nitrogen is deposited on the leaf surface as a result of atmospheric deposition. Ammonia is usually assimilated by leaf microbiota, but chemoautotrophic ammonia oxidizers may be present in the phyllosphere. Oxidized nitrogen species (nitrate and nitrite) are soluble in water and their supply changes during rain events. Bacterial nitrogen fixation can occur on the leaf surface, likely in pockets of moisture, due to the anaerobic restrictions of the fixation process. Leaf epiphytes are particularly vulnerable to UV radiation, and bacterial and fungal species isolated from the phyllosphere are usually more pigmented than those isolated from soil. This pigmentation will help to survive in the phyllosphere environment. Microorganisms in the phyllosphere may also produce enzymes that handle reactive oxygen species generated by solar radiation, as well as DNA defense proteins (Delmotte et al., 2009). Rapid expression of DNA repair mechanisms in response to UV exposure may be a key factor in phyllosphere survival. The majority of data on the effects of UV radiation come from research at the organismal or population level and little is understood about the effects of UV radiation on the microbial community as a whole. UV radiation has been shown to increase phyllosphere bacterial diversity. Day length may also be related to the growth and abundance of certain microorganisms since solar radiation has a diurnal cycle. Thus day lengths that change seasonally can affect the growth of microorganisms on the leaf surface; however, no studies have adequately disentangled such effects from other variables that also obey seasonal trends (i.e., temperature and relative humidity). Furthermore, since the majority of phyllosphere research focuses on agricultural systems, several studies are completed at harvest or at the end of the growing season, before seasonal patterns become apparent.

5.3 Phyllosphere microbiome: nature and composition

The microbiota of the phyllosphere includes a wide variety of microorganisms, but bacteria are usually the most prevalent. Bacterial assemblages in the phyllosphere are typically less diverse than those in the rhizosphere or soil (Knief et al., 2012). *Alphaproteobacteria* are abundant on the surface of leaves, and they play a variety of ecological roles (Innerebner et al., 2011). *Gammaproteobacteria* have also been included in phyllosphere bacterial population composition surveys (Redford et al., 2010). Proteobacteria are metabolically diverse and the phyllosphere bacteria that perform methyltrophy, nitrification, nitrogen fixation, or anoxygenic photosynthesis are normally characterized by this phylum. *Bacteroidetes* and *Actinobacteria* are the second most common bacterial lineages in phyllosphere cultures and both of these phyla are also well represented in the rhizosphere. In the phyllosphere, *Bacteroidetes* tend to belong to the *Cytophagaceae* or *Chitinophagaceae* families. These species are often pigmented and aerobic, indicating that they are well adapted to the plant's surface. Plant parasites, nitrogen-fixing symbionts, fungal antagonists, and decomposers are all known as members of the phylum *Actinobacteria*. Many of these functions have not been investigated in the phyllosphere but *Corynebacterium*, an *Actinobacteria* has been used as a foliar-applied plant growth promoter. Since archaea seem to make up a minor portion of the microbial population, less is known regarding their existence and distribution in the phyllosphere (Knief et al., 2012). Because of pathogenic fungal organisms, fungi are an essential component of the phyllosphere microbiota, which has been known for years. The fungal group is made up of species that play a wide range of ecological roles and whose population sizes vary according to the growing season and eventually leaf senescence (Kinkel et al., 2000). Before senescence, *Ascomycota* molds are often the dominant fungi on the leaf surface (Abdelfattah et al., 2015). Yeasts from the *Ascomycota* and *Basidiomycota* families are also important fungi. Following leaf senescence, filamentous fungi take over the fungal microbiome (Voriskova & Baldrian, 2013). Other microbial eukaryotes (protists) have not been studied extensively on plant leaves, but they primarily function as predators on the bacterial population, as they do in other systems. Next-generation sequencing of population metagenomes has shown that phyllosphere microorganisms have highly redundant functionality, implying that environmental factors (low nutrients, high UV, changing temperature, and humidity) select for consistent biological traits and low functionality at the community level (Lambais et al., 2017). In general, aerobic organoheterotrophs dominate the phyllosphere and metabolic diversity occurs mainly in the form of utilizable carbon compounds. Proteorhodopsin genes involved in anoxygenic photosynthesis were found to be taxonomically widespread in the phyllosphere of the tamarisk salt cedar (*Tamarix nilotica*) (Atamna-Ismaeel et al., 2012).

Proteorhodopsins are light-activated proton pumps that have been implicated in oligotrophic marine environments as starvation prevention mechanisms. Such adaptations seem to be particularly useful in the phyllosphere but further research is required to fully understand this phenomenon. While there is significant variability in microbiome composition even within a single plant species, plant species identification appears to be the most important factor affecting the composition of the phyllosphere microbial group (Redford et al., 2010). Leaf age and succession, in conjunction with environmental variation (Kinkel et al., 2000) and changes in the microbial composition of the atmosphere are likely to account for temporal variation in phyllosphere composition. Even on evergreen plants which would be expected to display less impact of leaf age, this variance can be dramatic and phyllosphere communities vary dramatically on seasonal time scales. The degree to which dispersal limitation or environmental gradients drive phyllosphere composition is uncertain and spatial variations in the phyllosphere of individual plant species are less well studied.

5.4 Bacterial diversity in the phyllosphere

Bacteria are the most abundant inhabitants of the leaves, with an average number of 10^6–10^8 cells/cm^2 in the phyllosphere. The population of epiphytic bacteria on the other hand varies depending on the plant species and its surroundings. The phyllosphere's physical and dietary factors account for the majority of the difference. Broad-leaf trees, rather than grasses or waxy broad-leaf plants have the largest number of bacteria (Kinkel et al., 2000). *Proteobacteria, Firmicutes, Bacteroides,* and *Actinobacteria* are the four main phyla of bacteria found in the phyllosphere. Methylotrophic bacteria are predominant in phyllosphere which includes genera such as *Methylobacterium, Methylophilus, Methylibium, Hyphomicrobium, Methylocella, Methylocapsa,* and *Methylocystis* (Krishnamoorthy et al., 2018). The genera *Methylobacterium* and *Sphingomonas* are the most common alphaproteobacterial genera found in plant phyllospheres (Delmotte et al., 2009). Unique assemblage regulations regulate the bacterial population organization on the phyllosphere. Normally factors such as soil type, plant genotype and organisms, plant immune system, age, climatic conditions, and geographic area force the bacterial population to form (Copeland et al., 2015). In *Arabidopsis thaliana* and maize, extensive studies on the soil and rhizosphere bacterial community on phyllosphere bacterial colonization are accessible (Peiffer et al., 2013). *Proteobacteria, Actinobacteria,* and *Bacteroidetes* are the most abundant phyla colonizing the leaf and root of *A. thaliana* (Delmotte et al., 2009). *Massilia, Flavobacterium, Pseudomonas,* and *Rathayibacter* are a prevalent bacterial genus in *A. thaliana, Deinococcus thermus* on tree phyllosphere (Redford et al., 2010), and *Bacillus* and *Pantoea* dominate the lettuce (Rastogi et al., 2012).

5.5 Fungal microbiota of phyllosphere

The phyllosphere fungi in the plant *Camellia japonica* are endophytic *Colletotrichum gloeosporioides* and *Colletotrichum acutatum,* as well as epiphytic *Pestalotiopsis* sp., *Aureobasidium pullulans, Phoma* sp., and *Ramichloridium* sp., according to Osono (2008). The abundance and diversity of fungi, on the other hand, varies by plant species and ecoclimatic conditions. Furthermore, seasonal and leaf age-dependent differences occur in the epiphytic and endophytic phyllosphere fungal assembly; for example, *Geniculosporium* sp. leaf age varies and *Cladosporium cladosporioides* leaf age varies (Osono, 2008). In the forest environment, phyllosphere fungi play an important role in mineral absorption and mineral recycling, specifically carbon, nitrogen, and phosphorus recycling. As a result, research into phyllosphere fungi and their physiology in relation to the host plant is important.

5.6 Actinomycetes diversity in phyllosphere

Actinobacteria have a strong presence in epiphytic and endophytic life forms in the phyllosphere, in addition to bacterial diversity. They are soil-dwelling saprophytic bacteria that have been extensively investigated for secondary therapeutic metabolites. This adaptable species has thrived in a number of environments, including the plant phyllosphere. Through their novel metabolites, some actinobacteria that form symbiotic associations in plant tissues have created enormous significance for the host and its climate. It has been widely published on the diversity and spread of endophytic actinobacteria from medicinal plants, fruit plants, and a variety of other terrestrial plants. Several species of actinobacteria have been reported from plants such as *Triticum aestivum, Lupinus termis, Lobelia clavatum, Acacia auriculiformis, Aquilaria crassna, Oryza sativa, Xylocarpus granatum,* and *Elaeagnus angustifolia* from various environments like arid, semiarid, and mangrove like *Actinoplane missouriensis, Actinoallomurus acacia, Actinoallomurus coprocola, Actinomadura glauciflava, Amycolatopsis tolypomycina, Actinoallomurus oryzae, Jishengella endophytica, Kribbella* sp., *Microbispora mesophila, Microbispora* sp., *Micromonospora* sp., *Nocardioides* sp., *Nocardia alba, Nonomuraea rubra, Micromonospora* sp. *Nonomuraea* sp., *Pseudonocardia* sp., *Planotetraspora* sp., *Pseudonocardia endophytica, Pseudonocardia halophobica, Streptomyces* sp., and *Streptomyces javensis* (Yadav & Yadav, 2018). The richness of actinomycetes in the phyllosphere is strong in tropical and temperate ecosystems, according to studies. Plant physiology and environmental factors also influence the actinobacterial relationship in plants, which aids in the establishment of endophytic life. Table 5.1 shows several examples of phyllosphere fungal endophytes.

5.7 Microbial interaction and phyllosphere

The phyllosphere's microbial diversity and abundance are determined by leaf physiology. It creates a microhabitat in which microorganisms change their physiology in order to survive in this environment. As a colonial type, epiphytic microbes were developed to protect the microorganisms from this harsh microhabitat (Lindow & Brandl, 2003). Bacteria typically establish a greater colonial association on the leaf surface, especially in veins as well as epidermal cell groves. Epidermal grooves are rich, particularly in sugar and water, in

TABLE 5.1 Phyllosphere fungal endophytes.

Endophytic fungi	Host plant	Type	References
Ramularia spp.	*Vitis riparia*	Wild	Kernaghan et al. (2017)
Alternaria alternate, *Cladosporium* sp., *Nigrospora oryzae*, *Botrytis cinerea*, *Epicoccum nigrum*, *Epicoccum* sp.	*Vitis vinifera*	Cultivar	
Absidia sp., *Aspergillus* sp., *Cladosporium* sp., *Cunninghamella* sp., *Fusarium* sp., *Nigrospora* sp., *Paecilomyces* sp., *Penicillium* sp., *Rhizopus* sp.,	*Meyna spinosa Roxb.*	Medicinal plant	Bhattacharyya et al. (2017)
Alternaria alternata, *Setosphaeria* sp., *Cochliobolus* sp., *Alternaria* sp., *Phoma herbarum*, *Davidiella tassiana*, *Botryosphaeria dothidea*, *Ulocladium alternariae*, *Phoma macrostoma* var. *incolorata*, *Phoma exigua* var. exigua, *Cladosporium cladosporioides* strain, *Botryosphaeria* sp., *Guignardia mangiferae*, *Pyrenophora tritici-repentis*, *Guignardia alliacea*, *Rhizopus oryzae*	*Catharanthus roseus*	Medicinal plant	
Ascomycetes: Trichoderma, *Penicillium*, *Fusarium*, and *Aspergillus*. Nonascomycetes: *Mucor* (Mucoromycota) and *Schizophyllum*	*Stanhopea tigrina*	Orchid	
Trichothecium sp., *Epicoccum nigrum*, *Alternaria alternaria*, *Alternaria arborescens*, *Nigrospora sphaerica*, *Epicoccum* sp., *Alternaria* sp., *Nigrospora* sp., *Colletotrichum gloeosporioides*, *Fusarium oxysporum*, *Trichothecium roseum*	*Vitis vinifera*	Fruit plant	
Aspergillus japonicus	*Euphorbia indica*	Wild plant	
Nigrospora sphaerica, *Acremonium alciforme*, *Allomyces arbuscula*, *Penicillium chrysogenum*, *Acrophialophora* sp., *Mycelia sterilia*	*Litsea cubeba*	Medicinal plant	
Colletotrichum destructivum, *Acremonium* sp.,	*Carex secalina*	Sedge	
Colletotrichum gloeosporioides f. sp. camelliae and *Pleosporales* sp.	*Camellia sinensis*	Tea	
Tremellales, *Davidiellaceae*, *Basidiomycota*, *Rhodotorula*, *Tremellales*, *Meria*, *Cryptococcus*, *Cladosporium*, *Acaromycetes*, *Erythrobasidium*, etc.	*Aegiceras corniculatum*, *Avicennia marina*, *Bruguiera gymnorrhiza*, *Kandelia candel*, *Rhizophora stylosa*, *Excoecaria agallocha*	Mangrove	

nutrients. The waxy cuticle on the leaf surface usually prevents bacteria from invading the phyllosphere by inhibiting permeability and moisture on the leaf surface (Burch et al., 2014). The waxy cuticle is diffused by the leaf surface water droplets, which improves the permeability of compounds diffused from the apoplast to the phyllosphere surface (Schreiber, 2005). The supply of nutrients for microorganisms is being hampered by leached compounds and water on the phyllosphere. Most commonly, the permeability and water availability of guard cells and their surface cuticles are improved by the flow of water from stomata (transpiration). Therefore greater cuticle layer permeation allows the bacteria to densely colonize. In addition, such compounds such as biosurfactants (syringe fact in developed by *Pseudomonas syringae*) can be produced by surface bacteria (Burch et al., 2014) which can alter and associate the leaf's cuticle surfaces. This will promote the supply of water and regulate the production of sugar that can enhance living conditions for the growth of epiphytic bacteria (Lindow & Brandl, 2003). Epiphytes such as *Pseudomonas* sp., *Stenotrophomonas* sp., and *Achromobacter* sp., increase lipophilic cuticle water permeability in *Hedera* and *Prunus*, which improves the supply of phyllosphere compounds to enhance epiphytic health on the surface of the leaves (Schreiber, 2005).

It has been experimentally proved in the bean phyllosphere containing fructose facilitates the growth of *Erwinia herbicola* and *Pantoea agglomerans* (Remus-Emsermann et al., 2012). Irregular fructose distribution, however, encourages the population of *Pseudonocardia eucalyptii* on bean leaves differently (Mercier & Lindow, 2000). These studies indicated that the epiphytic microorganisms rarely use the permeated carbon sources on the leaf surface for their development and multiplication. At the same time, the microbial community of the phyllosphere will affect the regulation of the leaf's physicochemical properties using both biotic and abiotic conditions (Quan & Liang, 2017). The phyllosphere's microbial diversity can also be influenced by the soil microbial community. Microbes, on the other hand, can create niches in the phyllosphere microhabitat where they can sustain and grow their population over time (Aguiar-Pulido et al., 2016). The existence of sugars and nutrients in this microhabitat will substantially alter the individual bacterial cells inside the microbial aggregates and direct physical associations with cell-to-cell interactions would be spatially developed (Tecon et al., 2018). On the basis of guided variables such as dispersal, microbe collection, diversification, and ecological drift, the population structure is coordinated. The community's fitness is attributed to the phyllosphere's internal (strain types) and external determinants (environment) (Schlechter et al., 2019). The internal population factors usually focus on the microbial interaction within the aggregates. The group structure is formed by both cooperative and mutualistic microbial relations and the formation of a greater colonial alliance comprising the maximum microbial population. Although commensals have poor group interactive partners, they are dispersed in the habitat at random. The interactive association within the organized group must not be affected by the commensals (Stubbendieck & Vargas-Bautista, 2016). Besides, in the population, antagonistic microbes have a hostile relationship, one will out compete the other and the ecosystem will destroy the responsive microbes. The effect on the establishment of the phyllosphere group system of mutual microbial interactions is not demonstrated (Schlechter et al., 2019). With heterogeneous communities, usually with shared and cooperative partners, bacteria may recognize the cell-to-cell contact mechanism and create a greater group structure. However, there may be some sort of mutualistic interactions between rapidly developing bacteria and pathogenic fungi, contributing to superficial infection of the host plant, which enhances the bacteria's nutritional accessibility to its rich population. *Dioszegia* sp. belongs to oomycete genus which is an example of fungal – fungal associations, which tends to decrease the bacterial population, *Albugo* sp., and compete on *A. thaliana* leaves with bacterial microbiota (Agler et al., 2016). In addition, competitive microbial interactions have detrimental impacts on at least one species in the ecosystem. As secondary metabolites, some competitive microbes develop toxic chemical substances (antibiotics and siderophores) that are harmful to their competitors. The best example of such an interaction is a Gram-negative *P. agglomerans* bacterium that inhibits the growth of *Erwinia amylovora*, an apple phytopathogen, via antibiotic activity (Peiffer et al., 2013; Pusey et al., 2011). In general, microbes compete primarily for food and space. The phyllosphere is a nutrient-limited world in which the competitive partner's development has been hampered by coexistence or exclusion from the location. Furthermore, the phyllosphere is heavily colonized by both oligotrophic and competitive microbes, both of which play important roles in the formation of population systems (Schlechter et al., 2019). The main variables of the group assemblage of the phyllosphere, however, remains actually unclear. Therefore further study is required to define the major variables deciding the assemblage of the phyllosphere population framework.

5.7.1 Metabolic dynamics of phyllosphere microbiota

Flavobacteria are abundant in the rhizosphere and phyllosphere of terrestrial plants like *A. thaliana*, where they are one of the most dominant leaf microbiota genera. From marine environments, *Flavobacteria* has been isolated which showed the carbohydrate metabolism adaptance with plant while on comparing with genomic sequences. The glycoside hydrolase encoding genes of the families GH78 and GH106 contained only genomes of *Flavobacteria* from terrestrial plant populations and not from aquatic groups responsible for the use of Rhamnogalacturonan, which is exclusively associated with hemicellulose terrestrial plants (Kolton et al., 2013). Microorganisms associated with the phyllosphere reside in a sunlight-exposed setting. In a nutrient-limited environment, photochemical conversion of this light resource into biomass and energy that can supplement the host plant's carbon supplies can be a critical advantage for growth. In phyllospheric cultures, metagenomic data revealed the existence of bacterial rhodopsin genes (Atamna-Ismaeel et al., 2012). It appears that certain epiphytic microorganisms have retinal-dependent rhodopsin proton pumps that can be light-activated by radiations that are different from the absorption spectrum of chlorophylls and carotenoids, which drive plant photosynthetic processes and thus the production of plant carbon resources that are eventually available to epiphytic microorganisms (Atamna-Ismaeel et al., 2012). With regard to the supply of carbon and nitrogen, it is likely that the carbohydrates and nitrogen state of the host plant could influence many metabolic pathways of epiphytic bacteria in the light-dependent process (Moran & Miller, 2007) and hence, fine variations in plant-light and photo-sap interaction. Manching et al. (2014) have recently identified global relations between leaf epiphytic bacterial species' balance of plant nitrogen and maize richness. In a free air CO_2 enrichment experiment, Ren et al. (2014) showed major changes in phyllosphere bacterial species in rice subject to different combinations of elevated CO_2 and differing degrees of nitrogen fertilization. In contrast, the enzymatic behavior of phyllospheric microorganisms is an important pyridoxal to pyridoxine (Huang et al., 2014). The potential presence of gene markers associated with aerobic phototrophic anoxygenic bacteria was explored in a parallel study (Atamna-Ismaeel et al., 2012). In the homologs of bchY encoding the Y subunit of chlorophyllide reductase and pufM encoding the M subunit of the photosynthetic reaction heart, five different phyllosphere microbiota from rice, soybean, tobacco, tamarix, clover were identified using metagenomics. Microscopic epifluorescence has detected the presence of distinctive pigments consistent with aerobic phototrophic anoxygenic bacteria. It was then observed that these bacteria make up 1%–7% of the total epiphetic population with the inclusion of the *Methyl bacterium* family, and more surprisingly with the involvement of an unidentified community of bacteria which appears to be phyllosphere-specific (Atamna-Ismaeel et al., 2012). The bacterial diversity of the phyllosphere is indeed very complicated than the microbiome of human diversity which were evaluated by using Rarefaction curves which imprints the ribosomal genes. The phyllosphere, on the other hand, tends to have much less bacterial diversity than the rhizosphere, soil, or marine habitats. The nutritional and functional mechanisms of microbial adaptation to life in the phyllosphere, such as the potential involvement of auxotrophic relationships and the potential dependence of phyllospheric microbial community structure on light availability, and thus foliage and canopy stratification, must also be addressed (Delmotte et al., 2009).

5.8 Omics approaches and future prospective

New insights into the microbiome phyllosphere's work and life histories are due to the use of high-throughput sequencing and "omics" techniques. Promising future research is possible through the knowledge of mechanisms for the microbe–plant and microbe–microbe interactions which decides phyllosphere's true functionality (Hunter et al., 2010). Metatranscriptomics has been used to prove that after entry of fungal pathogens, the rhizosphere bacterial population up regulates stress resistance genes, thus inducing the plants to express fungal antagonistic genes (Chapelle et al., 2016). A promising avenue for future study is the understanding of the molecular foundations for analogous collective defense mechanics in the phyllosphere. Based on the sequence length and guanine-cytosine content, fungal transcripts can be isolated from those of the plants. Delhomme et al. (2015) have demonstrated the capacity for coanalysis of both the transcriptome of the host plant and that of its fungal population. In elucidating relationship between pathogen and the host, as well as the wider microbiota, such techniques may be useful. An effort to relate metagenomics, metatranscriptomics, or metaproteomics to the evolving metabolomics area, which characterizes the chemical outputs of cell metabolism, will also be of interest (Aguiar-Pulido et al., 2016).

References

Abdelfattah, A., Nicosia, M. G. L. D., Cacciola, S. O., Droby, S., & Schena, L. (2015). Metabarcoding analysis of fungal diversity in the phyllosphere and carposphere of olive (Olea europaea). *PLoS ONE, 10*, e0131069.

Agler, M. T., Ruhe, J., Kroll, S., Morhenn, C., Kim, S. T., Weigel, D., & Kemen, E. M. (2016). Microbial hub taxa link host and abiotic factors to plant microbiome variation. *PLoS Biology, 14*(1). Available from https://doi.org/10.1371/journal.pbio.1002352.

Aguiar-Pulido, V., Huang, W., Suarez-Ulloa, V., Cickovski, T., Mathee, K., & Narasimhan, G. (2016). Metagenomics, metatranscriptomics, and metabolomics approaches for microbiome analysis. *Evolutionary Bioinformatics, 12*, 5–16. Available from https://doi.org/10.4137/EBO.S36436.

Atamna-Ismaeel, N., Finkel, O. M., Glaser, F., Sharon, I., Schneider, R., Post, A. F., Spudich, J. L., von Mering, C., Vorholt, J. A., Iluz, D., Béjà, O., & Belkin, S. (2012). Microbial rhodopsins on leaf surfaces of terrestrial plants. *Environmental Microbiology, 14*(1), 140–146. Available from https://doi.org/10.1111/j.1462-2920.2011.02554.x.

Beattie, G. A., & Lindow, S. E. (1999). Bacterial colonization of leaves: a spectrum of strategies. *Phytopathology, 89*, 353–359.

Bhattacharyya, L. H., Borah, G., Parkash, V., & Bhattacharyya, P. N. (2017). Fungal endophytes associated with the ethnomedicinal plant Meyna spinosa Roxb. *Current Life Sciences, 3*(1), 1–5.

Burch, A. Y., Zeisler, V., Yokota, K., Schreiber, L., & Lindow, S. E. (2014). The hygroscopic biosurfactant syringafactin produced by *Pseudomonas syringae* enhances fitness on leaf surfaces during fluctuating humidity. *Environmental Microbiology, 16*(7), 2086–2098. Available from https://doi.org/10.1111/1462-2920.12437.

Chapelle, E., Mendes, R., Bakker, P. A. H., & Raaijmakers, J. M. (2016). Fungal invasion of the rhizosphere microbiome. *ISME Journal, 10*(1), 265–268. Available from https://doi.org/10.1038/ismej.2015.82.

Copeland, J. K., Yuan, L., Layeghifard, M., Wang, P. W., & Guttman, D. S. (2015). Seasonal community succession of the phyllosphere microbiome. *Molecular Plant Microbe Interactions, 28*, 274–285.

Delhomme, N., Sundström, G., & Zamani, N. (2015). Serendipitous metatranscriptomics: The fungal community of Norway spruce (*Picea abies*). *PLoS One, 10*.

Delmotte, N., Knief, C., Chaffron, S., Innerebner, G., Roschitzki, B., Schlapbach, R., Von Mering, C., & Vorholt, J. A. (2009). Community proteogenomics reveals insights into the physiology of phyllosphere bacteria. *Proceedings of the National Academy of Sciences of the United States of America, 106*(38), 16428–16433. Available from https://doi.org/10.1073/pnas.0905240106.

Huang, S. H., Zhang, J. Y., Tao, Z., Lei, L., Yu, Y. H., & Huang, L. Q. (2014). Enzymatic conversion from pyridoxal to pyridoxine caused by microorganisms within tobacco phyllosphere. *Plant Physiology and Biochemistry, 85*, 9–13. Available from https://doi.org/10.1016/j.plaphy.2014.10.006.

Hunter, P. J., Hand, P., Pink, D., Whipps, J. M., & Bending, G. D. (2010). Both leaf properties and microbe-microbe interactions influence within-species variation in bacterial population diversity and structure in the lettuce (lactuca species) phyllosphere. *Applied and Environmental Microbiology, 76*(24), 8117–8125. Available from https://doi.org/10.1128/AEM.01321-10.

Innerebner, G., Knief, C., & Vorholt, J. A. (2011). Protection of Arabidopsis thaliana against leaf-pathogenic Pseudomonas syringae by Sphingomonas strains in a controlled model system. *Applied and Environmental Microbiology, 77*, 3202–3210.

Jumpponen, A., & Jones, K. L. (2009). Massively parallel 454 sequencing indicates hyperdiverse fungal communities in temperate Quercus macrocarpa phyllosphere. *New Phytol, 184*, 438–448.

Kernaghan, G., Mayerhofer, M., & Griffin, A. (2017). Fungal endophytes of wild and hybrid Vitis leaves and their potential for vineyard biocontrol. *Canadian Journal of Microbiology, 63*(7), 583–595.

Kinkel, L. L., Wilson, M., & Lindow, S. E. (2000). Plant species and plant incubation conditions influence variability in epiphytic bacterial population size. *Microbiol Ecology, 39*, 1–11.

Knief, C., Delmotte, N., Chaffron, S., Stark, M., Innerebner, G., Wassmann, R., von Mering, C., & Vorholt, J. A. (2012). Metaproteogenomic analysis of microbial communities in the phyllosphere and rhizosphere of rice. *ISME Journal, 6*(2012), 1378–1390.

Kolton, M., Sela, N., Elad, Y., & Cytryn, E. (2013). Comparative genomic analysis indicates that niche adaptation of terrestrial flavobacteria is strongly linked to plant glycan metabolism. *PLoS One, 8*(9). Available from https://doi.org/10.1371/journal.pone.0076704.

Krishnamoorthy, R., Kwon, S. W., Kumutha, K., Senthilkumar, M., Ahmed, S., Sa, T., & Anandham, R. (2018). Diversity of culturable methylotrophic bacteria in different genotypes of groundnut and their potential for plant growth promotion. *3 Biotech, 8*(6). Available from https://doi.org/10.1007/s13205-018-1291-2.

Lambais, M. R., Barrera, S. E., Santos, E. C., Crowley, D. E., & Jumpponen, A. (2017). Phyllosphere metaproteomes of trees from the Brazilian Atlantic forest show high levels of functional redundancy. *Microbial Ecology, 73*(1), 123–134. Available from https://doi.org/10.1007/s00248-016-0878-6.

Lindow, S. E., & Brandl, M. T. (2003). Microbiology of the phyllosphere. *Applied and Environmental Microbiology, 69*(4), 1875–1883. Available from https://doi.org/10.1128/AEM.69.4.1875-1883.2003.

Manching, H. C., Balint-Kurti, P. J., & Stapleton, A. E. (2014). Southern leaf blight disease severity is correlated with decreased maize leaf epiphytic bacterial species richness and the phyllosphere bacterial diversity decline is enhanced by nitrogen fertilization. *Frontiers in Plant Science, 5*. Available from https://doi.org/10.3389/fpls.2014.00403.

Mercier, J., & Lindow, S. E. (2000). Role of leaf surface sugars in colonization of plants by bacterial epiphytes. *Applied and Environmental Microbiology, 66*(1), 369–374. Available from https://doi.org/10.1128/AEM.66.1.369-374.2000.

Moran, M. A., & Miller, W. L. (2007). Resourceful heterotrophs make the most of light in the coastal ocean. *Nature Reviews Microbiology, 5*(10), 792–800. Available from https://doi.org/10.1038/nrmicro1746.

Morris, C. E., & Kinkel, L. L. (2002). Fifty years of phylosphere microbiology: significant contributions to research in related fields. In S. E. Lindow, E. I. Hecht-Poinar, & V. Elliott (Eds.), *Phyllosphere microbiology* (pp. 365–375). St. Paul, Minn: APS Press.

Osono, T. (2008). Endophytic and epiphytic phyllosphere fungi of Camellia japonica: seasonal and leaf age-dependent variations. *Mycologia, 100*, 387–391.

Peiffer, J. A., Spor, A., Koren, O., Jin, Z., Tringe, S. G., Dangl, J. L., Buckler, E. S., & Ley, R. E. (2013). Diversity and heritability of the maize rhizosphere microbiome under field conditions. *Proceedings of the National Academy of Sciences, 110*, 6548–6553.

Pusey, P. L., Stockwell, V. O., Reardon, C. L., Smits, T. H. M., & Duffy, B. (2011). Antibiosis activity of Pantoea agglomerans biocontrol strain E325 against Erwinia amylovora on apple flower stigmas. *Phytopathology, 101*, 1234–1241.

Quan, M., & Liang, J. (2017). The influences of four types of soil on the growth, physiological and biochemical characteristics of *Lycoris aurea* (L' Her.) Herb. *Scientific Reports, 7*. Available from https://doi.org/10.1038/srep43284.

Rastogi, G., Sbodio, A., Tech, J. J., Suslow, T. V., Coaker, G. L., & Leveau, J. H. J. (2012). Leaf microbiota in an agroecosystem: Spatiotemporal variation in bacterial community composition on field-grown lettuce. *ISME Journal, 6*(10), 1812–1822. Available from https://doi.org/10.1038/ismej.2012.32.

Redford, A. J., Bowers, R. M., Knight, R., Linhart, Y., & Fierer, N. (2010). The ecology of the phyllosphere: Geographic and phylogenetic variability in the distribution of bacteria on tree leaves. *Environmental Microbiology, 12*(11), 2885–2893. Available from https://doi.org/10.1111/j.1462-2920.2010.02258.x.

Remus-Emsermann, M. N. P., Tecon, R., Kowalchuk, G. A., & Leveau, J. H. J. (2012). Variation in local carrying capacity and the individual fate of bacterial colonizers in the phyllosphere. *ISME Journal, 6*(4), 756–765. Available from https://doi.org/10.1038/ismej.2011.209.

Ren, G., Zhang, H., Lin, X., Zhu, J., & Jia, Z. (2014). Response of phyllosphere bacterial communities to elevated CO_2 during rice growing season. *Applied Microbiology and Biotechnology, 98*(22), 9459–9471. Available from https://doi.org/10.1007/s00253-014-5915-0.

Schlechter, R. O., Miebach, M., & Remus-Emsermann, M. N. P. (2019). Driving factors of epiphytic bacterial communities: A review. *Journal of Advanced Research, 19*, 57–65. Available from https://doi.org/10.1016/j.jare.2019.03.003.

Schreiber, L. (2005). Polar paths of diffusion across plant cuticles: New evidence for an old hypothesis. *Annals of Botany, 95*(7), 1069–1073. Available from https://doi.org/10.1093/aob/mci122.

Stubbendieck, R., & Vargas-Bautista, C. (2016). Straight PD bacterial communities: Interactions to scale. *Front Microbiol, 7*.

Tecon, R., Ebrahimi, A., Kleyer, H., Levi, S. E., & Or, D. (2018). Cell-to-cell bacterial interactions promoted by drier conditions on soil surfaces. *Proceedings of the National Academy of Sciences of the United States of America, 115*(39), 9791–9796. Available from https://doi.org/10.1073/pnas.1808274115.

Vorholt, J. A. (2012). Microbial life in the phyllosphere. *Nature Review, 10*, 828–840.

Voriskova, J., & Baldrian, P. (2013). Fungal community on decomposing leaf litter undergoes rapid successional changes. *ISME Journal, 7*, 477–486.

Yadav, A. N., Kumar, V., Prasad, R., Saxena, A. K., & Dhaliwal, H. S. (2018). Microbiome in Crops: Diversity, distribution and potential role in crops improvements. In R. Prasad, S. S. Gill, & N. Tuteja (Eds.), *Crop Improvement through Microbial Biotechnology* (pp. 305–332). USA: Elsevier.

CHAPTER

6

Fungal association in hotspot of rhizosphere

Manju Shree Shakya Hada[1], *Reshma Tuladhar*[2], *Anima Shrestha*[1], *Sarita Manandhar*[1] and *Anjana Singh*[2]

[1]Department of Microbiology, Tri-Chandra Multiple College, Tribhuvan University, Kathmandu, Nepal, [2]Central Department of Microbiology, Tribhuvan University, Kirtipur, Kathmandu, Nepal

6.1 Introduction

A small soil volume of the root where microbial population is stimulated by root activities is termed as rhizosphere and is also considered as a hotspot of microbial activities (Chandrashekar et al., 2014). Peter Farb in 1959 first stated that "the soil is the placenta of life." The major role played by microorganisms in soil fertility was not clearly understood until 1953 when Troug mentioned that a one foot-acre of soil may contain as much as 8 tons of living microorganisms (Feiro, 1978). The soil bears diverse organisms which couple with the root systems to present a complex picture. The most common examples of microbial relationship to plants are Rhizobium- legume association and the mycorrhiza in gymnosperms (Bergersen & Costin, 1964).

Lorenz Hiltner in 1904 first coined the term "rhizosphere" to describe the interaction between legume root and bacteria. Later it was further categorized as "endorhizosphere" and "ectorhizosphere" which represent the interaction within and outside the root. Thus rhizosphere can be referred as a part of the soil ecosystem which involved the interaction among soil, soil microbes, and plant roots (Lynch et al., 2002).

Microbial Diversity and Ecology in Hotspots
DOI: https://doi.org/10.1016/B978-0-323-90148-2.00016-X

The roots deposit mainly two types of exudates by the process known as rhizodeposition: water-soluble and water-insoluble exudates. Plants deposit up to 40% of the photosynthetically fixed carbon through rhizodeposition making it available for microorganisms residing in rhizosphere. This increases the microbial density of some distinct microbes by 10- to 100-fold from the surrounding bulk soil considering it as microbial hotspot (Chandrashekar et al., 2014). Such hotspot occupy only a small portion of soil volume (usually between 1% and 5%, or 0.2% in subsoil) and the process rates are up to two orders of magnitude faster than in bulk soil (Kuzyakov & Blagodatskaya, 2015). So, the rhizosphere comprises an important habitat for a diverse microorganisms including both prokaryotes and eukaryotes that interact with each other (Bender et al., 2016; Hinsinger et al., 2009). The biological processes that takes place in rhizosphere determine various activities within ecosystems as turnover of organic matter, symbiotic and nonsymbiotic atmospheric nitrogen fixation, soil aggregation, nutrient availability, cycling of nutrients, and solubility of toxic elements etc. Fungi in rhizosphere are mostly beneficial in increasing soil fertility and plant growth. Therefore fungal association may assist in world hunger management to some extent by increasing in agricultural production and developing strategies for plant stress management.

6.2 Fungal association in rhizosphere

Rhizosphere is considered as soil microenvironment where numerous microorganisms like fungi, bacteria, etc. coexist with plants. So, it exhibits the highest levels of biological diversity of any environment (Bender et al., 2016). Numerous studies were carried out to understand the interaction between plants and microbes in rhizospheric soil, considering it as a hotspot for interactions. It is considered that for the growth of healthy plant, underground root–soil–microbe interactions plays an important part (Zhang et al., 2017). Soil provides water and nutrients to plants through plant roots and plants provide rhizodeposits as nutrients to microbes. The microbes then take the advantage of rhizodeposits and produce vast range of metabolites which can either stimulate or inhibit plant growth (Lynch et al., 2002). So, this complex system results active involvement of root–soil, soil–microbe, and microbe–microbe interactions in rhizosphere. The complex system of rhizosphere focusing fungal association is presented as the interaction among the plant root, fungi, soil, and its environment stimuli, and change any of these components may affect in their association.

6.2.1 Plant root and rhizodeposition

In rhizosphere, root is the main part associated with the plant where "rhizodeposition" plays an important role. The chemical characteristics and rate of exudates flow may lead to a proliferation of microbes in and outside the root (Jones et al., 2004). Rhizodeposition occurs by two different processes: leakage of minor organic compounds (less than 5% of carbon fixed by photosynthesis daily) over which the plant exerts little control and exudation of major organic compound with a specific function (obtained from root respiration or dying root cells) over which the plant exerts control (Lambers et al., 2009). The exudates may be soluble compounds as sugars, organic acids, amino acids, vitamins, hormones etc., and insoluble materials as root debris, sloughed-off materials, etc. It is estimated that there may occur the loss of up to 40% of plant's primary carbon production through rhizodeposition and it depends on the plant species, plant age, root architecture, membrane permeability, supply of carbon from shoot to root, and environmental conditions (Lynch et al., 2002).

The damage in membrane integrity due to biotic or abiotic stresses, like drought and nutrient deficiency, can increase the rate of exudation from root and diffuse in the soil. But fungi can utilize the exudates effectively showing significant reduction of diffused exudates in the soil (Jones et al., 2004). Different organic acids as rhizodeposits depend upon the plant species and its environment. There are various examples like cultivars of rice (*Oryza sativa*) release oxalate in the presence of lead (Yang et al., 2000), while *Triticum aestivum* release malate and citrate in the presence of copper (Clemens et al., 2002). Such released organic acids may chelate toxic metals like aluminum and prevent the plants from its harmful effects. Similarly, sometimes shoot regulate the rate of sugar supply to root which may act as signaling molecule for the regulation of root architecture (Freixes et al., 2002). For example, strigolactones as rhizodeposits from *Lotus japonicus* induces extensive hyphal branching in the arbuscular mycorrhizal fungus (AMF) during spore germination (Akiyama et al., 2005).

As significant amount of carbon flow through rhizodeposition does not take part in dry matter production, it is considered as a cost for the plant. This cost is balanced by endophytes like mycorrhizal fungi which utilize 4%–20% of the total carbon assimilated by host plant.

6.2.2 Fungi in rhizosphere

The fungi are a diverse group of organisms ranging from single celled yeasts to large macrofungi mushrooms. Hundred thousand species fungi described so far spend some stage of their life-cycle in the soil environment (Bridge & Spooner, 2001). In terrestrial ecosystems, especially in the rhizosphere, fungi play a major role for plant growth and soil health through different ecological processes (Gams, 2007). They are involved in activities like decomposition, mineralization, building of organic matter, biological control, and nitrogen and phosphorus cycling in soil. Fungi through mycorrhizal associations increase crop production and also help in carbon sequestration particularly in the forest soils (Clemmensen et al., 2013).

The well-studied beneficial rhizosphere fungi are endophytes such as mycorrhizal fungi and root-associated fungi such as *Trichoderma* and *Gliocladium* from the *Deuteromycetes,* and *Piriformospora* from the Sebacinales order (Kogel et al., 2006; Qiang et al., 2012) and deleterious rhizosphere fungi are pathogenic oomycetes fungi (Mendes et al., 2013).

6.2.2.1 Mycorrhizal fungi

The word mycorrhiza itself explain it as the symbiotic association between plant root and fungi since this word is derived from the Greek word "mikos" meaning fungi and "rhiza" meaning roots (Bonfante & Anca, 2009; Pathan et al., 2020). Mycorrhiza is classified into two groups as ectomycorrhiza and endomycorrhiza on the basis of their anatomical location in plant root. Ectomycorrhiza remains extracellular colonizing in intercellular spaces of epidermal and cortical cells of root forming hartig net associate mainly with woody plants like trees and shrubs, while endomycorrhiza is intracellular penetrating the hyphae inside the root cells forming hyphopodia (or appressoria), coils, and arbuscule. The endomycorrhiza is further classified into three subgroups as AMF, ericoid, and orchid. The AMF is most predominant associated with more than 80% of land plants including most agricultural crops (Akiyama et al., 2005; Daniell et al., 2001; Garcia et al., 2015), while ericoid mycorrhiza is restricted to the order Ericales and orchid to the family Orchidaceae. The extensive studies have been done primarily on two types of mycorrhiza: ectomycorrhizal (ECM) and arbuscular mycorrhizal (AM) due to their ecological and economic importance (Garcia et al., 2015).

Their functions include beneficial effects in establishment of plant, nutrient uptake, against biotic and abiotic stresses, as biocontrol agent, and the improvement of soil structure (Barea et al., 2005; Ocón et al., 2007).

Mycorrhiza colonize plant tissues through several steps like host recognition, spore germination, penetration of the epidermis, and tissue colonization (Petrini, 1991). They invade plant tissues with extensive branching in the vicinity of host root and penetrate the cuticle and epidermal cell wall. The infection may be inter or intracellular limiting to a single or multiple cells around the penetration site (Akiyama et al., 2005).

The sporulation and hyphal growth of mycorrhiza is initiated by signal molecules of root exudates. For example, Rutin, a flavonoid signal, in *Eucalyptus globulus* spp. *bicostata* root exudates initiates the ECM hyphal growth (Lagrange et al., 2001). Similarly, abietic acid in *Pinus sylvestris* root exudates induces germination of *Suillus* spp. spores (Garcia et al., 2015). Mycorrhiza also produces symbiotic signals like Myc factors that helps it to proliferate in diverse families like Fabaceae, Asteraceae, and Umbelliferae of the plants. Such discovery of Myc factors open the door for molecular and cellular dissection of plant root endosymbiosis (Maillet et al., 2011).

These fungal symbionts form a bridge between soil and plant roots termed "wood-wide web" which create network among different plants resulting horizontal nutrient movement (Simard et al., 1997). These networks promote aggregation and stability of soil that significantly affects soil quality and increase plant productivity (Rillig & Mummey, 2006). Many achlorophyllous heterotrophic plants obtain organic carbon through this network for their growth.

6.2.2.2 *Mycorrhiza like fungi*—Piriformospora indica

Piriformospora indica is an endophytic root-colonizing fungus, having similar characteristics of a typical AMF. But unlike AMF, it can be grown in pure culture, so also considered as a model organism (Varma et al., 1999). It has a broad host spectrum and can colonize in plant like *Arabidopsis* that is not colonized by AMF (Deshmukh et al., 2006). It bears chlamydospores which initiates colonization, germination, and finally forms an inter- and intrahyphal network in the root. This promotes the growth of a broad range of plants, increases biomass and yield of the crops, so it is utilized as a good alternative candidate to improve commercial plant production (Varma et al., 1999). Like in case of *Arabidopsis*, it enhances seed yield, reduces the time for seed ripening, and increases tolerance to abiotic stress (Peškan-Berghöfer et al., 2004). Besides it act as a potential biological agent against numerous root pathogens and also as plant protector by relieving stress conditions caused by acidity, desiccation, and heavy metal toxicity (Tripathi et al., 2008). For example, in *Hordeum vulgare* (barley), it induces local and systemic resistance to fungal diseases and abiotic stress (Deshmukh et al., 2006).

6.2.2.3 *Root-associated fungi*—Trichoderma *spp*

Trichoderma spp., unlike mycorrhizal fungi, are free-living root-associated fungi and common in root ecosystems. They are cosmopolitan in nature colonizing woody and herbaceous plant material. The temperate and tropical soils contain 10^1-10^3 culturable propagules per gram of soil (Harman et al., 2004). Most of *Trichoderma* spp. are fast growers, spore producers, and powerful antibiotic producers. These properties make them a strong invader that can reside in agricultural to desert soils of all climatic zones (including tundra, Antarctica, and tropics), and is ecologically abundant in diverse habitats such as in lake water, dead plant material to living roots of virtually any plant species (Woo et al., 2006). Their colonization in the plants has many advantages like they promote growth, develop resistant to soilborne diseases, and decrease susceptibility to abiotic stresses such as lack of nutrients, drought, and xenobiotic pesticides (Woo et al., 2006). There exist many strains of *Trichoderm*a which can produce over 200 types of antibiotics highly toxic to pathogenic microorganism (Woo et al., 2006).

Trichoderma spp. show dual characteristics as being plant symbionts and mycoparasite. Being opportunistic avirulent plant symbionts, they compete with other microorganisms for space and root exudates as nutrients. Such competent strains can colonize entire root surfaces for several weeks to months. They can degrade essential enzymes like pectinase for plant-pathogenic fungi to invade plants. And as mycoparasites, they can parasitize wide range of pathogenic fungi with the involvement of about 20−30 known genes, proteins, and metabolites (Harman et al., 2004). This mechanism follows certain steps like detection of other fungi, expression of cell wall degrading enzymes (cellulose, chitinase), expression of fungitoxic agents (endochitinases), attachment to host fungi (coils and appressoria), production of fungitoxic cell wall degrading enzymes, and dissolution of host cell wall, and finally death of host cell (Harman et al., 2004).

Several studies showed that *Trichoderma* spp. induces systemic disease resistance by producing signal molecules easily sensed by the plant. They can protect different dicots and monocots like Graminaceae, Solanaceae, and Cucurbitaceae from invasion by pathogenic fungi (*Rhizoctonia solani, Botrytis cinerea, Colletotrichum* spp., *Magnaporthe grisea, Phytophthora* spp., *Alternaria* spp., etc.), bacteria *(Xanthomonas* spp., *Pseudomonas syringae*, etc.), and even viruses (*Cucumber mosaic virus*) (Harman et al., 2004).

Trichoderma propagules are culturable in laboratory and can be produced in both liquid with water or oil based and dry as granules in cheaper price in industrial level. It can be stored for several months without losing their beneficial activities. Hence, *Trichoderma* spp. are widely used as efficient biocontrol agents as well as for the production of useful enzymes in agricultural and industrial sector.

6.2.2.4 Pathogenic fungi in rhizosphere

Root diseases are caused by various soilborne pathogenic fungi. These phytopathogens are ubiquitous in nature and difficult to control. The most common pathogens include *Fusarium culmorum, Fusarium pseudograminearum, Gaeumannomyces graminis* var. *Tritici, Bipolaris sorokiniana,* and *Alternaria* spp. in Ascomycota; *Rhizoctonia, Rhizopus oryzae, Rhizoctonia solani*, and *Penicillium* spp. in Basidiomycota; *Pythium* spp. in Oomycota; and *Curvularia* spp. in Deuteromycota (Zhang et al., 2017).

Fusarium culmorum causes head blight, and *Blumeria graminis* f. spp. *hordei*. causes powdery mildew in barley (Deshmukh et al., 2006). They can easily reside in plant debris or in soil in the form of chlamydospores. So, they are responsible for significant loss of grain worldwide (Chaudhary et al., 2018).

6.2.3 The soil and surrounding environment

The rhizosphere as microhabitat represents movement and storage of elements and water as well as the ability of roots and associated microorganisms to explore the soil under the influence of environmental factors like oxygen, pH, temperature, light intensity, etc.

6.2.3.1 Soil and its characteristics

Soil is the largest reservoir of biodiversity on the Earth which provide habitats for all forms of life ranging from unicellular prokaryotes to eukaryotes including protozoa and nematodes to mites, insects, and earthworms (Hinsinger et al., 2009). Nunan et al. (2003) stated that 1 g of soil may contain about 10^7-10^{12} bacteria, 10^4 protozoa, 10^4 nematodes, 5–25 km of fungal hyphae, and this may form hotspot of activities (Nunan et al., 2003).

In nature, the resources like water, air, nutrients present in soil is difficult to obtain by soil fauna, microbes, and roots. To facilitate the intake of resources, soil structure stability plays an important part. Soil structure is defined as the arrangement of soil particles bound together in aggregates of different sizes and building pores. The ability of the soil to maintain such an arrangement after being exposed to environmental stresses is called soil structure stability (Oades, 1984). Mycorrhiza plays an important role in the formation and stabilization of soil aggregates that ensures appropriate soil tilth (Barea et al., 2005) by either secreting the fungal products into the environment or releasing the inner contents via hyphal turnover and decomposition. Involvement of mycorrhizal fungi in soil aggregation can influence soil at different levels starting from fungal mycelium itself to host plant root and plant communities (Rillig & Mummey, 2006). Microorganisms that contribute to such soils include species of *Trichoderma, Fusarium, Streptomyces, Bacillus,* and *Actinomyces* spp.

The existence of disease-suppressive soils is based on the presence of specific microbial populations like *Trichoderma, Fusarium, Streptomyces, Bacillus,* and *Actinomyces* spp. These organisms suppress the pathogens by following the mechanisms like antagonism via secondary metabolites (e.g., antibiotics) production; competition for space and nutrients; hyperparasitism and elicitation of induced systemic resistance (ISR).

One of the examples is suppression of pathogenic *Fusarium* spp. by competing for carbon by nonpathogenic *Fusarium* spp. in case of Fusarium wilt (Pieterse et al., 2014).

6.2.3.2 Influence of temperature

Temperature plays a major role on root exudation influencing the growth of both plant and the microbes including fungi. The plant exudation response to temperature varied among species of plants. For example, *Solanum lycopersicum* (tomato) and *Trifolium subterraneum* (subterranean clover) roots secrete high amount amino acid asparagine at high temperature, while strawberry roots secrete higher amount of organic acids at low temperature (Pathan et al., 2020). Temperature affects enzyme kinetics which interfere with the spatial distribution of microbial hotspots. In a study of enzyme kinetics of phosphatase, cellobiohydrolase and leucine-aminopeptidase responsible for phosphorous, carbon, and nitrogen cycle in *Zea mays* (maize) plant, Vmax was found to increase at 10°C–30°C and decrease at 40°C and above. This shows that soil warming (at least up to 30°C) increases enzyme activity in hotspot areas (Ma et al., 2017).

6.2.3.3 Oxygen

Fungi are aerobic organisms requiring oxygen for their growth. In well-drained soils, air penetrates readily and oxygen concentrations are usually relatively high, whereas in water-logged soil, limited oxygen is present in dissolved form in soil water. The limited oxygen is used by fungi as well as plant root creating anaerobic condition in the rhizosphere. This results in proliferation of anaerobic microbes and deposition of nitrogen dioxide, manganese, iron, hydrogen peroxide along with different organic acids like acetic and butyric acids produced during microbial metabolism. The higher concentration of these products may damage root and ultimately affect the fungal population.

6.2.3.4 Water availability

Water acts as solvent for metabolic activities of plants and microbes in natural systems. The water availability in soil depends upon various factors such as soil structure and texture, soil organic matter content, rainfall, drainage, and plant cover (Lynch et al., 2002). Soil structure is influenced by soil water content which varies with the magnitude and frequency of wetting and drying cycles affecting plant growth (Rillig & Mummey, 2006). Water deficit cause the delay in plant growth, but some of the plants which can grow even in drought are equipped with special fungi. For example, plants associated with vesicular arbuscular mycorrhizae (VAM) can resist against drought by a number of mechanisms such as increased root hydraulic conductivity, stomatal regulation, hyphal water uptake, and osmotic adjustment (Ruiz-Lozano & Azcon, 1995).

6.2.3.5 Influence of pH

The pH of soil influences the life of below-ground. Extremes of pH impose major constraints on the root growth due to the toxicities of ions like in the acidic range by Al^{3+}, Mn^{2+}, and H^{+} ions (Marschner & Römheld, 1983) and in the alkaline range by HCO^{3-} ion. For example, at high pH 9, *Lypopersicum esculentum*, *Lactuca sativa*, and *Cynodon dactylon* showed the reduction of root growth (Tang et al., 1993). The nutrient deficiencies like iron in calcareous soils or phosphate in both acidic and alkaline soils were also observed (Richardson et al., 2009). In comparison to bulk soil, rhizosphere pH has been reported to be lesser than or greater than 1.0–2.0 pH units due to the physiological activities of plants and microbes.

The application of chemical fertilizers can also change the pH of rhizosphere. If ammonium-based fertilizer is used, hydrogen ions are released, while application of nitrate-based fertilizer release hydroxyl ions which results in the shift of soil pH to either acidic or basic, affecting the microflora as well as rhizodeposition For example, the iron is found predominantly in highly oxidized ferric form in aerated soil. This form is soluble at pH 7.4, but if there is an increase in pH, decrease in solubility occurs. This causes the concentration of available iron in soil to be too low to support the growth of microorganisms as well as plants.

6.2.3.6 Light intensity

The rate of photosynthetic carbon fixation by plants is directly related to light intensity and the carbon source present in root exudates is directly linked to amount of carbon fixation which affects the growth of rhizospheric microbes. Pathan et al. (2020) reported that the type of amino acid secretion also depends upon the intensity of light. It is mentioned that in case of trefoils, serine, glutamic acid, and alanine were released in high concentration in day light, while aspartic and glutamic acid, phenylalanine, and leucine were inhibited in shadow (Pathan et al., 2020). Similarly, different studies showed an increase in component of root exudate with increase in the intensity of light like catechin in spotted knapweed (Duke et al., 2009) and citrate in white tulips (Amiri et al., 2018).

6.2.3.7 Nutrient availability

Nutrient availability in soil depends upon different factors like soil chemistry, plant genotype, and nutrient supply (Jones et al., 2004). In rhizosphere, plants provide organic matter through rhizodeposition and the availability of organic matter drives the heterotrophic microorganisms (Jones et al., 2004; Wardle et al., 2004). The predominant microbial communities in rhizosphere are the most nutrient-efficient microorganisms which can compete for available nutrients in soils (Hinsinger et al., 2009). For example, in alpine and arctic soils where the availability of inorganic nitrogen is low, other communities than nitrate utilizers will predominate (Chapin et al., 1993). Besides competition, facilitation and complementarity activity may be shown by certain plants. For example, a model created by Raynaud et al. (2008) showed that plants having high rate of rhizodeposition can facilitate nutrients to neighboring plant roots with poor exudation potential (Raynaud et al., 2008). So, facilitation may act as one of the major driving forces of ecosystem productivity (Gross et al., 2007). This applies for mutualistic relationships between the host plant and the microsymbiont in mycorrhizal associations.

The soil nutrient content varies with season. It was high during winter as the preceding rainy season solubilized minerals available in soil layers enhancing soil nutrient availability, and thus favoring fungal growth. The availability of mineral nutrients in soil plays a crucial role in the occurrence of fungal communities in the rhizosphere. The flowering in *Ischaemum ciliare, Digitaria bicornis,* and *Heteropogon contortus* during the winter coincided with the species richness of fungal communities. Besides, the fibrous root system of grass species during winter might increase the root exudate content in soil which may also promote fungal communities in the rhizosphere (Vasanthakumari & Shivanna, 2011).

6.2.3.8 *Application of organic fertilizers*

The soil fungal population highly depends upon organic carbon status of soil. The application of different organic manure positively affect the fungal population (13×10^4 CFU/g soil) compared to the control soil (4×10^4 CFU/g) soil where organic manure was not applied (Das & Dkhar, 2011). The organic fertilizers like plant compost, vermicompost, integrated plant compost, and farmyard manure increase the rhizosphere microbial population and soil physicochemical properties than inorganic fertilizers. The fungal population for different treatments follows the order as farmyard manure followed by plant compost, then integrated compost, vermicompost, and control (Das & Dkhar, 2011). Recently, application of biochar in agricultural soils to improve soil properties and plant performance showed AMF abundance (Güereña et al., 2015).

6.3 Outcomes of fungal rhizosphere association

The outcomes of fungal rhizosphere association majorly depends upon the type of fungi associated with it along with rhizodeposition, soil properties, and environment stimuli. Generally, beneficial endophytes like mycorrhiza and root-associated fungi like *Trichoderma* spp. are responsible for positive outcomes and deleterious fungi for negative outcome.

6.3.1 Positive outcomes

6.3.1.1 Biological control

The term "biological control" in plant pathology is used to define the purposeful utilization of microbial antagonists to suppress diseases as well as the use of host-specific pathogens to control weed populations (Pal & Gardener, 2006). The biocontrol agents act upon pathogens through different mode of interaction preferably by mutualism, competition, parasitism, and predation (Odum & Barrett, 1971). If fungi can suppress the activities of other plant pathogens without harming host plant, such fungi can be used as biocontrol agents in nature. Fungi contributing to the disease control can be classified as competitive saprophytes, facultative and obligate plant symbionts, and facultative hyperparasites (Pal & Gardener, 2006).

The competitive saprophytes include avirulent fungi like *Fusarium oxysporum* which is phylogenetically very similar to plant pathogens. They can cause multiple infections in plants and suppress virulent pathogens, via the induction of host defenses. The facultative symbioants like *P. indica* can block the mycelial development of plant pathogens like *Gaeumannomyces graminis* and *Fusarium* spp. The obligate symbionts like AMFs can prevent root infections through various modes like competing for the access sites, stimulation of host defense, physical protection, and chemical interactions (Fitter & Garbaye, 1994) while ECM fungi prevent through antibiosis, synthesis of fungistatic compounds, and a physical barrier of the fungal mantle (Garcia-Garrido et al., 2000). For example, AM fungi *Glomus fasciculatum* and *Glomus macrocarpum* can suppress phytotoxic myxomycetes, causative agent of apple replant disease while ECM fungi like *Paxillus involutus* can suppress *F. oxysporum* and *F. moniliforme* responsible for root rot in red pine. The facultative hyperparasites can directly attach plant pathogen and kill it. For example, powdery mildew pathogens can be parasitized by hyperparasites like *Acremonium alternatum, Acrodontium crateriforme,* and *Gliocladium virens* (Kiss, 2003).

The use of biocontrol agent is preferred these days because of its advantage over chemical control for being ecological and eco-friendly, nontoxic, and does not show any detrimental effect in the long term (Tripathi et al., 2008).

6.3.1.2 Soil nutrient cycling and natural fertilizer

Soil nutrient cycling in the soil–plant system of crops relies on the effects of the agronomical practices on soil conditions. The soil microbes are responsible for various biogeochemical cycles and the fate of important elements like carbon, nitrogen, and phosphorous depend upon microbial activities (Lambers et al., 2009). AMFs provide the host plant with mineral nutrients and water, in exchange for photosynthetic products (Smith et al., 2000). AM support nitrogen fixation process in legumes and involves in absorption, transportation, and utilization of NH^{4+}. In both ECM and AM fungi, genes involved in organic and inorganic uptake of nitrogen have been identified and found that plant nitrogen transporters were activated during mycorrhization. These transporters help in release of a substantial amount of nitrogen to their hosts plant (Barea et al., 2005). Due to these beneficial characteristics, they are potential natural fertilizers with a view of sustainable agriculture (Berruti et al., 2016).

6.3.1.3 Fungi as decomposer

Fungi play major role as decomposers of organic matter. A study was carried out to detect the major functional groups of fungi through depth profile using 454 pyro sequencing technique which showed that free-living saprophytes predominate in upper layer of soil while mycorrhizal and root-associated fungi predominate with increase in soil depth. This suggests that regulation of organic matter in soil depends upon type of fungi present in soil (Clemmensen et al., 2013).

6.3.1.4 Phosphate solubilization and intake

In nature, phosphorous is found in insoluble forms in bone meals, rock phosphates, and chemically fixed soil phosphorous compounds. So, to make phosphorous available for the plants, phosphate solubilization process plays an important part. This process is generally carried out by fungi of *Penicillium*, *Aspergillus*, and *Fusarium* genera and potential phosphorous solubilizers are *Aspergillus awamori*, *Penicillium digitatum*, *Aspergillus niger*, and *Schwanniomyces occidentalis* (Pindi & Satyanarayana, 2012). Such solubilized phosphate will be efficiently uptake by the plants with the help of AM fungi. (Barea et al., 2005). Different studies showed that AM fungi possess active phosphate transporters that take up inorganic phosphate (Pi) from the soil and deliver to the plants and such fungi can transport even from low phosphorous content soil (Daniell et al., 2001). The plants are usually Mycorrhiza specific and help in the expression of Pi transporter exclusively during AM symbiosis to acquire Pi delivered by the AM fungus (Bonfante & Anca, 2009).

6.3.1.5 Induced systemic resistance

The term "induced resistance" is defined as the induced state of resistance in plants to protect other nonexposed plant parts against pathogenic microbes and herbivorous insects attack. It is developed by plants through exposure to a pathogen, herbivorous insects and beneficial microbes as biological inducers as well as to chemicals as chemical inducers. This is expressed by the activation of latent defense mechanisms against those agents locally at the site of induction as well as systemically in plant parts that are spatially separated from the inducer and is known as ISR. ISR is regulated by a network of interconnected signaling pathways with plant hormones as major regulatory molecules. The signaling components are usually shared by beneficial microbes with pathogens and protect the plants from their detrimental effects. A wide variety of root-associated mutualists, including *Trichoderma*, *P. indica*, and Mycorrhiza species sensitize the plant immune system for enhanced defense without directly activating costly defenses (Pieterse et al., 2014).

6.3.1.6 Abiotic stress

Microorganisms can cope with different environmental stress. During unfavorable condition, stress induces a cellular stage which leads to the synthesis of specific proteins and metabolites to adapt and survive till favorable condition return (Ocón et al., 2007). The plants associated with AM fungi show more tolerance to environmental stress such as drought, salinity, and cold by the accumulation of sugars and other compatible solutes as osmoprotectants acting as stabilizing biomolecules (Voit, 2003). For example, trehalose is found at high concentrations in resting cells and survival forms, like spores in most of fungi. Auxin is another phytohormone which is involved in biotic and abiotic stress responses and plays a major role in signaling between plant and fungi during ECM establishment (2012). The development of plants with high productivity and survival rate under drought conditions helps to fulfill the main goal of crop breeding programs (Balestrini et al., 2018).

6.3.2 Negative outcome

Besides the majority of beneficial fungi, rhizosphere bear phytopathogens in its microbiome community. The studies revealed that rhizosphere can be considered as an infectious courtyard where soilborne pathogens, including bacteria, fungi, oomycetes, and nematodes are in association with the plant (Chaudhary et al., 2018). Generally, interactions between plant and phytopathogens are host specific. Plants secrete particular classes of molecules like amino acids, sugars, and other exudate components through root exudate which directly stimulate phytopathogens (Lynch et al., 2002), while pathogens emit particular classes of effectors into the apoplast to infect in the root tissues (Chaudhary et al., 2018). The continuous cropping of the same crop can lead to decreased yields due to an increase in specific plant pathogen numbers in the soil, so crop rotation is one of the best employed method to solve this problem (Lynch et al., 2002).

Wheat is one of the three major staple foods in the world, grown in Asia (China), Australia, Europe, North America, and South America (Duffy, 2000). All cultivars of wheat are vulnerable to attack by several soilborne fungal pathogens that cause root diseases resulting in tremendous yield loss. These root diseases are difficult to control due to capability of phytopathogens to form durable chlamydospores in the soil with or in the absence of growing hosts and can evade plant defenses. These phytopathogens can cause huge loss of nutritionally valuable plants like potato, tomato, and eggplants as well as industrially valuable like cotton and some olive tree species. Root infection is the initial phase for a number of different diseases like decay, rot, damping off, and wilt. The destruction of plant roots affects plant water and nutrient uptake causing nutrient deficiency symptoms, reduced size, and drought stress etc. (Pathan et al., 2020; Raaijmakers et al., 2009).

6.4 Application of fungal rhizosphere association

The Population Division of United Nations had stated that world population will continue to grow with an estimated 7.7 billion global population in 2019 could grow to around 8.5 billion in 2030, 9.7 billion in 2050, and 10.9 billion by 2100 (World Population Prospects, 2019). Global Report on Food Crises (GRFC) 2020 mentioned that in 2019, the number of people in crisis (IPC/CH Phase 3 or above) was the highest in the four years of the GRFC's existence (Food Security Information Network, 2020). To meet the demand of increasing world population, abundant food and feed need to be produced (Pal & Gardener, 2006) which can be achieved by applying fertilizers and pesticides.

Despite chemical fertilizers and pesticides were in use from 19th century to meet world population demand, development of pesticide resistance and reduction of beneficial rhizospheric microbial population have instigated researchers to develop biofertilizers and biocontrol agents as the alternatives (Berruti et al., 2016; Woo et al., 2006). The beneficial fungi like *Trichoderma* spp. and AMFs are the potent source of both biofertilizers and biocontrol agents.

Trichoderma spp. are marketed as biopesticides, biofertilizers, growth enhancers, and stimulants of natural resistance (Woo et al., 2014). More than 50 different *Trichoderma*-based agricultural products are available and applied to protect and improve yield of vegetable, ornamentals, and fruit trees (Woo et al., 2006). The genes *Taabc2* (from *Trichoderma atroviride*) are responsible for resistance to pathogen-secreted metabolites, mycotoxins, and some fungicides (Ruocco et al., 2009) and *Hytlo1* (from MK1 strain of *Trichoderma longibrachiatum*) are responsible for stimulation of root formation and growth as well as antifungal activity (Ruocco et al., 2015).

AMFs are widely used biofertilizers in agriculture industry to replace chemical fertilizer because they are formed early in the development of the plants, represent nearly ubiquitous root, and assist plants in the uptake of nutrients (especially phosphorus and micronutrients) (Berruti et al., 2016; Pal & Gardener, 2006). AMF inoculation in the field has proven to be effective and give better performance on plant production (by more than 20%) in both controlled and open-field conditions (Berruti et al., 2016; Mäder et al., 2011). *P. indica* is another potent biofertilizer since they are culturable in laboratory (Deshmukh et al., 2006). So, application of rhizospheric fungi to improve plant performance is a widely applied technique in many countries.

6.5 Conclusion

The governing bodies of most of the countries are implementing legislative mandates in agricultural sector with the objective of minimizing the application of pesticides and chemical fertilizers for environmental safety, so biocontrol agents and biofertilizers are emerging as potent agents for their replacement. With *Trichoderma* and Mycorrhiza considered as the prominent biocontrol agents and biofertilizers, the research on rhizospheric fungi are getting more attention and promoted by organization like International Organisation for Biological Control. But European and Mediterranean Plant Protection Organization had prepared the guidelines regarding import, release, and application procedures of nonindigenous biological control agents to minimize the risks to natural ecosystems. Hence, with restrictive regulations for their implementation, but at the same time need to meet the demand of food for the ever-increasing world population, comprehensive studies of fungal interactions in rhizospheric hotspots are even more urgent to support crop production.

References

Akiyama, K., Matsuzaki, K. I., & Hayashi, H. (2005). Plant sesquiterpenes induce hyphal branching in arbuscular mycorrhizal fungi. *Nature*, *435*(7043), 824–827. Available from https://doi.org/10.1038/nature03608.

Amiri, A., Kafi, M., Kalate-Jari, S., & Matinizadeh, M. (2018). Tulip response to different light sources. *Journal of Animal and Plant Sciences, 28*(2), 539–545. Available from http://thejaps.org.pk/docs/v-28-02/23.pdf.

Balestrini, R., Chitarra, W., Antoniou, C., Ruocco, M., & Fotopoulos, V. (2018). Improvement of plant performance under water deficit with the employment of biological and chemical priming agents. *Journal of Agricultural Science, 156*(5), 680–688. Available from https://doi.org/10.1017/S0021859618000126.

Barea, J. M., Azcón, R., & Azcón-Aguilar, C. (2005). Soil biology. In F. Busco, & A. Varma (Eds.), *Interactions between mycorrhizal fungi and bacteria to improve plant nutrient cycling and soil structure* (Vol. 3, pp. 195–212). Springer Science and Business Media LLC. Available from https://doi.org/10.1007/3-540-26609-7_10.

Bender, S. F., Wagg, C., & van der Heijden, M. G. A. (2016). An underground revolution: Biodiversity and soil ecological engineering for agricultural sustainability. *Trends in Ecology and Evolution, 31*(6), 440–452. Available from https://doi.org/10.1016/j.tree.2016.02.016.

Bergersen, F. J., & Costin, A. B. (1964). Root nodules on *Podocarpus lawrencei* and their ecological significance. *Australian Journal of Biological Sciences, 17*, 44–48. Available from https://doi.org/10.1071/bi9640044.

Berruti, A., Lumini, E., Balestrini, R., & Bianciotto, V. (2016). Arbuscular mycorrhizal fungi as natural biofertilizers: Let's benefit from past successes. *Frontiers in Microbiology, 6*. Available from https://doi.org/10.3389/fmicb.2015.01559.

Bonfante, P., & Anca, I. A. (2009). Plants, mycorrhizal fungi, and bacteria: A network of interactions. *Annual Review of Microbiology, 63*, 363–383. Available from https://doi.org/10.1146/annurev.micro.091208.073504.

Bridge, P., & Spooner, B. (2001). Soil fungi: Diversity and detection. *Plant and Soil, 232* (1–2), 147–154. Available from https://doi.org/10.1023/A:1010346305799.

Chandrashekar, M. A., Pai., & Raju, N. S. (2014). Fungal diversity of rhizosphere soils in different agricultural fields of Nanjangud Taluk of Mysore District. *International Journal of Current Microbiology and Applied Sciences, 3*(5), 559–566.

Chapin, F. S., Moilanen, L., & Kielland, K. (1993). Preferential use of organic nitrogen for growth by a non-mycorrhizal arctic sedge. *Nature, 361*(6408), 150–153. Available from https://doi.org/10.1038/361150a0.

Chaudhary, R., Balhara, M., Dangi, M., Dangi, M., & Chhillar, A. K. (2018). Rhizosphere microbiome: An emerging frontier in causing and curing infectious diseases. *Asian Journal of Pharmaceutical and Clinical Research, 11*(6), 65–69. Available from https://doi.org/10.22159/ajpcr.2018.v11i6.25299.

Clemens, S., Palmgren, M. G., & Krämer, U. (2002). A long way ahead: Understanding and engineering plant metal accumulation. *Trends in Plant Science, 7*(7), 309–315. Available from https://doi.org/10.1016/S1360-1385(02)02295-1.

Clemmensen, K. E., Bahr, A., Ovaskainen, O., Dahlberg, A., Ekblad, A., Wallander, H., Stenlid, J., Finlay, R. D., Wardle, D. A., & Lindahl, B. D. (2013). Roots and associated fungi drive long-term carbon sequestration in boreal forest. *Science, 340*(6127), 1615–1618. Available from https://doi.org/10.1126/science.1231923.

Daniell, T. J., Husband, R., Fitter, A. H., & Young, J. P. W. (2001). Molecular diversity of arbuscular mycorrhizal fungi colonising arable crops. *FEMS Microbiology Ecology, 36* (2–3), 203–209. Available from https://doi.org/10.1016/S0168-6496(01)00134-9.

Das, B. B., & Dkhar, M. S. (2011). Rhizosphere Microbial Populations and Physico Chemical Properties as Affected by Organic and Inorganic Farming Practices, Am-Euras. *Journal of Agriculture and Environmental Sciences, 10*(2), 140–150.

Deshmukh, S., Hückelhoven, R., Schäfer, P., Imani, J., Sharma, M., Weiss, M., Waller, F., & Kogel, K. H. (2006). The root endophytic fungus *Pirifomospora indica* requires host cell for proliferation during mutualistic symbiosis with barley. *Proceedings of the National*

Academy of Sciences of the United States of America, 103(49), 18450–18457. Available from https://doi.org/10.1073/pnas.0605697103.

Duffy, B. (2000). Combination of pencycuron and Pseudomonas fluorescens strain 2-79 for integrated control of rhizoctonia root rot and take-all of spring wheat. *Crop Protection, 19*(1), 21–25. Available from https://doi.org/10.1016/S0261-2194(99)00078-2.

Duke, S. O., Blair, A. C., Dayan, F. E., Johnson, R. D., Meepagala, K. M., Cook, D., & Bajsa, J. (2009). Is (-)-catechin a novel weapon of spotted Knapweed (*Centaurea stoebe*)? *Journal of Chemical Ecology, 35*(2), 141–153. Available from https://doi.org/10.1007/s10886-008-9587-z.

Feiro, A. D. (1978). The rhizosphere. *The American Biology Teacher, 64*, 50–52.

Fitter, A. H., & Garbaye, J. (1994). Interactions between mycorrhizal fungi and other soil organisms. *Plant and Soil, 159*, 123–132. Available from https://doi.org/10.1007/bf00000101.

Food Security Information Network, 2020. Global Report on Food Crisis 2020, FSIN, Italy.

Freixes, S., Thibaud, M. C., Tardieu, F., & Muller, B. (2002). Root elongation and branching is related to local hexose concentration in arabidopsis thaliana seedlings. *Plant, Cell and Environment, 25*(10), 1357–1366. Available from https://doi.org/10.1046/j.1365-3040.2002.00912.x.

Gams, W. (2007). Biodiversity of soil-inhabiting fungi. *Biodiversity and Conservation, 16*(1), 69–72. Available from https://doi.org/10.1007/s10531-006-9121-y.

Garcia, K., Delaux, P. M., Cope, K. R., & Ané, J. M. (2015). Molecular signals required for the establishment and maintenance of ectomycorrhizal symbioses. *New Phytologist, 208*(1), 79–87. Available from https://doi.org/10.1111/nph.13423.

Garcia-Garrido, J. M., Tribak, M., Rejon-Palomares, A., Ocampo, J. A., & Garcia-Romera, I. (2000). Hydrolytic enzymes and ability of arbuscular mycorrhizal fungi to colonize roots. *Journal of Experimental Botany, 51*(349), 1443–1448. Available from https://doi.org/10.1093/jxb/51.349.1443.

Gross, N., Suding, K. N., Lavorel, S., & Roumet, C. (2007). Complementarity as a mechanism of coexistence between functional groups of grasses. *Journal of Ecology, 95*(6), 1296–1305. Available from https://doi.org/10.1111/j.1365-2745.2007.01303.x.

Güereña, D. T., Lehmann, J., Thies, J. E., Enders, A., Karanja, N., & Neufeldt, H. (2015). Partitioning the contributions of biochar properties to enhanced biological nitrogen fixation in common bean (*Phaseolus vulgaris*). *Biology and Fertility of Soils, 51*(4), 479–491. Available from https://doi.org/10.1007/s00374-014-0990-z.

Harman, G. E., Howell, C. R., Viterbo, A., Chet, I., & Lorito, M. (2004). Trichoderma species—Opportunistic, avirulent plant symbionts. *Nature Reviews Microbiology, 2*(1), 43–56. Available from https://doi.org/10.1038/nrmicro797.

Hinsinger, P., Bengough, A. G., Vetterlein, D., & Young, I. M. (2009). Rhizosphere: Biophysics, biogeochemistry and ecological relevance. *Plant and Soil, 321*(1–2), 117–152. Available from https://doi.org/10.1007/s11104-008-9885-9.

Jones, D. L., Hodge, A., & Kuzyakov, Y. (2004). Plant and mycorrhizal regulation of rhizodeposition. *New Phytologist, 163*(3), 459–480. Available from https://doi.org/10.1111/j.1469-8137.2004.01130.x.

Kiss, L. (2003). A review of fungal antagonists of powdery mildews and their potential as biocontrol agents. *Pest Management Science, 59*(4), 475–483. Available from https://doi.org/10.1002/ps.689.

Kogel, K. H., Franken, P., & Hückelhoven, R. (2006). Endophyte or parasite—what decides? *Current Opinion in Plant Biology, 9*(4), 358–363. Available from https://doi.org/10.1016/j.pbi.2006.05.001.

Kuzyakov, Y., & Blagodatskaya, E. (2015). Microbial hotspots and hot moments in soil: Concept and review. *Soil Biology and Biochemistry, 83*, 184–199. Available from https://doi.org/10.1016/j.soilbio.2015.01.025.

Lagrange, H., Jay-Allgmand, C., & Lapeyrie, F. (2001). Rutin, the phenolglycoside from eucalyptus root exudates, stimulates *Pisolithus hyphal* growth at picomolar concentrations. *New Phytologist, 149*(2), 349–355. Available from https://doi.org/10.1046/j.1469-8137.2001.00027.x.

Lambers, H., Mougel, C., Jaillard, B., & Hinsinger, P. (2009). Plant-microbe-soil interactions in the rhizosphere: An evolutionary perspective. *Plant and Soil, 321*(1–2), 83–115. Available from https://doi.org/10.1007/s11104-009-0042-x.

Lynch, J. M., Brimecombe, M. J., & De Leij, F. A. (2002). Rhizosphere, In: eLS. John Wiley & Sons, Ltd: Chichester. *eLS*. Available from https://doi.org/10.1038/npg.els.0000403

Ma, X., Razavi, B. S., Holz, M., Blagodatskaya, E., & Kuzyakov, Y. (2017). Warming increases hotspot areas of enzyme activity and shortens the duration of hot moments in the root-detritusphere. *Soil Biology and Biochemistry, 107*, 226–233. Available from https://doi.org/10.1016/j.soilbio.2017.01.009.

Mäder, P., Kaiser, F., Adholeya, A., Singh, R., Uppal, H. S., Sharma, A. K., Srivastava, R., Sahai, V., Aragno, M., Wiemken, A., Johri, B. N., & Fried, P. M. (2011). Inoculation of root microorganisms for sustainable wheat-rice and wheat-black gram rotations in India. *Soil Biology and Biochemistry, 43*(3), 609–619. Available from https://doi.org/10.1016/j.soilbio.2010.11.031.

Maillet, F., Poinsot, V., André, O., Puech-Pagés, V., Haouy, A., Gueunier, M., Cromer, L., Giraudet, D., Formey, D., Niebel, A., Martinez, E. A., Driguez, H., Bécard, G., & Dénarié, J. (2011). Fungal lipochitooligosaccharide symbiotic signals in arbuscular mycorrhiza. *Nature, 469*(7328), 58–64. Available from https://doi.org/10.1038/nature09622.

Marschner, H., & Römheld, V. (1983). In vivo Measurement of Root-induced pH Changes at the Soil-Root Interface: Effect of Plant Species and Nitrogen Source. *Z. Pjlanzenphysiol. Bd., 111*, 241–251. Available from https://doi.org/10.1094/PHI-A-2006-1117-02

Mendes, R., Garbeva, P., & Raaijmakers, J. M. (2013). The rhizosphere microbiome: Significance of plant beneficial, plant pathogenic, and human pathogenic microorganisms. *FEMS Microbiology Reviews, 37*(5), 634–663. Available from https://doi.org/10.1111/1574-6976.12028.

Nunan, N., Wu, K., Young, I. M., Crawford, J. W., & Ritz, K. (2003). Spatial distribution of bacterial communities and their relationships with the micro-architecture of soil. *FEMS Microbiology Ecology, 44*(2), 203–215. Available from https://doi.org/10.1016/S0168-6496(03)00027-8.

Oades, J. M. (1984). Soil organic matter and structural stability: Mechanisms and implications for management. *Plant and Soil, 76*, 319–337. Available from https://doi.org/10.1007/bf02205590.

Ocón, A., Hampp, R., & Requena, N. (2007). Trehalose turnover during abiotic stress in arbuscular mycorrhizal fungi. *New Phytologist, 174*(4), 879–891. Available from https://doi.org/10.1111/j.1469-8137.2007.02048.x.

Odum, E. P., & Barrett, G. W. (1971). *Fundamentals of ecology* (5th ed.). Thomson, Brooks/Cole.

Pal, K. K., & Gardener, B. M. (2006). Biological Control of Plant Pathogens, *The Plant Health Instructor*, 1–25. Available from https://doi.org/10.1016/S0044-328X(83)80083-X.

Pathan, S. I., Ceccherin, M. T., Sunseri, F., & Lupini, A. (2020). Rhizosphere as hotspot for plant-soil-microbe interaction. In R. Datta, R. S. Meena, S. I. Pathan, & M. T. Ceccherini (Eds.), *Carbon and nitrogen cycling in soil* (pp. 17–44). Springer Nature Singapore Pte Ltd. Available from https://doi.org/10.1007/978-981-13-7264-3_1.

Peškan-Berghöfer, T., Shahollari, B., Giong, P.H., Hehl, S., Markert, C., Blanke, V., Kost, G., Varma, A., & Oelmüllera arm R. (2004). Association of *Piriformospora indica* with *Arabidopsis thaliana* roots represents a novel system to study beneficial plant–microbe interactions and involves early plant protein modifications in the endoplasmic reticulum and at the plasma membrane, *Physiologia Plantarum, 122*, 465–477. Available from https://doi.org/10.1111/j.1399-3054.2004.00424.x.

Petrini, O. (1991). *Fungal endophytes of tree leaves. Microbial ecology of leaves* (pp. 179–197). Springer-Verlag New York Inc.

Pieterse, C. M. J., Zamioudis, C., Berendsen, R. L., Weller, D. M., Van Wees, S. C. M., & Bakker, P. A. H. M. (2014). Induced systemic resistance by beneficial microbes. *Annual Review of Phytopathology, 52*, 347–375. Available from https://doi.org/10.1146/annurev-phyto-082712-102340.

Pindi, P. K., & Satyanarayana, S. D. V. (2012). Liquid microbial consortium—A potential tool for sustainable soil health. *Journal of Biofertilizers & Biopesticides, 3*(4), 1–9. Available from https://doi.org/10.4172/2155-6202.1000124.

Qiang, X., Weiss, M., Kogel, K. H., & Schäfer, P. (2012). Piriformospora indica—A mutualistic basidiomycete with an exceptionally large plant host range. *Molecular Plant Pathology, 13* (5), 508–518. Available from https://doi.org/10.1111/j.1364-3703.2011.00764.x.

Raaijmakers, J. M., Paulitz, T. C., Steinberg, C., Alabouvette, C., & Moënne-Loccoz, Y. (2009). The rhizosphere: A playground and battlefield for soilborne pathogens and beneficial microorganisms. *Plant and Soil, 321*(1–2), 341–361. Available from https://doi.org/10.1007/s11104-008-9568-6.

Raynaud, X., Jaillard, B., & Leadley, P. W. (2008). Plants may alter competition by modifying nutrient bioavailability in rhizosphere: A modeling approach. *American Naturalist, 171*(1), 44–58. Available from https://doi.org/10.1086/523951.

Richardson, A. E., Barea, J. M., McNeill, A. M., & Prigent-Combaret, C. (2009). Acquisition of phosphorus and nitrogen in the rhizosphere and plant growth promotion by microorganisms. *Plant and Soil, 321*(1–2), 305–339. Available from https://doi.org/10.1007/s11104-009-9895-2.

Rillig, M. C., & Mummey, D. L. (2006). Mycorrhizas and soil structure. *New Phytologist, 171*(1), 41–53. Available from https://doi.org/10.1111/j.1469-8137.2006.01750.x.

Ruiz-Lozano, J. M., & Azcon, R. (1995). Hyphal contribution to water uptake in mycorrhizal plants as affected by the fungal species and water status. *Physiologia Plantarum, 95*, 472–478. Available from https://doi.org/10.1034/j.1399-3054.1995.950320.x.

Ruocco, M., Lanzuise, S., Lombardi, N., Woo, S. L., Vinale, F., Marra, R., Varlese, R., Manganiello, G., Pascale, A., Scala, V., Turrà, D., Scala, F., & Lorito, M. (2015). Multiple roles and effects of a novel Trichoderma hydrophobin. *Molecular Plant-Microbe Interactions, 28*(2), 167–179. Available from https://doi.org/10.1094/MPMI-07-14-0194-R.

Ruocco, M., Lanzuise, S., Vinale, F., Marra, R., Turrà, D., Woo, S. L., & Lorito, M. (2009). Identification of a new biocontrol gene in *Trichoderma atroviride*: The role of an ABC transporter membrane pump in the interaction with different plant-pathogenic fungi. *Molecular Plant-Microbe Interactions, 22*(3), 291–301. Available from https://doi.org/10.1094/MPMI-22-3-0291.

Simard, S. W., Perry, D. A., Jones, M. D., Myrold, D. D., Durall, D. M., & Molina, R. (1997). Net transfer of carbon between ectomycorrhizal tree species in the field. *Nature, 388* (6642), 579–582. Available from https://doi.org/10.1038/41557.

Smith, F. A., Jakobsen, I., & Smith, S. E. (2000). Spatial differences in acquisition of soil phosphate between two arbuscular mycorrhizal fungi in symbiosis with *Medicago truncatula*. *New Phytologist, 147*(2), 357–366. Available from https://doi.org/10.1046/j.1469-8137.2000.00695.x.

Tang, C., Kuo, J., Longnecker, N. E., Thomson, C. J., & Robson, A. D. (1993). High pH causes disintegration of the root surface in *Lupinus angustifolius* L. *Annals of Botany, 71* (3), 201–207. Available from https://doi.org/10.1006/anbo.1993.1025.

Tripathi, S., Kamal, S., Sheramati, I., Oelmuller, R., & Varma, A. (2008). *Mycorrhizal fungi and other root endophytes as biocontrol agents against root pathogens. Mycorrhiza: State of the art, genetics and molecular biology, eco-function, biotechnology, eco-physiology, structure and systematics* (3rd ed., pp. 281–306). Springer-Verlag Berlin Heidelberg. Available from https://doi.org/10.1007/978-3-540-78826-3_14.

Varma, A., Verma, S., Sudha., Sahay, N., Bütehorn, B., & Franken, P. (1999). *Piriformospora indica*, a cultivable plant-growth-promoting root endophyte. *Applied and Environmental Microbiology, 65*(6), 2741–2744. Available from https://doi.org/10.1128/aem.65.6.2741-2744.1999.

Vasanthakumari, M. M., & Shivanna, M. B. (2011). Fungal assemblages in the rhizosphere and rhizoplane of grasses of the subfamily Panicoideae in the Lakkavalli region of Karnataka, India. *Microbes and Environments, 26*(3), 228–236. Available from https://doi.org/10.1264/jsme2.ME10163.

Voit, E. O. (2003). Biochemical and genomic regulation of the trehalose cycle in yeast: Review of observations and canonical model analysis. *Journal of Theoretical Biology, 223*(1), 55–78. Available from https://doi.org/10.1016/S0022-5193(03)00072-9.

Wardle, D. A., Bardgett, R. D., Klironomos, J. N., Setälä, H., Van Der Putten, W. H., & Wall, D. H. (2004). Ecological linkages between aboveground and belowground biota. *Science, 304*(5677), 1629–1633. Available from https://doi.org/10.1126/science.1094875.

Woo, S. L., Ruocco, M., Francesco, V., Nigro, M., Marra, R., Lombardi, N., Pascale, A., Lanzuise, S., Manganiello, G., & Lorito, M. (2014). Trichoderma-based products and their widespread use in agriculture. *The Open Mycology Journal, 8*, 71–126. Available from https://doi.org/10.2174/1874437001408010071.

Woo, S. L., Scala, F., Ruocco, M., & Lorito, M. (2006). The molecular biology of the interactions between *Trichoderma* spp., phytopathogenic fungi, and plants. *Phytopathology, 96*(2), 181–185. Available from https://doi.org/10.1094/PHYTO-96-0181.

World Population Prospects (2019). New York: United Nations.

Yang, Y. Y., Jung, J. Y., Song, W. Y., Suh, H. S., & Lee, Y. (2000). Identification of rice varieties with high tolerance or sensitivity to lead and characterization of the mechanism of tolerance. *Plant Physiology, 124*(3), 1019–1026. Available from https://doi.org/10.1104/pp.124.3.1019.

Zhang, R., Vivanco, J. M., & Shen, Q. (2017). The unseen rhizosphere root–soil–microbe interactions for crop production. *Current Opinion in Microbiology, 37*, 8–14. Available from https://doi.org/10.1016/j.mib.2017.03.008.

CHAPTER

7

Diversity of actinomycetes in Western Ghats

Aparna Gunjal[1] *and Devidas S. Bhagat*[2]

[1]Department of Microbiology, Dr. D.Y. Patil, Arts, Commerce & Science College, Pune, India, [2]Department of Forensic Chemistry and Toxicology, Government Institute of Forensic Science, Aurangabad, India

7.1 Introduction

7.1.1 Western Ghats, the hotspot of biodiversity

Western Ghats of India is the main hotspot for many plants, animals, and microorganisms (bacteria, actinomycetes, fungi, algae, and yeasts) (Gautham et al., 2012). The Western Ghats is termed as "gene pool" due to diversity of trees and crops, animals, birds, reptiles, insects, fish, and microorganisms. This variety of plants, animals, and microorganisms in the Western Ghats is immensely unique which needs to be conserved. The conservation of biodiversity in the Western Ghats is by the establishment of different protected areas, national parks, and sanctuaries (Nampoothiri et al., 2013). The Western Ghats ranks eighth among the hotspots for biodiversity. The Western Ghats is peninsular India's highest peak. The Western Ghats covers various states in India viz., Maharashtra, Gujarat, Goa, Karnataka, Kerala, and Tamil Nadu (Nampoothiri et al., 2013). The area of Western Ghats is around 180,000 km^2 (Nampoothiri et al., 2013), which are ecologically sensitive areas. The Western Ghats are the reservoirs of important metabolites from many microorganisms, especially from the actinomycetes group. The forest part of Western Ghats for the diversity of actinomycetes is not studied (Siddharth et al., 2020). There is a report on diversity of classification and primary and secondary substances produced by actinobacteria (Buedenbender et al., 2017). The geographical location of Western Ghats is shown in Fig. 7.1.

DOI: https://doi.org/10.1016/B978-0-323-90148-2.00007-9

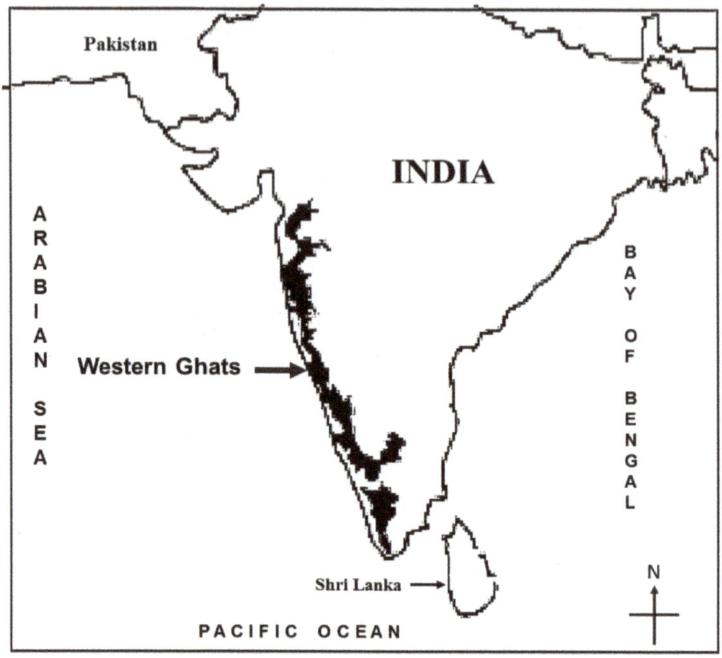

FIGURE 7.1 Plant growth-promoting substances by solid state fermentation using *Streptomyces* sp.

There are very few reports on the diversity of actinomycetes from Western Ghats (Dezfully & Ramanayaka, 2015; Radhakrishnan et al., 2010). The chapter focuses on the diversity and application of actinomycetes from the Western Ghats.

7.2 Actinomycetes

The actinomycetes group includes filamentous Gram-positive bacteria (Devadass et al., 2016). The Western Ghats include the diversity of actinomycetes that produce different antibiotics (Akshatha, Manjunatha, Divyashree, et al., 2013; Watve et al., 2001). The actinomycetes have high G + C content (Mesta et al., 2017). Their growth is compact, colonies are very hard, and the surface is flat or elevated (Tiwari et al., 2019). The actinomycetes have nonseptate, sporulation mycelium branched or nonbranched, straight, or spiral-shaped. They produce spores with varying shapes viz., spherical, cylindrical, or oval (Chamikara, 2016). The morphological features of actinomycetes are similar to fungi and cell wall contents are the same as Gram-positive bacteria. Because of these reasons, the actinomycetes are put in a separate category other than the usual bacteria. The color of the aerial mycelium varies viz., white, creamy white, chalky, powdery, brown, gray to pinkish and violet.

The color of substrate mycelium can be brown, yellow to orange (Salim et al., 2017). Some actinomycetes show the formation of concentric rings (Tiwari et al., 2019). They also known to produce pigments such as melanin (Mohseni et al., 2013). The genus *Streptomyces* is the most major actinomycete and is reported to produce more than 75% of primary and secondary metabolites (Milind & Charushila, 2007; Prashith et al., 2015; Ravi Ranjan et al., 2015; Sharma et al., 2016). There are 843 species of *Streptomyces* studied from various environments (Parte, 2018). This genus *Streptomyces actinobacteria, Acidimicrobiia, Coriobacteriia, Nitriliruptoria, Rubrobacteria*, and *Thermoleophilia* (Tiwari et al., 2019) and 16 orders. The diversity of actinomycetes from the Western Ghats is usually ignored and therefore needs to be given attention. There are acidophilic (able to survive where $pH < 6.0$), alkaliphilic (able to survive where $pH > 7.0$), psychrophilic (cold-loving), thermophilic (tolerate high temperature), halophilic (salt-loving), and xerophilic actinomycetes.

7.3 Habitat of actinomycetes

The actinomycetes are mostly found in the soil (Tiwari et al., 2019), ocean (Tiwari et al., 2019), marine (Mohseni et al., 2013), and mangrove ecosystem (Zainal Abidin et al., 2016), composts (Workie & Abate, 2016), deep sediments, food material, etc. The actinomycetes can grow on media containing macro and micronutrients. The earthy smell of soil during the rainy season is due to the actinomycetes.

7.3.1 Importance of diversity of actinomycetes in the Western Ghats

The diversity of actinomycetes in Western Ghats has gained importance because they are a special group of microorganisms due to diverse metabolites and plant growth substances produced by them. The diversity of actinomycetes in Western Ghats helps to understand their morphology and characteristics. This diversity also helps to explore the novel aspects and research related to actinomycetes. The diversity of actinomycetes in Western Ghats seen depends on the quality and composition of soil, isolation method, growth media, temperature, and pH.

7.3.2 Actinomycetes diversity in morphology and pigment production isolated from the Western Ghats

There is a report where actinomycetes viz., *Streptomyces* (50%), *Actinomadura* (32.2%), *Spirillospora* (6.32%), *Actinocorallia* (5.08%), *Arthrobacter* (3.3%), *Saccharopolyspora* (0.8%), *Nonomuraea*, and *Nonomuraea antimicrobica* were isolated from the rhizosphere region of coffee plant in Western Ghats.

There is a report on the diversity of *Streptomyces* sp. with differences in color of spores (white, gray, blue, red, and pink); colonies (discrete, lichenoid, leathery, powdery, and butyrous); different color of aerial and substrate (white, gray, and blue); and arrangement of the spores (rectus, flexibilis, retinaculum aperatum—open loops, hooks, and spira—simple spirals, short, and compact spirals). The color of the pigments produced by *Streptomyces* sp. was dark red, green, yellow, and brown. There is a study on the diversity of endophytic actinobacteria isolated from *Combretum latifolium* Blume, medicinal plant from the Western Ghats of India. From these endophytic actinomycetes, about 35% included genus *Streptomyces*, 17% *Nocordiopsis*, and 13% *Micromonospora*. The *Streptomyces* sp. from the Western Ghats of the Karnataka region, India viz., Kodachadri, Nagara, Hosanagara, Ripponpet, Agumbe, Tirthahalli, and Chikmagalore have shown diversity in spore arrangement like open looks, hooks, spirals, closed spirals, straight, and flexous (Shobha & Onkarappa, 2014). A wide diversity of actinomycetes has been reported which are isolated from the Western Ghats of Tamil Nadu region viz., Kanyakumari, Thirunelveli, Dindigul, Nilgiri (Valan et al., 2012), and Karnataka (Siddharth et al., 2020). The wide diversity of actinomycetes is observed in tropical rainforests region (Nalini & Prakash, 2017). The diversity of actinomycetes is also studied from the Western Ghats of different regions of Tamil Nadu, India (Padmadhas & Ragunathan, 2010). Deivanai and Nambu (2018) have reported the diversity of actinomycetes in the Western Ghats of Tamil Nadu region viz., Courtallum, Palakad, and Theni. Around 161 actinomycetes isolated from the Western Ghats of Tamil Nadu, India, 27 isolates showed diversity in pigment production (difference in pigment concerning color) and these were *Streptomyces* sp. Actinobacteria are well known to produce pigments with diverse color viz., pink, red, brown, yellow, orange, and greenish-brown. The *Streptomyces* sp. isolated from Western Ghats of Karnataka, India showed diversity in colony characters and spore color (Gautham et al., 2012). There is a report on the study of the diversity of actinomycetes from the Western Ghats of Maharashtra (Gunjal et al., 2014).

7.3.3 Diverse group of actinomycetes

The main groups of actinomycetes are *Nocardioform*, *Streptomycetes*, *Actinoplanetes*, *Maduromycetes*, *Thermomonospora*, and related genera and *Thermoactinomycetes*.

7.3.3.1 Nocardioform actinomycetes

Nocardia actinomycetes are aerobic and 0.5–1.2 μm in diameter. They are abundantly found in soil and may be either rod or cocci shape. These actinomycetes have aerial hyphae and are Gram-positive in nature. The colonies of *Nocardia* are chalky, irregular, and in different color viz., brown, pink, orange, red, purple, gray, white. The cell wall is made up of mycolic acids and also glycan moiety which shows the presence of N-glycolic residues.

7.3.3.2 Streptomycetes

The *Streptomyces* are aerobic, Gram-positive, abundantly found in soil, and 0.5–2.0 μm in diameter. They show the presence of vegetative hyphae and branched mycelium. Aerial mycelium is floccose, granular, powdery, or velvety and shows the presence of spores arranged in chains. The spores are nonmotile. The colonies of *Streptomyces* are smooth. They form diffusible pigments that have industrial applications. The sclerotia, pycnidial, sporangia, and synnemata structures are observed in some species. The major composition of the cell wall is L-diaminopimelic acid, but mycolic acids are absent. The ideal temperature for the growth of *Streptomyces* is 25°C–35°C and pH 6.5–8.0. They produce catalase enzyme and also show the ability to reduce nitrates to nitrites. *Streptomyces* also have the ability to use organic compounds as major carbon source.

7.3.3.3 Actinoplanetes

Actinoplanes are Gram-positive, and show the presence of branched mycelium and hyphae. The arrangement of hyphae is in palisade form. The colonies of *Actinoplanes* show immense diversity in color. They produce spores within sporangia. The shape of the spores may be either spherical or irregular, with size (3–20 × 6–30 μm). Polar flagella are found in motile spores. The cell wall is mainly made up of meso-diaminopimelic acid (meso-DAP) and glycine. The best temperature required for the growth of *Actinoplanes* is 15°C–35°C. They are chemoorganotrophic.

7.3.3.4 Maduromycetes

This group has two subgroups, viz., *Streptosporangium* and related taxa; and *Actinomadura*.

7.3.3.5 *Streptosporangium and related taxa*

These are aerobic and Gram-positive actinomycetes. They have branched and nonfragmented vegetative mycelium. The aerial mycelium is present which differentiates either into arthrospores or spore vesicles. The spores may be either motile or nonmotile. The *Streptosporangium* and related taxa groups are chemoorganotrophic and thermophilic. The cell wall peptidoglycan is of A1γ type. The major component of cell wall is *N*-acetylated muramic acid and glucosamine.

7.3.3.6 Actinomadura

Actinomadura is aerobic, Gram-positive, and chemoorganotrophic. They are present in soil. They have branched hyphae and substrate mycelium. Aerial mycelium is also seen with long chain of arthrospores.

The color of aerial mycelium can be blue, brown, gray, green, pink, red, white, or yellow. The spore chains are straight or irregular spirals. The surface of spores may be irregular, smooth, spiny, or warty. The temperature required for the growth of *Actinomadura* is 10°C–60°C. The peptidoglycan of the cell wall is made of meso-DAP and *N*-acetylated muramic acid. The mycolic acids are absent.

7.3.3.7 Thermomonospora *and related genera*

Thermonospora actinomycetes are aerobic and form spores. The vegetative mycelium is present and cell wall is made up of meso-DAP. Mycolic acids are absent.

7.3.3.8 Thermoactinomycetes

Thermoactinomycetes are aerobic, Gram-positive, and chemoorganotrophs. They both have substrate and aerial mycelium. Single spores are present on aerial and vegetative hyphae. The spores are true endospores and contain dipicolinic acid. They form spores that can tolerate heat. The cell wall peptidoglycan contains meso-DAP.

The genus *Thermoactinomyces* falls under this category. *Thermoactinomyces* have septate, branched substrate mycelium (0.4–0.8 μm diameter), and aerial mycelium (0.5–1.0 μm diameter). They form spores which can tolerate heat.

7.4 Growth of actinomycetes on different media

Actinomycetes can grow on nutrient agar (NA) and starch casein agar (SCA) media at 37°C in 15 days. The colonies of actinomycetes are hard and chalky and vary in color on NA and SCA. The advanced microscopy studies such as phase contrast, scanning electron microscope, and transmission electron microscope will help to understand their structure (i.e., aerial and substrate mycelia, branching, spores and their arrangement, hyphae, etc.) and identify them (Passari et al., 2015). In nutrient and starch casein broth, the growth is seen as filamentous biomass after incubation on a rotary incubator shaker at 37°C for 2 weeks. The biomass formed can be harvested and used for various applications (e.g., bioremediation).

7.5 Diversity in enzyme production by actinomycetes isolated from the Western Ghats

There is a report where actinomycetes isolated from coffee plantation soil of Western Ghats showed immense diversity in the production of enzymes viz., cellulase, pectinase, xylanase, amylase, and protease. About 127 actinomycetes were isolated from the Western Ghats of Kerala region, India viz., Wayandu, Munnar, Chinnar, Marayoor, Anamala, Neryamangalam,

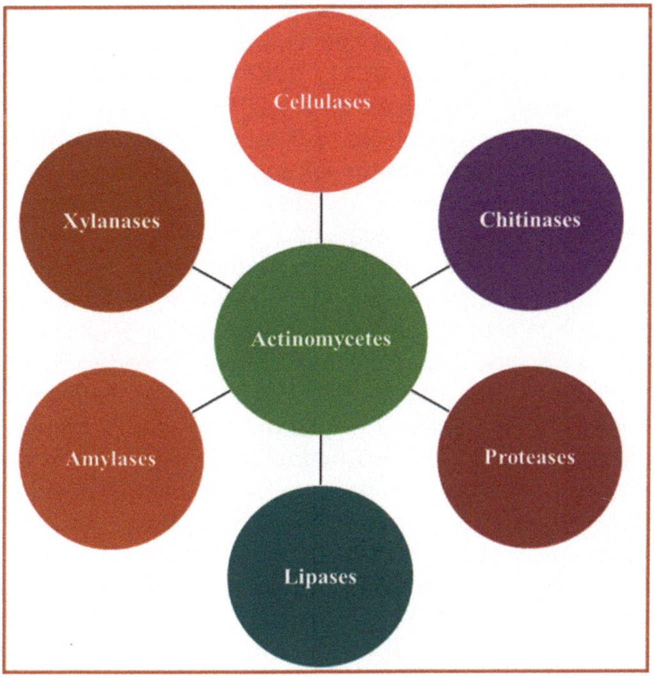

FIGURE 7.2 Pigment extraction from *Streptomyces* sp.

Nelliyampathi, Agasthyarkoodum, Palode, and Kulathuppuzha. From these 127 actinomycetes, around 107 strains (83%) and 79 strains (62%) produced exo-β-1,4-glucanase and endo-β-1,3-glucanase enzyme, respectively (Edison & Pradeep, 2020). Also, there is a report on diversity in the production of β-glucanase enzyme by actinomycetes isolated from the Western Ghats in the region of Kerala (Edison et al., 2017). There is a report where actinomycetes isolated from the fallen leaves of medicinal tree *Emblica officinalis* in Western Ghats of Southern India have shown diversity in enzyme production with amylase (80%), lipase (60%), and cellulase (60%). There is a report on production of enzyme L-asparaginase by *Streptomyces* sp. isolated from the Western Ghats of Agumbe, Karnataka, India (Salimath & Onkarappa, 2016). This L-asparaginase from *Streptomyces* sp. has the anticancer potential. The diverse enzymes viz., cellulases, lipases, chitinases, amylases, proteases, and xylanases produced by actinomycetes are represented in Fig. 7.2.

7.6 Antimicrobial diversity of actinomycetes isolated from the Western Ghats

Actinomycetes group produces many antibiotics which include streptomycin, penicillin, tetracycline, cephalosporins, etc. There is a study on variation in the antimicrobial activity, where 30% showed inhibition of Gram-negative bacteria, 5% inhibited Gram-positive bacteria, fungi, and yeasts, respectively (Gautham et al., 2012). The inhibition zones were between 0 and 40 mm (Gautham et al., 2012). There is a report where *Streptomyces* sp. from various places of Western Ghats of Shivamogga (district), Karnataka has shown antifungal activity against yeasts and molds (Prashith et al., 2015; Shobha et al., 2014). This *Streptomyces* sp. can be used for the development of insecticidal agents. The *Streptomyces* sp. from the Western Ghats of Karnataka region, India viz., Kodachadri, Nagara, Hosanagara, Ripponpet, Agumbe, Tirthahalli, and Chikmagalore has shown antimicrobial activity against *Salmonella typhi, Staphylococcus aureus, Pseudomonas fluorescence, Escherichia coli, Proteus vulgaris, Bacillus subtilis, Klebsiella pneumoniae, Streptococcus* sp., *Vibrio cholerae, Cryptococcus neoformens, Candida albicans, Candida lipolytica,* and *Saccharomyces cerevisiae* (Shobha & Onkarappa, 2014). Also, actinobacteria from the Western Ghats of Karnataka have shown diversity in the production of novel metabolites against Gram-negative and Gram-positive bacteria. Fathima et al. (2018) have also isolated actinomycetes from the Western Ghats of Karnataka, which have shown production of bioactive natural products with chemical diversity as well as broad spectrum of antimicrobial activity against Gram-positive (*S. aureus* and *B. subtilis*) and Gram-negative bacteria (*P. aeruginosa* and *E. coli*). There is a report on antibacterial activity by *Streptomyces* sp. SRDP-07 isolated from the Western Ghats of Thirthahalli, Shivamogga (district), Karnataka (Prashith Kekuda et al., 2013). There is a study on *Streptomyces aurantiacus* actinomycete isolated from the Western Ghats which has shown immense diversity in antimicrobial activity (Nampoothiri et al., 2013). The *Streptomyces* sp. G0S2 has shown diversity in antimicrobial activity against various test organisms with the zone of inhibition ranging from 21 to 27 mm (Gautham et al., 2016). The actinomycetes isolated from the Western Ghats of the Kodagu district of Karnataka, India have shown diversity in antimicrobial activity (Nooshin, 2012). About 36 (26%) and 15 isolates (11%) showed antibacterial activity against Gram-positive and Gram-negative bacteria, respectively, and 87 isolates (63%) showed activity against both Gram-positive and Gram-negative bacteria. Total 30 (22%) isolates showed antifungal activities (Nooshin, 2012). Around 367 actinobacteria were isolated from the Western Ghats of Tamil Nadu, India. From these 367 actinomycetes, 140 isolates showed activity against *B. subtilis* (Valan et al., 2012). Siddharth et al. (2020) have also found actinomycetes from the Western Ghats of Tamil Nadu, India to exhibit antimicrobial activity against various phytopathogens. There is a study on the antibacterial activity of actinobacteria isolated from the Western Ghats of Tamil Nadu, India. Diversity in the antimicrobial activity has been reported by *Streptomyces chartreusis* isolated from the Western Ghats Kodagu (district), Karnataka, India. The zone of inhibition was between 10 and 36 mm (Dezfully & Hanumanthu, 2018). *Streptomyces* isolated from the Western Ghats soil sample of Kanyakumari district, India has been reported to show antimicrobial activity. Siva Kumar et al. (2011) have studied the antibacterial and antifungal activity of actinomycetes isolated from the Western Ghats of Tamil Nadu, India.

7.7 Biotechnological applications of actinomycetes from the Western Ghats

The diverse products produced by actinomycetes are represented in Fig. 7.3.

7.7.1 Enzymes from actinomycetes

The enzymes from actinomycetes can be used in food, agriculture, brewery, pharmaceutical industries, and also biomedical (Saima, 2012). The production of endoglucanases, cellulases, xylanase, chitinase, amylase, etc. from actinomycetes will be eco-friendly, easy, and economical.

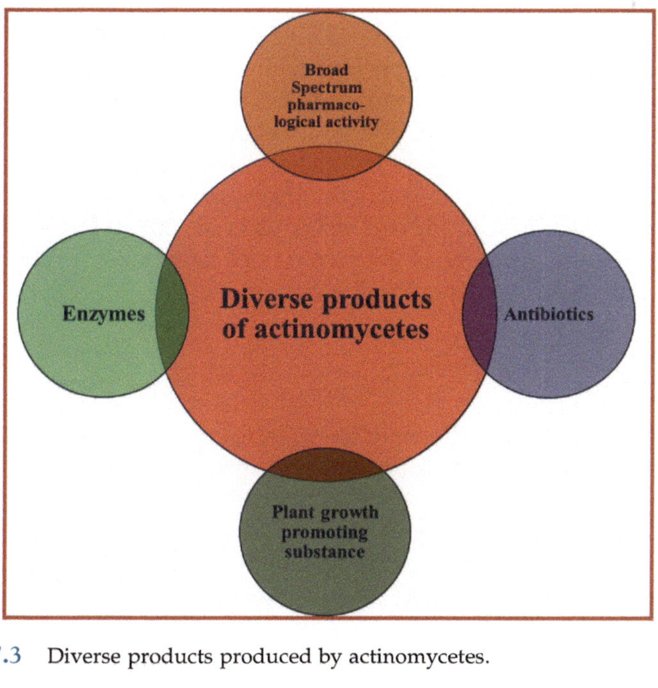

FIGURE 7.3 Diverse products produced by actinomycetes.

Lipases have use in various detergent, food, and medical industries, and also in the diagnostic field (Milind & Charushila, 2007; Prashith et al., 2015; Ravi Ranjan et al., 2015; Sharma et al., 2016).

Cellulases enzymes from actinomycetes have applications in detergent and cloth and paper manufacturing (Azzeddine et al., 2013).

Proteases from *Streptomyces* and *Nocardiopsis* sp. are used in processing various agroindustrial and plant wastes (Mukhtar et al., 2017) and in leather, textile, and dairy industries (Gohel & Singh, 2012), respectively.

Amylases from actinomycetes have immense demand in the market with important applications in the making of detergent formulation (Shigeri et al., 2009).

Xylanases from various actinomycetes are used in paper and pulp industries (Priya et al., 2012).

Chitinases from *Streptomyces* and *Microbispora* sp. are used to hydrolyze chitin oligosaccharides. The chitin oligosaccharides are used as antioxidant and anticancer agents (Mukhtar et al., 2017).

The study is done on the production of L-asparaginase by actinomycetes from Western Ghats of Karnataka, India (Akshatha, Manjunatha, Pooja, et al., 2013; Watve et al., 2001).

7.7.2 Antioxidants

The diversity of actinomycetes is important in bioprospecting other important biomolecules. The antioxidants from actinomycetes inhibit the production of radicals and thus prevent damage to tissues due to these radicals (Lobo et al., 2010), thus protecting human beings against various diseases (Narendhran et al., 2014). The antioxidants from actinomycetes can be used as therapeutics which are nontoxic and economical. There is a report on antioxidant activity of *Streptomyces variabilis* which is isolated from the Gulf of Khambhat, Gujarat, India (Dholakiya et al., 2017). The study is done on the antioxidant property of *Streptomyces cinnamonensis* VITNS1 isolated from Serkadu Region, Vellore, Tamil Nadu, India.

7.7.3 Pigments

Natural pigments are alternate to chemical dyes in various industries. *Streptomyces coelicolor* and *Streptomyces violaceoruber* produce red-blue antitoxin and α, β, and ε actinorhodin called as Actinorhodin-related blue pigments. These pigments have applications in the medical field. The pigments from actinobacteria are also found to have antimicrobial activity against the pathogens (Parmar et al., 2016; Parmar & Singh, 2018). These pigments also have application in food, cosmetics, and pharmaceutical industries (Parmar et al., 2016). They are also used in

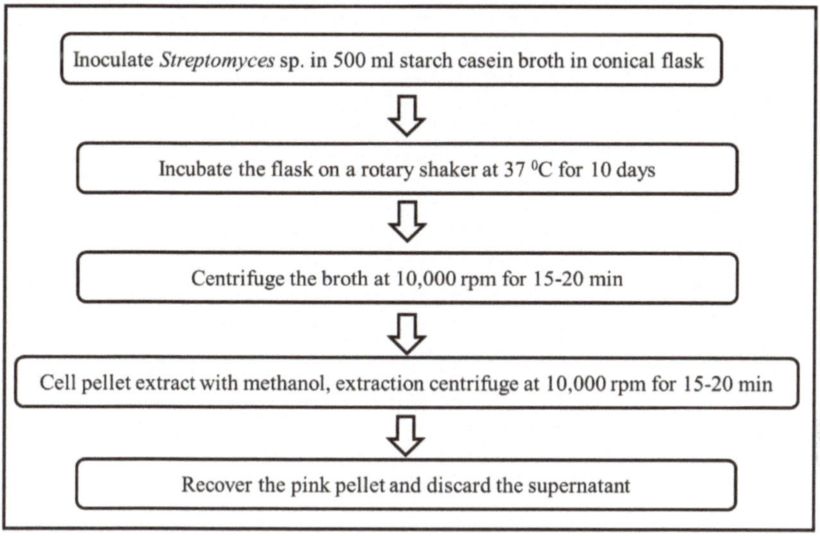

FIGURE 7.4 Diverse enzymes produced by actinomycetes.

biopolymers. Carotenoids are yellow-to-orange red pigments produced by actinomycetes. Carotenoids have applications in nutraceuticals and cosmetics products and also to increase immunity. The extraction of pigment from *Streptomyces* sp. is shown in Fig. 7.4. For the extraction of pigment from *Streptomyces* sp., it is inoculated in starch casein broth and then incubated the flask on a rotary shaker at 37°C for 10 days. After this, the broth is centrifuged at 10,000 rpm for 15–20 min and then the pellet is extracted using the solvent methanol. Again, it is centrifuged at 10,000 rpm for 15–20 min. The extra solvent is evaporated and the pink color pigment is recovered and the supernatant is discarded (Udhyakumar et al., 2017).

7.7.4 Actinomycetes for the plant growth

The plant growth hormones produced by actinomycetes from the Western Ghats have applications in agriculture to increase the growth of plants and many crops. The antimicrobial compounds produced by actinomycetes can be used to inhibit the phytopathogens. The role of actinobacteria in enhancement of plant growth has been reported (Sharma et al., 2014). There is a report on the control of stem rot disease and enhancement in the growth of chili plant by actinomycete (Pattanapipitpaisal & Kamlandharn, 2012). Also, there is a study on actinomycetes for sustainable production and preservation of leguminous plants (Sathya et al., 2017). *Streptomyces* has been reported to produce

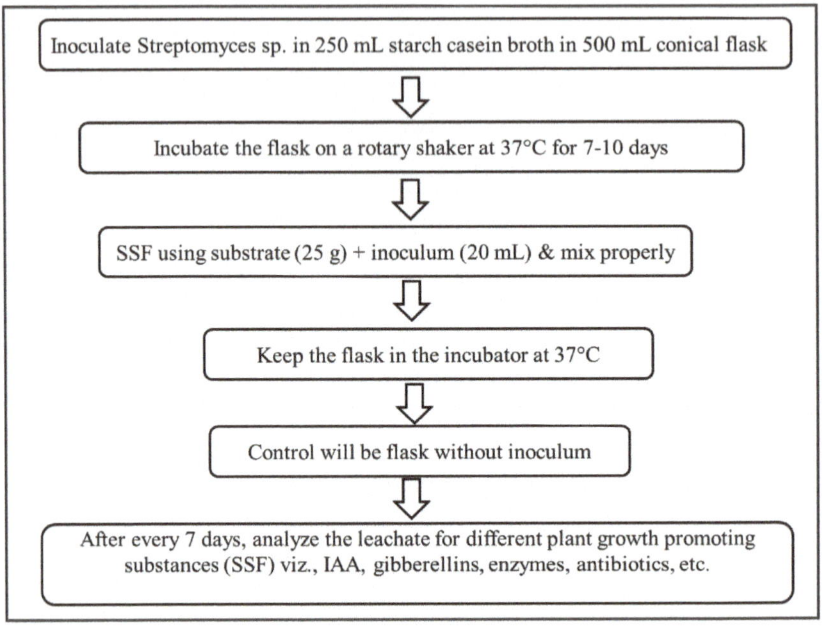

FIGURE 7.5 Geographical location of Western Ghats.

variety of metabolites, antibiotics, and enzymes. Actinomycetes can convert inorganic form of phosphorus to organic form and thus make it easily available to the plants. This will help in the growth of plants (Anwar et al., 2016). *Streptomyces pseudovenezuelae* due to plant growth-promoting traits has shown to enhance maize growth under drought stress (Chukwuneme et al., 2020).

Using solid state fermentation (SSF), the production of plant growth enhancing substances viz., IAA, gibberellins, cytokinins, ammonia (NH_3), enzymes, etc. can be obtained from *Streptomyces* sp. In SSF, different carriers or substrates viz., wheat straw, rice husk, sawdust, peanut shell, bagasse, pressmud, coconut husk, corn cob, etc. are used. The SSF can be used to obtain various important value-added products (Gunjal et al., 2018). The plant growth-promoting substances by SSF using *Streptomyces* sp. is represented in Fig. 7.5.

7.8 Conclusion

The actinomycetes will have many applications in agriculture, medical as well as pharmaceutical industries. The diversity study of actinomycetes will help to understand new species along with their applications. They are very important group of microorganisms. The various groups of actinomycetes show immense diversity in morphology characteristics, colony, and pigment production. The actinomycetes also have antibacterial and antifungal activity. The diverse metabolites produced by actinomycetes have significant applications or roles. The production of plant growth-promoting substances by SSF using *Streptomyces* sp. will help in the growth of many plants. The actinomycetes tolerate wide range of temperature, pH, and stress which is very important. The metabolites produced by actinomycetes must be developed on a large scale by fermentation. The actinomycetes will play an important role in new drug discovery because of the diverse antibiotics produced by them. The novel products can be obtained from the actinomycetes which will be very economical, easy, and eco-friendly. This will help to conserve the microbial world which requires attention. This diversity of actinomycetes will also help to understand the new applications which will benefit various industries and also agriculture.

7.9 Future prospects

The research on the molecular study of actinomycetes concerning to morphology must be carried.

The advanced microscopy studies of actinomycetes for identification and to understand their structure needs to be given attention.

The novel applications and structure of metabolites produced by actinomycetes must be studied.

Ecological studies of actinomycetes.

Characterization of all secondary metabolites produced by actinomycetes.

References

Akshatha, M., Manjunatha, B., Divyashree, B., Fathima, A., & Priya, J. (2013). Screening for L-asparginase producing actinomycetes from Western Ghats of Karnataka, India. *Indian Journal of Applied Research*, *3*, 6–8.

Akshatha, M. D., Manjunatha, B. K., Pooja, R., Umesh, T. M., & Sreevijeth, R. (2013). Screening for novel antibiotic producing actinomycetes from Western Ghats of Karnataka state, India. *Biotechnology*, *2*, 11–13.

Anwar, S., Ali, B., & Sajid, I. (2016). Screening of rhizospheric actinomycetes for various in-vitro and in-vivo plant growth promoting (PGP) traits and for agroactive compounds. *Frontiers in Microbiology*, *7*. Available from https://doi.org/10.3389/fmicb.2016.01334.

Azzeddine, B., Abdelaziz, M., Estelle, C., Mouloud, K., Nawel, B., Nabila, B., Francis, D., & Said, B. (2013). Optimization and partial characterization of endoglucanase produced by streptomyces SP. B-PNG23. *Archives of Biological Sciences*, *65*(2), 549–558. Available from https://doi.org/10.2298/ABS1302549A.

Buedenbender, L., Carroll, A. R., Ekins, M., & Kurtböke, D. I. (2017). Taxonomic and metabolite diversity of actinomycetes associated with three Australian ascidians. *Diversity*, *9*(4). Available from https://doi.org/10.3390/d9040053.

Chamikara, P. (2016). Advanced study on selected taxonomic groups of bacteria and archaea actinomycetes (pp. 1–9). http://www.academia.edu/26067743/Actinomycetes.

Chukwuneme, C. F., Babalola, O. O., Kutu, F. R., & Ojuederie, O. B. (2020). Characterization of actinomycetes isolates for plant growth promoting traits and their effects on drought tolerance in maize. *Journal of Plant Interactions*, *15*(1), 93–105. Available from https://doi.org/10.1080/17429145.2020.1752833.

Deivanai, K., & Nambu, M. R. (2018). Antibacterial activity of actinomycetes isolated from a few selected soil samples of Southern region of Western Ghats. *International Research Journal of Pharmacy and Medical Sciences*, *1*, 26–28.

Devadass, B., Paulraj, M., & Ignacimuthu, S. (2016). Antimicrobial activity of soil actinomycetes isolated from Western Ghats in Tamil Nadu, India. *Journal of Bacteriology & Mycology: Open Access*, *3*(2), 224–232.

Dezfully, N. K., & Hanumanthu, N. (2018). *Streptomyces chartreusis* strain ACTM-8 from the soil of Kodagu, Karnataka state (India): Isolation, identification and antimicrobial activity. *International Journal of Pharmaceutical, Chemical and Biological Sciences*, *8*, 187–194.

Dezfully, N. K., & Ramanayaka, J. G. (2015). Isolation, identification and evaluation of antimicrobial activity of *Streptomyces flavogriseus* strain ACTK2 from soil sample of Kodagu, Karnataka State (India). *Journal of Microbiology*, *8*, 1–8.

Dholakiya, R. N., Kumar, R., Mishra, A., Mody, K. H., & Jha, B. (2017). Antibacterial and antioxidant activities of novel actinobacteria strain isolated from Gulf of Khambhat, Gujarat. *Frontiers in Microbiology*, *8*. Available from https://doi.org/10.3389/fmicb.2017.02420.

Edison, L. K., & Pradeep, N. S. (2020). Functional screening of β-glucanase producing actinomycetes strains from Western Ghats ecosystems of Kerala, India. BioRXiv. https://doi.org/10.1101/2020.04.11.036731.

Edison, L. K., Shiburaj, S., & Pradeep, N. S. (2017). Exploration of Western Ghats regions in Kerala for isolating beta-glucanase producing actinomycetes strains. *Trends in Biosciences*, *10*, 8453–8456.

Fathima, A., Belgami, D., Saba, S. S., Aurora, S., Manjunatha, B. K., & Divakara, R. (2018). Screening of antibacterial and antioxidant activities of *Streptomyces* species isolated from Western Ghats region of Karnataka. *International Journal of Creative Research Thoughts*, *6*, 393–399.

Gautham, S. A., Mesta, S. C., & Onkarappa, R. (2016). Characterization and antimicrobial spectrum of a potent *Streptomyces* sp. GOS2 isolated from Western Ghats of Karnataka, India. *International Journal of Pharmacy and Pharmaceutical Sciences*, *8*(9), 310–312. Available from https://doi.org/10.22159/ijpps.2016.v8i9.12547.

Gautham, S. A., Shobha, K. S., Onkarappa, R., & Prashith, K. T. R. (2012). Isolation, characterisation and antimicrobial potential of streptomyces species from western ghats of Karnataka, India. *Research Journal of Pharmacy and Technology*, *5*(2), 233–238.

Gohel, S. D., & Singh, S. P. (2012). Purification strategies, characteristics and thermodynamic analysis of a highly thermostable alkaline protease from a salt-tolerant alkaliphilic actinomycete, *Nocardiopsis alba* OK-5. *Journal of Chromatography B: Analytical Technologies in the Biomedical and Life Sciences*, *889–890*, 61–68. Available from https://doi.org/10.1016/j.jchromb.2012.01.031.

Gunjal, Rajput, N., & Kapadnis, B. (2014). Microbial biodiversity in Western Ghats of Maharashtra. In *Proceedings of perspectives on biodiversity of India, Indian biodiversity congress conference* (pp. 30–37).

Gunjal, A. B., Kapadnis, B. P., & Pawar, N. J. (2018). PressMud, a lignocellulosic waste as potential carrier for in-situ production of plant growth promoting substances by

Bacillus circulans. *Journal of Solid Waste Technology and Management*, *44*(3), 281–287. Available from https://doi.org/10.5276/JSWTM.2018.281.

Lobo, V., Patil, A., Phatak, A., & Chandra, N. (2010). Free radicals, antioxidants and functional foods: Impact on human health. *Pharmacognosy Reviews*, *4*(8), 118–126. Available from https://doi.org/10.4103/0973-7847.70902.

Mesta, S., & Onkarappa, R. (2017). Characterization and antimicrobial potential of marine actinomycetes from estuaries of Uttara Kannada district, Karnataka. *Journal of Drug Delivery and Therapeutics*, *7*, 27–37.

Milind, W., & Charushila, K. (2007). *Streptomyces* sp. as predators of bacteria. *Nature Precedings*. Available from https://doi.org/10.1038/npre.2007.1263.1.

Mohseni, M., Norouzi, H., Hamedi, J., & Roohi, A. (2013). Screening of antibacterial producing actinomycetes from sediments of the Caspian sea. *International Journal of Molecular and Cell Medicine*, *2*, 64–71.

Mukhtar, S., Zaheer, A., Ayesha, D., Malik, K. A., & Mehnaz, S. (2017). Actinomycetes: A source of industrially important enzymes. *Journal of Proteomics and Bioinformatics*, *10*, 316–319.

Nalini, M. S., & Prakash, H. S. (2017). Diversity and bioprospecting of actinomycete endophytes from the medicinal plants. *Letters in Applied Microbiology*, *64*(4), 261–270. Available from https://doi.org/10.1111/lam.12718.

Nampoothiri, K. M., Ramkumar, B., & Pandey, A. (2013). Western ghats of india: Rich source of microbial biodiversity. *Journal of Scientific and Industrial Research*, *72*(9–10), 617–623. Available from http://nopr.niscair.res.in/bitstream/123456789/20956/1/JSIR%2072%289-10%29%20617-623.pdf.

Narendhran, S., Rajiv, P., Vanathi, P., & Sivaraj, R. (2014). Spectroscopic analysis of bioactive compounds from *Streptomyces cavouresis* KUV39: Evaluation of antioxidant and cytotoxicity activity. *International Journal of Pharmacy and Pharmaceutical Sciences*, *6*(7), 319–322. Available from http://innovareacademics.in/journals/index.php/ijpps/article/download/2009/9983.

Nooshin, K. D. (2012). Diversity of antibiotic producing actinomycetes from soil samples of Kodagu district—Karnataka (India). In *International conference on biodiversity and sustainable energy development*. https://doi.org/10.4172/2157-7625.S1.009.

Padmadhas., & Ragunathan, R. (2010). Isolation, characterization and identification of novel actinomycetes collected from Western Ghats region of India. *Journal of Pharmaceutical and Biomedical Sciences*, *1*, 1–7.

Parmar, R., Singh, C., Saini, P., & Kumar, A. (2016). Isolation and screening of antimicrobial and extracellular pigment producing actinomycetes from Chambal territory of Madhya Pradesh region, India. *Asian Journal of Pharmaceutical and Clinical Research*, *9*, 157–160.

Parmar, R. S., & Singh, C. (2018). A comprehensive study of eco-friendly natural pigment and its applications. *Biochemistry and Biophysics Reports*, *13*, 22–26. Available from https://doi.org/10.1016/j.bbrep.2017.11.002.

Parte, A. C. (2018). LPSN - List of prokaryotic names with standing in nomenclature (Bacterio.net), 20 years on. *International Journal of Systematic and Evolutionary Microbiology*, *68*(6), 1825–1829. Available from https://doi.org/10.1099/ijsem.0.002786.

Passari, A. K., Mishra, V. K., Saikia, R., Gupta, V. K., & Singh, B. P. (2015). Isolation, abundance and phylogenetic affiliation of endophytic actinomycetes associated with medicinal plants and screening for their in vitro antimicrobial biosynthetic potential. *Frontiers in Microbiology*, *6*. Available from https://doi.org/10.3389/fmicb.2015.00273.

Pattanapipitpaisal, P., & Kamlandharn, R. (2012). Screening of chitinolytic actinomycetes for biological control of *Sclerotium rolfsii* stem rot disease of chilli. *Songklanakarin Journal of Science and Technology*, *34*(4), 383–389. Available from http://rdo.psu.ac.th/sjstweb/journal/34-4/0353-3345-34-4-387-393.pdf.

Prashith, T. R., Onkarappa, R., Gautham, S. R., Sunita, C. M., & Raghavendra, H. L. (2015). Antimicrobial, antioxidant and cytotoxic activity of *Streptomyces* sp. from Western Ghats soils of Karnataka, India. *Science Technology and Arts Research Journal*, *4*, 164–180.

Prashith Kekuda, T. R., Dileep, N., Junaid, S., Rakesh, K. N., Mesta, S. C., & Onkarappa, R. (2013). Biological activities of *Streptomyces* species SRDP-07 isolated from soil of Thirthahalli, Karnataka, India. *International Journal of Drug Development and Research, 5* (3), 268–285. Available from http://www.ijddr.in/Dacuments/10/33.pdf.

Priya, B. S., Stalin, T., & Selvam, K. (2012). Efficient utilization of xylanase and lipase producing thermophilic marine actinomycetes (*Streptomyces albus* and *Streptomyces hygroscopicus*) in the production of eco-friendly alternative energy from waste. *African Journal of Biotechnology, 11*, 14320–14325.

Radhakrishnan, M., Suganya, S., Balagurunathan, R., & Kumar, V. (2010). Preliminary screening for antibacterial and antimycobacterial activity of actinomycetes from less explored ecosystems. *World Journal of Microbiology and Biotechnology, 26*(3), 561–566. Available from https://doi.org/10.1007/s11274-009-0198-9.

Ravi Ranjan, K., Vasantba, J., Bhoomi, M., Bonisha, T., & Bhumika, C. (2015). Antibacterial potentials of actinomycetes isolated from Gujarat. *International Journal of Pharmaceutical Sciences Review and Research, 30*(1), 78–83. Available from http://globalresearchonline.net/journalcontents/v30-1/15.pdf.

Saima, A. (2012). Cloning and expression of *endo*-1,4–β-glucanase gene from *Bacillus licheniformis* ATCC 14580 into *Escherichia coli* BL21 (DE 3). *African Journal of Biotechnology, 11*, 2846–2854.

Salim, F. M., Sharmili, S. A., Anbumalarmathi, J., & Umamaheswari, K. (2017). Isolation, molecular characterization and identification of antibiotic producing actinomycetes from soil samples. *Journal of Applied Pharmaceutical Science, 7*(9), 69–75. Available from https://doi.org/10.7324/JAPS.2017.70909.

Salimath, M. B., & Onkarappa, R. (2016). Screening of Streptomycetes for L-asparaginase, therapeutic agent of lymphocytic leukemia from Western Ghats of Karnataka, India. *International Journal of Drug Development and Research, 8*, 23–29.

Sathya, A., Vijayabharathi, R., & Gopalkrishnan, S. (2017). Plant growth promoting actinobacteria: A new strategy for enhancing sustainable production and protection of grain legumes. *Biotech, 3*, 1–10.

Sharma, M., Dangi, P., & Choudhary, M. (2014). Actinomycetes: Source, identification, and their applications. *International Journal of Current Microbiology and Applied Sciences, 3*, 801–832.

Sharma, P., Kalita, M. C., & Thakur, D. (2016). Broad spectrum antimicrobial activity of forest-derived soil actinomycete, *Nocardia* sp. PB-52. *Frontiers in Microbiology, 7*. Available from https://doi.org/10.3389/fmicb.2016.00347.

Shigeri, Y., Matsui, T., & Watanabe, K. (2009). Decomposition of intact chicken feathers by a thermophile in combination with an acidulocomposting garbage-treatment process. *Bioscience, Biotechnology and Biochemistry, 73*(11), 2519–2521. Available from https://doi.org/10.1271/bbb.90398.

Shobha, K. S., & Onkarappa, R. (2014). Actinomycetes of Western Ghats Karnataka. India. *Research and Reviews in Biosciences, 8*, 172–178.

Shobha, K. S., Onkarappa, R., & Gautham, S. A. (2014). Antifungal potential of pigmented *Streptomyces* ksro4 from Agumbe region of Western Ghats. *Journal of Pharmaceutical and Scientific Innovation, 3*, 43–46.

Siddharth, S., Ravishankar Rai, V., Wink, J., & Steinert, M. (2020). Diversity and bioactive potential of actinobacteria from unexplored regions of western Ghats, India. *Microorganisms, 8*(2). Available from https://doi.org/10.3390/microorganisms8020225.

Siva Kumar, J., Santhanam, P., & Masilamani Selvam, M. (2011). Antimicrobial activity of actinomycetes isolated from the western ghats of Tamilnadu. *International Journal of Pharma and Bio Sciences*, 2(1), 42–49. Available from http://www.ijpbs.net/volume2/issue1/biological/_4.pdf.

Tiwari, D., Bhati, P., Das, P., & Shouche, S. (2019). Potential of actinomycetes as bioremediating and biocontrolling agent. *Paripex - Indian Journal of Research, 8*, 36–40.

Udhyakumar, K., Ramalingam, S., Saravanan, R., & Dheeba, B. (2017). Extraction of actinomycetes (*Streptomyces* sp.) pigment and evaluation of its anticancer property on HeLa cell line. *Der Pharma Chemica, 9*, 106–113.

Valan, A. M., Ignacimuthu, S., & Agastian, P. (2012). Actinomycetes from Western Ghats of Tamil Nadu with its antimicrobial properties. *Asian Pacific Journal of Tropical Biomedicine, 2*(2), S830–S837. Available from https://doi.org/10.1016/S2221-1691(12)60320-7.

Watve, M. G., Tickoo, R., Jog, M. M., & Bhole, B. D. (2001). How many antibiotics are produced by the genus *Streptomyces*? *Archives of Microbiology, 176*(5), 386–390. Available from https://doi.org/10.1007/s002030100345.

Workie, M., & Abate, D. (2016). Screening of bioactive compounds from actinomycetes isolated from compost prepared for mushroom growth against *Candida albicans*. *International Journal of Innovative Pharmaceutical Sciences and Research, 4*, 887–889.

Zainal Abidin, Z. A., Abdul Malek, N., Zainuddin, Z., & Chowdhury, A. J. K. (2016). Selective isolation and antagonistic activity of actinomycetes from mangrove forest of Pahang, Malaysia. *Frontiers in Life Science, 9*(1), 24–31. Available from https://doi.org/10.1080/21553769.2015.1051244.

CHAPTER

8

Microbial diversity at the polluted sites

Suneeta Gireesh Panicker

Department of Microbiology, D.Y. Patil Arts Commerce and Science College, Pune, India

8.1 Introduction

Contaminants that have entered into nature and passed the accepted limits due to human activities are called pollutants. These pollutants have harmful effects to all the inhabitants in that area. They can be in different forms like chemicals, metals, energies, etc. Human interventions like mining, chlorination, use of radioactive substances, heavy metals in an ecosystem introduce various materials which are not natural to the ecosystem leading to pollution of that site. In addition, volcanic eruptions, atmospheric fallouts, sea salt particles, etc. also contaminate the area. Furthermore, strenuous utilization of natural resources and lighting technological advancements for industrialization has added more pollutants. The pollutants like hydrocarbon, organophosphates and organochlorine compounds, pest and herb killers, radioisotopes, etc., are of genuine health and environmental worries.

These polluted sites could be all three forms, land, water, and air. Such sites which contain numerous pollutants are designated as "pollution hotspots." Environmental and social awareness in modern days has evoked a considerable control on the pollutants disposal, but a notable number still exists and additions are constantly happening.

Different types of pollution is proving to be a great threat to human life globally in recent years. The polluted environment which include dirty atmosphere, water bodies, and ground are most leading causes resulting in premature deaths which reports about nine million deaths

Microbial Diversity and Ecology in Hotspots
DOI: https://doi.org/10.1016/B978-0-323-90148-2.00011-0

around the globe in 2015 (Landrigan et al., 2018). Uncontrolled and unlimited usage of all the spheres of earth by the growing human population has significantly contributed to the pollution of atmosphere, water bodies, and land.

The radioactive hotspot detection and localization can be done using robots' detectors viz., undirected (e.g., scintillation detectors of NaI) and directed (e.g., a gamma camera). The undirected detectors are to plot environments with respect to radiation intensity or to check stationary or mobile sources. Fixed radiation sensors can search radioactivity mobile sources against background radiation. The directional detectors include CZT Compton camera. Cameras with double fastened and encoded crevices were utilized to determine the non-linear coordinates belonging to the radioactive source by using triangulation protocols. A detection system which consists of detectors of 18 CZT type and positioning apparatus, not like that of static detectors, was the most used system to locate the sources of radioactive elements. It was able to locate minimum amount of radioisotopes in an area of 10 cm around the main source within 4.5 min. It was also capable of detecting and quantifying the activity of the radioisotopes by rare and on-site functioning (Ardiny et al., 2019).

Environments may be polluted with varied pollutants. The polluted systems are becoming highly complex to understand them and to predict their effects on the environment. It requires various methods to analyze this complexity and to generate a huge data set related to the physicochemical properties of the polluted areas.

Various human activities like powdering of nuclear fuel, checking of nuclear weapon and periodic nuclear disaster similar that occurred in Chernobyl mining have resulted in introduction of highly dangerous nuclides in the marine bodies (Garcia-Leon, 1990; Shaw & Bell, 1994). Radioisotopes are extensively used in various research activities, medical field and in the factory areas, resulting in huge piling of various radioactive wastes.

Nuclear power stations contribute almost 95% of this radioactive wastes produced from different origins (Ahier & Tracy, 1995; Tamponnet & Declerck, 2008). Radionuclide also known as radioisotope is a chemical element in an unstable form which degrades radioactively causing the release of nuclear emission. Dosages of radiation are measured in gray (Gy). Exposure to just 1 Gy of radiations causes the initial symptoms of radiation illness in human. Around 50% of all the population coming in contact of 4.5 Gy radiations die, whereas radiations of 8 Gy or more are always deadly to humans. These radionuclides have a long half-life. Even small concentrations of radionuclides are known to have influence for an elongated period and hence its impact is increasing with clock hour. The often-confronted radioactive nuclides include cobalt-60 (60Co), plutonium-239 (239Pu), radium-226 (226Ra), radon-222

(222Rn), technetium-99 (99Tc), thorium-232 (232Th), and uranium-238 (238U). Particularly, those that turned out from the nuclear reactors through cleaving of basic atoms are iridium-192 (192Ir), thallium-201 (201Tl), strontium-90 (90 Sr), and cesium-137 (137Cs) take a remarkably longer period to degrade (Kurnaz et al., 2007). The presence of radioactive nuclides in the recent environment is due to the nuclear weapons' checking that happened in the 1950s and 1960s. Also, due to some accidental contamination like Chernobyl catastrophe, it has resulted in accumulation of these radionuclides in the nature. Moreover, industrial activities involving nuclear power, add to the burden of anthropogenic radioactivity with managed release of their effluents. These pollutants happen to launch into the food web and eventually results in "biomagnification" (Dua et al., 2002; Naso et al., 2003; Chiuchiolo et al., 2004). Moreover, 238U decomposes to form 226Ra, having a half-life of 1600 years. It is also observed that 235U, after thermal fission generates transuranic elements Np, Pu, and Am products which are nearly in the scale of 72Zn to 161Tb with bigger or smaller half-lives. Corrosion of fuel cladding or assemblies or activation, caused in some nuclear wastes, leads to the emission of dangerous radionuclides (60Co, 59Ni, 93Zr). All the nuclides that lie between the 15 metallic elements that is from atomic number 89 (actinium) to atomic number 103 (lawrencium) are actinides. These are known to be highly unstable, radioactive, heavier nuclides and do not occur naturally. In the actinides produced and the fission products produced, there are various important radionuclides which if released in the environment would be of serious environmental concern (Lloyd & Renshaw, 2005a, 2005b). After disposal of these radionuclide wastes in the near-surface environments, there are high chances that these get dispersed in the groundwater. As a result of this, such contaminants widespread to many nearby sites, and the toxic aqueous uranium widespread dispersion are a big example of such situations (Wronkiewicz & Buck, 1999). The oxidized form of uranium (UVI) is soluble in water and thus enters groundwater (Liu et al., 2005).

Bacterial count varies widely and in general, it is estimated that 10 g soil contains about 10^7 different bacterial species (Gans et al., 2005). In the soil, microorganisms can reside in several different habitats viz., decaying organic matter, bulk soil, rhizosphere, micropores, and macropores. The diverseness of a domain and the phylogenetic heterogeneity of bacteria are found to be directly proportional. The spatial variation and flexibility of the resources increase when the average quality of these domains increase. The bacterial communities were found to be less varied in saturated subsurface soil than the unsaturated soils, which may be because of higher heterogeneity of the unsaturated soils (Zhou et al., 2002). The diversity and the efficacy of a niche can increase if the resources in it increase potentially and vice versa. This indicates that decrease in the microbial community due to pollution in any region leads to a basic loss of that ecological system. Thus the first step toward the remediation of polluted sites, microorganism deficient ecosystems is to maintain the biodiversity of such ecosystems.

Microorganisms are omnipresent and our biosphere is ruled by an extensively diversified mass of microorganisms. The environmental factors that influence their distribution are vegetation (Gans et al., 2005; Joner et al., 2001), soil form and features (Hamamura et al., 2005) and regional climate (Maila et al., 2006). The microbial community varies crucially with change in the contaminant at a given site. Those with specified machinery and mechanisms usually outgrow the less/nontolerant species in such polluted sites and the natural selection caused by the force of contaminants make them supreme (Head et al., 2006).

There is immense change in the diversity of microorganisms viz., tuberculosis *Bacilli*; hepatitis, Herpes, and tobacco mosaic viruses, Cytomegalovirus; soil micromycetes, etc., near the Chernobyl site contaminated with the radioactive pollutants (Yablokov et al., 2009). The radiations effect on aquatic life close to Fukushima Daiichi nuclear power plant has been reported before and after the incident. The effect of radiocesium on the microorganisms is still seen including those in the hotspots region (Battle, 2015). Diverse microorganisms inhabit largely nuclear waste and accident sites (e.g., Hanford, Chernobyl, Fukushima). There are diverse microorganisms which are reported to divide in extreme gamma-CIR dose rates of 13–126 Gy/hour (Shuryak et al., 2017).

Microbial catalytic diversity is mainly due to the enzyme genes included in utilization of different pollutants as carbon source. Many genes like amoA (ammonia oxidation), nirK and nirS (denitrification), nifH (nitrogen fixation), dsrA (sulfate reduction), etc., have been studied to look into the key processes applied by the microorganisms for exploitation of a wide variety of carbon sources (Hansel et al., 2008). Various environmental pollutants are degraded by the microorganisms without producing toxic intermediates.

Radionuclides in the oxidized forms are known to be hugely soluble in aqueous medium due to which they become mobile in the groundwater and thus launch into the food chain or adversely affect the inhabitants of that area. As opposite to this, the reduced form of radionuclides are highly insoluble and usually tend to precipitate out. Microbes have great ability of this process. They contain the enzymes, which can reduce the radionuclides and convert them into insoluble nontoxic forms.

One way to prevent or decrease the diffusion of the pollutants in the environment is by the use of actively growing bacteria that are near to contamination sources, where the amount of the radionuclide are heavily elevated. The microorganisms are known to carry out metal reduction or immobilization and degradation of the toxic organic compounds.

Bacterial genera like *Serratia, Geobacter, Hymenobacter metalli* sp., *Shewanella, Deinococcus, Kineococcus radiotolerans*, etc., can tolerate and also can act on the radionuclides present in the radioactive waste at a high concentration. These microbes are known to reduce their capacity to become soluble, their mobility, and bioavailability (Chung et al., 2010; Lloyd & Gadd, 2011).

Different microorganisms found adhered to the surfaces of the pool where the nuclear materials were stored in the Spanish nuclear plant were isolated and characterized (Chicote et al., 2005). Six bacteria belonging to the beta-*Proteobacteria, Actinomycetales* and *Bacillus, Staphylococcus* groups and a fungal species of *Aspergillus* were obtained from these sites. In 1962, sediments near the waste tank of USA DOE Hanford Site, Washington State which had huge doses of aluminate, alkali, nitrate, chromate, ^{137}Cs and ^{99}Tc got contaminated due to a leak in that tank (Fredrickson et al., 2004). Various bacterial isolates were found in this site, of which Gramþ bacteria closely associated with *Arthrobacter* sp. were usual, but *Rhodococcus* and *Nocardia* were also found. Active bacteria were also obtained from the samples containing higher concentration (> 10 mCi 137Cs g1) of radioactive material. Few isolates characterized from samples with >20 mCi of 137Cs g1 could withstand radiations (upto w20 kGy). They were found to be closely associated with *D. radiodurans* and were greatly resistant to desiccation. A canned meat which had received 4000 Gy of ionizing radiation showed the presence of active *D. radiodurans* (Anderson et al., 1956). This bacterium is listed in the *Guinness Book of World Records* as "the world's toughest bacterium." It is known as an extremophile as it can survive in presence of less nutrients, drought conditions and high radiations. It is correctly named as "strange berry" which indicates that it withstands radiation. This organism has a unique DNA repair system which makes it so tough and capable of resisting the high doses of radiations by repairing its DNA in few hours. It carries nearly four to ten copies of its genome and the repair system. Due to the advantage of this unusual system, *D. radiodurans* is capable of repairing its complete genome (at least one copy) that has been harmed by the radiations. After sequencing the whole genome of this microbe, it was found that it has less genes for DNA repair than the *Escherichia coli* bacteria which is highly sensitive to radiations. A number of studies have focused on the proteins and pathways that intricate in stitching a shattered DNA together again. These studies revealed that capability of *D. radiodurans* to rebuilt a normally working genome depends on the extended synthesis-dependent strand annealing (ESDSA) pathway and through homologous recombination pathways mediated by RecFOR and RecA (Bentchikou et al., 2010). Another mechanism which is also known to be involved in the DNA repair is RecA-independent single-strand annealing (SSA) process which requires DdrB. This is a single strand DNA binding protein possessing a SSA capacity (Sugiman-Marangos et al., 2016).

Another bacteria able to resist considerably huge amounts of metals and alkali and extreme doses of ionizing radiations was initially isolated from a waste cell with high doses of radioisotopes at the Savannah River Site in Aiken, South Carolina (Phillips et al., 2002). It was aerobic, produced orange color pigment, and belonged to the Actinobacteria phylum. It was also found to exhibit resistance to radiation similar to that of *D. radiodurans* (Phillips et al., 2002). Both these organisms belonged to different phyla but were found to have same level of resistance.

Bioremediation via microorganisms is a striking substitute for scooping out the pollutants from the soil or water. A wide variety of microorganisms possess various mechanisms for excavating the contaminants from polluted areas. They do this mainly by adding or removing electrons, making the transformed radionuclides soluble and mobile so that they can be simply removed from the nature (Amachi et al., 2010). This type of microorganism assisted biological alteration provide chances of biological treatments for clearing the radioisotopes from the nature. *Rhodanobacter* sp. and *Desulfuromusa ferrireducens* have such ability to interact with different pollutants (Green et al., 2012).

In order to control the subsurface expansion of uranium contamination, the conversion of soluble UVI to insoluble UIV must be reduced as this UIV can be precipitated and locked to one place. Lovley et al. (1996) and microorganisms are shown to participate in this process through the bioreduction of uranium (U) (Lovley, 1991). They are also known to be associated with the biosorption of U and other potentially toxic metals on bacterial membranes, the bioaccumulation of U inside bacteria and the bioprecipitation of U(VI)-phosphate phases (Merroun & Selenska-Pobell, 2008).

Uranium is used as a model to study the interaction between the microbes and the radionuclides. There are some studies on the microbial interaction with the major radioactive products like ^{137}Cs and ^{90}Sr. Wide ranges of new diverse organisms capable of removing uranium have been isolated from the polluted sites.

Sediments at the Midnite mine (Seilacher, 1992) are found to be acidic uranium-contaminated and harbor few heterotrophic bacteria along with *D. radiodurans*, which demonstrated uranium deposition and resistance to the actinide. A close relative of *Arthrobacterilicis*, with large G + C content demonstrated accumulation of uranium in the form of intracellular precipitates accompanied with polyphosphate granules. This could be a possible way of detoxification method in this bacterium. Another U(VI)-reducing organism studied was *Salmonella subterranea* sp. nov., which was found to utilize nitrate as the electron acceptor, hydrogen as the electron donor, and acetate as the carbon source (Shelobolina et al., 2004).

8.2 Uranium

Among the native living beings near the uranium-contaminated sites, *Geobacteraceae* is commonly found. These bacteria are able to fix atmospheric nitrogen. There are various nonbiological methods of UVI reduction which involve sulfides, FeII, or hydroquinones (Fredrikson et al., 2000), but they appear to be insubstantial when compared to the reduction methods adopted by the microbial world. Many microorganisms are known to utilize UVI as an electron acceptor (e.g., *Shewanella putrefaciens*). Sulfate-reducing and FeIII-reducing microorganisms are reported to remove uranium-contaminated waters by precipitating it (Lovley, Giovannoni, et al., 1993; Lovley, Widman, et al., 1993).

8.3 Thorium IV

Occurrence of Thorium IV is wide in the environment. Though the bioreduction of Thorium at valency IV is difficult, biosorption by microorganisms is a possible way to occur and there are several reports of ThIV biosorption by fungal biomass. *Mycobacterium smegmatis* at pH 1.0 was shown to biosorb thorium significantly (Andres et al., 1992). Some organisms were also reported to be capable of removing thorium as an insoluble biogenic phosphate (Yong & Macaskie, 1998).

8.4 Neptunium-237

An alpha-emitting radionuclide, Neptunium-237 is a major component yielded in large quantities in the nuclear reactors. This contaminant remains in the waste for nearly 104–107 years. It is very difficult to remove this highly mobile and biologically toxic pollutant by physicochemical or biotechnological methods. However, *S. putrefaciens* can reduce Np^{V} to Np^{IV} along with concurrent precipitation as phosphate, formed by *Citrobacter* (Lloyd et al., 2000).

8.5 Plutonium

Plutonium (Pu) is highly radiotoxic and shows complex redox chemistry. Thus its geochemistry remains to be a lot more challenging than the other actinides. Pu(IV) is the most stable oxidation state, but it also occurs in Pu(III), Pu(V), Pu(VI) states in the environment. Fe(III)-reducing bacteria are reported to reduce the Pu(IV) to Pu(III), which can solubilize the sediment-bound Pu(IV). Thus produced trivalent actinide can react with different microbial ligands (Lloyd et al., 2000). Although few bacteria such as *Rhizopus arrhizus, Pseudomonas aeruginosa, Citrobacter* are demonstrated to have the ability to regulate the solubility of Pu in a wide variety of habitats, the exact mode of action of Pu-microbe reaction is still not clear (Lloyd & Renshaw, 2005a, 2005b).

In a varied set of environmental conditions, Americium remains stable in its trivalent oxidation state. Different microorganisms are reported to efficiently biosorb 241Am. Biosorption of 241Am by *R. arrhizus* (Dhami et al., 1998) and removal of 90% of 241Am with *Candida utilis* (Lloyd & Renshaw, 2005a) have been reported. Phosphate biodeposition of 241Am that is an enzymatically mediated biodeposition by *Citrobacter* sp. has been reported by Macaskie et al. (1994).

Tc(*VII*) radionuclide can also be reduced and converted into insoluble low-valency oxide TC(IV) by metal-reducing microorganisms (Pignolet et al., 1989). Such reduction of Tc(VII) by *S. putrefaciens* and *Geobacter metallireducens* enzymes has been reported by Lloyd and Macaskie (1996). Different studies using phosphor imager technique, X-ray absorption spectroscopy (Wildung et al., 2000), etc., have confirmed the formation of insoluble oxidation state after Tc(VII) reduction by *Geobacter sulfurreducens* and *S. putrefaciens*. *Desulfovibrio desulfuricans* and some other organisms were shown to use formate as an alternate electron donor for reduction of Tc(VII) (Lloyd et al., 1999). The primary formate hydrogenlyase (FHL) complex is shown to be present in the periplasm of few microorganisms. The presence of periplasmic Ni-Fe hydrogenase has been shown to have a role in Tc(VII) reduction relative in the N subclass of *Desulfovibrio fructosovorans* (De Luca et al., 2001). Moreover, the iron-reducing bacteria viz., *Geobacter* sp. and *Rhodoferax ferrireducens* are reported to possess the capacity to enzymatically reduce the radionuclides (Kim et al., 2012).

Desulfovibrio vulgaris has tetraheme cytochrome c3 protein, which can reduce U(VI) using U(VI) reductase. Hydrogenase enzyme was used as a physiological electron donor (Lovley & Phillips, 1994) in this reaction. *D. desulfuricans* (G20) also demonstrated the possession of a homolog and established the role of cytochrome. c3.

Same findings were demonstrated by Elias et al. (2004), who showed that *D. vulgaris* was capable of reducing Fe(III), U(VI) and sulfate due to the presence of cytochrome c3, using hydrogen, lactate or pyruvate as the electron donor. The study proposed a usual pathway which involves the coupling of organic electron donors, for this type of reduction reaction. The results indicated that there are different mechanisms dependent on hydrogen for this reduction process. Moreover, the study made an interesting observation that though the *Desulfosporosinus* sp. lacked cytochrome c3, still were found to reduce U(VI). Few other well-known metal-reducing bacteria like *Geobacter* and *Shewanella* which showed the reduction of actinides and other toxic metals/radionuclides by different mechanisms need to be extensively studied (Lovley et al., 1989).

The important radionuclide Technetium 99Tc, which has a long half-life (2.1×10^5 years) is a fission product of 235U and is extensively studied. This radionuclide is present in the pertechnetate anion TcO4 [Tc (VII)] form under oxic conditions. In this form, Tc(VII) is largely soluble and motile. Microbial enzymes are known to convert into an insoluble Tc(IV) form (Dexter-Dyer et al., 1984). One such haloalkaliphilic bacterium with the capacity to reduce Tc(VII) enzymatically was isolated from the soda-lake environments (Schidlowski, 1988).

8.6 Mechanisms involved in bioremediation

Bacterial resistance to radiations is also attributed to a repair pathway carried out by an exonuclease enzyme (Misra et al., 2006). This kind of resistance **is** linked to deposition of Mn(II) in the bacterium (Daly et al., 2004). Even the presence of numerous amino acid synthesis pathways are been identified for this type of extensive resistance (Venkateswaran et al., 2000).

Various bacterial genus like *Pseudomonas, Bacillus, Serratia, Arthrobacter,* and *Rhodococcus* are found to be resistant to U, Ni, Co, Cu, and Cd. This resistance is due to their inherent properties such as intracellular sequestration of U, bioprecipitation and biomineralization (Choudhary et al., 2012). Recently a bacteria isolated from spent nuclear fuel pond demonstrated significant resistance to cobalt. This tolerance was associated with a highly organized efflux pump mechanism which is similar to Ni2þ and Zn2þ (Dekker et al., 2014).

P. aeruginosa J007 and *Arthrobacter nicotianae* showed resistance to uranium whereas, *Serratia* is tolerant to Cs137, Sr90, and Co60 (Paterson-Beedle et al., 2006). Similarly, *E. coli* was found to be resistant to nickel and cobalt due to the presence of novel YohM protein rcnA (Rodrigue et al., 2005). Gamma-radiation resistance has been credited to S-layer proteins (Gerbino et al., 2015). Studies performed on disastrous calamity in Chernobyl have demonstrated that radiations positively affected the fungal spore germination in the nuclear power plant regions. This effect was then named as "radiostimulation" (Dadachova & Casadevall, 2008). They also found that these fungal strains expressed radiotropism and revealed their capability to use such radiations as energy source.

Interaction between bacteria and radionuclides occur in different ways like bioaccumulation, biosorption, biotransformation, biosolubilization and bioprecipitation.

8.7 Biosorption

Biosorption is a metabolically independent, physicochemical process. It is based on various mechanisms including absorption, adsorption, ion exchange, surface complexation, and precipitation. Biosorption processes are greatly supreme in the nature and traditional biological treatment. This method of bioremediation results in aggregation of the radionuclides on the surface or interface. Dead biomass or even living cells can be used for biosorption. However, dead biomass is more effective for the removal of metals or metalloids than live biomass. Electrostatic forces of attraction between the metal cations and the negatively charged cell surface create an electrostatic binding. This type of binding is responsible for the adsorption of radionuclides onto the cell surface leading to extracellular accumulation.

Slime layers (S-layers) or the capsule present on the organism also significantly affect the process of bioaccumulation. Thus adsorption properties are attributed to the surface characteristics of the bacterial cell wall (Jiang et al., 2004). However, Gram-positive and Gram-negative bacteria exhibited less or no considerable difference in their bioaccumulation capability.

Electronegative charge on the membrane is due to the phosphate moieties present in it. Thus the positive cations of metals and radionuclides are attracted toward the membrane causing biosorption or adsorption of the nuclides on the cell surface. The ligands like amine, carboxyl, phosphate, hydroxyl, etc., help the binding of nuclides to the cell wall (Lukyanova et al., 2011) and this process is known as passive uptake process.

Rhodococcus opacus and *Rhodococcus rhodochrous* are shown to be a good source of extracellular polymeric substance (EPS) that can be utilized for Cd(II), Pb(II), Ni(II), Co(II), and Cr(VI) ions sorption leading to the immobilization of these radionuclides. Presence of carboxyl in the cell wall of *Citrobacter freudii* is as an effective biosorption site (Xie et al., 2008). The main flaw of this method is rapid saturation with the adsorbed molecules and competing desorption due to the cations onto the cells that are not involved in the binding (Jabbar & Wallner, 2015). Thus this process can be restricted for the elimination of smaller concentration of radionuclides from various effluents (Merroun & Selenska-Pobell, 2008).

8.8 Bioaccumulation

Retention and concentration of a substance within an organism is termed as "bioaccumulation" (Tabak et al., 2005). In other words, (Jorgensen, 2010) it is "an enhancement of contaminants in organisms respective to that habitat." Accumulation can be intracellular or extracellular.

Various stable molecules and radionuclides can penetrate inside the cell with the help of transport mechanism that is used for the transport of the ions. Inside the cells, these toxic substances are segregated by the cystein-rich metallothioneins (Turner & Robinson, 1995) or by getting in the vacuole which happens in the fungal cell (Gadd & Fomina, 2011). This type of transport mechanism is poor for the actinides like heavy radionuclides. Another mechanism for the bioaccumulation of such toxic heavy radioactive elements is also reported (Suzuki & Banfield, 1999). It states that the increased permeability of the membrane may be accountable for the movement of these elements inside the cell. Sulfate-reducing bacteria which are known to produce sulfide are found ubiquitously (Postgate, 1979). These bacteria perform a crucial role in avoiding the spread of toxic radionuclides in a wide variety of environments. These bacteria are known to liberate H_2S gas and are responsible to hasten the precipitation or reduction of the high-valence element.

Other bacterial species that can tolerate gamma-radiation doses higher than 25×10^3 Sv are *Deinococcus* and *Rubrobacter*. Even some species of *Chroococcidiopsis* are reported to be resistant to 15×10^3 Sv radiations. Some resistance is also reported by genus *Micrococcus* (Brim et al., 2000). A bioluminescent bacterium, *Photobacterium phosphoreum*, when exposed to radionuclide ^{241}Am, had a chronic effect on its growth. At low temperatures, the growth of the bacteria was found to decrease, but the length and the intensity of its luminescence were found to increase. The reason for such an effect was ascribed to the development of peroxides in the medium solution. These peroxides were formed as a result of radiolysis of water (Alexandrova et al., 2011). It was also shown that ^{241}Am radionuclide accumulated in the genome of the organism. Apart from radiation, these microorganisms have also displayed tolerance to ultraviolet radiation, oxidative agents and desiccation. The genetic constitution of these bacteria which are capable to overcome such environmental stress can be studied by the metagenomic approach.

Radionuclide transporters are responsible for intracellular accumulation. The presence of radionuclide transporter will decide the entry of the radionuclides into the cell. These transporters affect the nuclide porosity of the cell membrane (Merroun & Selenska-Pobell, 2008). Such intracellular deposition of U(VI) in *Halomonas* sp. and *Citrobacter* sp. has been studied by Francis et al. (2004). They demonstrated that the U(VI) accumulated in the form of phosphates such as hydroxophosphate or polyphosphates or as uranium hydrogen phosphate. *Bacillus subtilis* cell wall, when grown in more presence of glucose, showed increased accumulation of iodide (Amachi et al., 2010). In the same way, U(VI) at 5 mg/L concentration was bioaccumulated by *Bacillus vallismortis* after 72 hours of inoculation. The nuclides upon accumulation are shown to be acted on by polyphosphate bodies. Thus the metal and radionuclide tolerance can be achieved by such intracellular chelation mechanisms (Keasling & Hupf, 1996). The process of bioaccumulation takes place through a quick interaction between the nuclide and the anionic groups available in constituents of cell surface. Molecular properties of the nuclides; characteristics and properties of the bacteria mainly affect these types of interactions. The cell wall lipid content and the size of the organism also affects bioaccumulation. Smaller the size, lesser will be the less surface area for accumulation (Jorgensen, 2010).

8.9 Biotransformation

Biotransformation is a process in which the radionuclide gets chemically transformed by the bacterial ability to cause a redox reaction. These redox reactions create an alteration in the pH which changes the solubility of the nuclides (Francis & Nancharaiah, 2015). The main chemical reactions produced by the bacteria leads to the development of coprecipitates, oxides, ionic, organic or inorganic complexes of radionuclides. Actively growing aerobic as well as anaerobic bacteria have the capacity to generate various redox reactions that cause biotransformation. The nonsorptive nuclides are commonly transformed enzymatically or nonenzymatically. *S. putrefaciens* strain CN-32, *G. sulfurreducens* strain PCA and *Anaeromyxobacter dehalogenans* strain K are few examples of bacteria and archaea which are sulfate-reducing bacteria or dissimilatory Fe(II) (DSRB and DIRB) are capable of biotransformation.

The biotransformation (bioreduction and bioprecipitation) finally causes immobilization of the radionuclide (Rui et al., 2013). The bacteria viz., *Shewanella oneidensis*, *S. putrefaciens*, *G. metallireducens* are shown to be largely involved in enzymatic reduction of metals and radionuclides. Technetium (Tc) reduction under different pH conditions along with H_2O formation in the metabolic reaction has been demonstrated (Jabbar & Wallner, 2015). Plutonium IV [Pu(IV)] is anaerobically reduced to Pu (III) by *Bacillus* sp., *Serratia marcescens* and *Clostridium* sp. which was shown by Francis (2007). Similarly, largely nonsorptive radionuclide Np (V) was widely reduced by *D. desulfuricans* (Icopini et al., 2007). It was also found that riboflavin addition enhanced the reduction rates of Tc (VI), Np(V) and Pu(IV) by *S. oneidensis* (Cherkouk et al., 2016).

8.10 Biosolubilization

Autotrophic metabolism of radionuclides that are leached out from their solid matrices leads to biosolubilization. The autotrophic, mesophilic, and acidophilic bacteria in the process of obtaining energy from reduced Fe or S compounds indirectly carry out the solubilization of these radionuclides and metals (Ghauri & Johnson, 1991). They require moisture, acidic pH and oxygen to oxidize Fe, S, and leach out metals. Metal solubilization has been reported by *Acidithiobacillus ferrooxidans, Aciditiobacillus thiooxidans, Leptospirillum ferrooxidans,* and *Sulfolobus* (Bosecker, 1997). Bacterial metabolite like citrate was shown to be responsible for the increased solubility of nuclides for a longer time period. Nevertheless, greater salinity and phosphate presence cause the detachment of uranyl citrate complex that brings about the precipitation of uranium and citric acid. Thus the process of biosolubilization is highly affected by citric acid (Francis et al., 1992). It has been also observed that citrate can complex with the radionuclide and the complexes formed can be sometimes degraded by the microorganisms. Such degradation of the radionuclide citrate complex results in radionuclide precipitation (Francis et al., 1994) and was noticed with *P. aeruginosa* and *Pseudomonas putida* (Thomas et al., 2000).

8.11 Bioprecipitation

When a soluble form of a nuclide is converted into an insoluble form, precipitation occurs. Metals and radionuclides chiefly precipitate as carbonates or hydroxides. Bioprecipitation mainly occurs due to the oxidation and valence state of the radionuclide. Microbes carry out precipitation on a site called as "nucleation site." This microbial precipitation depends on the ligand concentration created by the cell. Bioprecipitation of the radionuclides is highly determined by the microbial ligands and biogenic formation of minerals (Jabbar & Wallner, 2015). The metabolism and secretions of bacteria produce a variation in the pH and Eh in the localized area, thus leading to an alteration in pH near the area neighboring the metal (Merroun & Selenska-Pobell, 2008). A concept of coprecipitation in which during the precipitation reaction, amalgamation of elements in metal oxide minerals occur is reported (Tabak et al., 2005).

8.12 Chelation

Microorganisms are known to produce chelating biopolymers such as pyomelanin, melanin, etc. These chelating agents can significantly alter the mobility of the radionuclides (Tamponnet & Declerck, 2008). Siderophores produced by *Pseudomonas fluorescens* and *Pseudomonas stutzeri* in presence of radionuclides 147Pm, 234Th, and 241Am can highly chelate these nuclides (Arlinger et al., 2004).

8.13 Complexation

Interaction between various groups results in complexation. Teichoic acid, organic phosphate groups, and radionuclides interact with each other to form a complex. Ethylene diaminetetraacetate, fulvic acids, oxalate, diethylenetriaminepentaaceticacid, citrate, humic acid, bicarbonates, etc., are few known complexing agents. Hydroxyapatite was reported to be produced by *Serattia* sp. on sequestration of Sr (Handley-Sidhu et al., 2011).

8.14 Conclusion

Radionuclides contamination into the surrounding from different sources that involve nuclear weapons checking and nuclear power plants is a matter of serious worry. Enumerable microorganisms interact with these main important radionuclides converting them into insoluble compounds. These products can be easily separated and removed through bioremediation processes. The diverse population of microbes in these hotspot polluted areas have a tremendous potential to deal with the contaminants present at that site. They have crucial and unique metabolic processes which help or contribute a vital part in the sustainability of other lives on this planet. These naturally occurring microorganisms with all these beneficial properties can be directly utilized for the bioremediation of nuclear wastes. Bioremediation of radionuclides or radioactive wastes is an inevitable need of the current situation. Recently, their ability to undergo genetic manipulations has been greatly explored. The capabilities of these indigenous microbes of the polluted areas need to be explored in depth, so that they can be used either in isolation or in combination to remove the higher concentration of the radioactive waste. The major advantage of applying microorganisms for this work is that they can be reused.

References

Ahier, B. A., & Tracy, B. L. (1995). Radionuclides in the Great Lakes basin. *Environmental Health Perspectives, 103*(9), 89–101. Available from https://doi.org/10.1289/ehp.95103s989.

Alexandrova, M., Rozhko, T., Vydryakova, G., & Kudryasheva, N. (2011). Effect of americium-241 on luminous bacteria. Role of peroxides. *Journal of Environmental Radioactivity, 102*(4), 407–411. Available from https://doi.org/10.1016/j.jenvrad.2011.02.011.

Amachi, S., Minami, K., Miyasaka, I., & Fukunaga, S. (2010). Ability of anaerobic microorganisms to associate with iodine: 125I tracer experiments using laboratory strains and enriched microbial communities from subsurface formation water. *Chemosphere, 79*(4), 349–354. Available from https://doi.org/10.1016/j.chemosphere.2010.02.028.

Anderson, W., Nordon, H. C., Cain, F., Parrish, G., & Dugan, D. (1956). Studies on a radio-resistant micrococcus. I. Isolation, morphology, cultural characteristics and resistance to g-radiation. *Food Technology, 10*, 575–578.

Andres, Y., MacCordick, H. J., & Hubert, J. C. (1992). Bacterial biosorption and retention of thorium and uranyl cations by *Mycobacterium smegmatis*. *Journal of Radioanalytical and Nuclear Chemistry Letters, 166*(5), 431–440. Available from https://doi.org/10.1007/BF02167788.

Ardiny, H., Witwicki, S., & Mondada, F. (2019). Autonomous exploration for radioactive hotspots localization taking account of sensor limitations. *Sensors (Switzerland), 19*(2). Available from https://doi.org/10.3390/s19020292.

Arlinger, J., Oskarsson, A., Albinsson, Y., Andlid, T., & Pedersen, K. (2004). Mobilisation of radionuclides by ligands produced by bacteria from the deep subsurface. in: V.M. Oversby, L.O. Werme (Eds.), *Materials Research Society symposium proceedings*. Materials Research Society, Warrendale, (Vol. 807, pp. 823–828).

Battle, J. V. I. (2015). Exposures and effects in the marine environment after the Fukushima accident. *Annals of the ICRP, 44*, 331–346. Available from https://doi.org/10.1177/0146645315576099.

Bentchikou, E., Servant, P., Coste, G., & Sommer, S. (2010). A major role of the RecFOR pathway in DNA double-strand-break repair through ESDSA in Deinococcus radiodurans. *PLoS Genetics, 6*(1).

Bosecker, K. (1997). Bioleaching: Metal solubilization by microorganisms. In FEMS microbiology reviews (Vol. 20, Issues 3–4, pp. 591–604). https://doi.org/10.1016/S0168-6445(97)00036-3.

Brim, H., McFarlan, S. C., Fredrickson, J. K., Minton, K. W., Zhai, M., Wackett, L. P., & Daly, M. J. (2000). Engineering *Deinococcus radiodurans* for metal remediation in radioactive mixed waste environments. *Nature Biotechnology, 18*(1), 85–90. Available from https://doi.org/10.1038/71986.

Cherkouk, A., Law, G. T. W., Rizoulis, A., Law, K., Renshaw, J. C., Morris, K., Livens, F. R., & Lloyd, J. R. (2016). Influence of riboflavin on the reduction of radionuclides by *Shewanella oneidenis* MR-1. *Dalton Transactions, 45*(12), 5030–5037. Available from https://doi.org/10.1039/c4dt02929a.

Chicote, E., García, A. M., Moreno, D. A., Sarró, M. I., Lorenzo, P. I., & Montero, F. (2005). Isolation and identification of bacteria from spent nuclear fuel pools. *Journal of Industrial Microbiology and Biotechnology, 32*(4), 155–162. Available from https://doi.org/10.1007/s10295-005-0216-3.

Chiuchiolo, A., Dickhut, M., Michele, C., & Hugh, D. (2004). Persistent organic pollutants at the base of the antarctic marine food web. *Environmental Science & Technology, 38*, 3551–3557.

Choudhary, S., Islam, E., Kazy, S. K., & Sar, P. (2012). Uranium and other heavy metal resistance and accumulation in bacteria isolated from uranium mine wastes. *Journal of Environmental Science and Health - Part A Toxic/Hazardous Substances and Environmental Engineering, 47*(4), 622–637. Available from https://doi.org/10.1080/10934529.2012.650584.

Chung, A. P., Lopes, A., Nobre, M. F., & Morais, P. V. (2010). *Hymenobacter perfusus* sp. nov., *Hymenobacter flocculans* sp. nov. and *Hymenobacter metalli* sp. nov. three new species isolated from an uranium mine waste water treatment system. *Systematic and Applied Microbiology, 33*(8), 436–443. Available from https://doi.org/10.1016/j.syapm.2010.09.002.

Dadachova, E., & Casadevall, A. (2008). Ionizing radiation: How fungi cope, adapt, and exploit with the help of melanin. *Current Opinion in Microbiology, 11*(6), 525–531. Available from https://doi.org/10.1016/j.mib.2008.09.013.

Daly, M. J., Gaidamakova, E. K., Matrosova, V. Y., Vasilenko, A., Zhai, M., Venkateswaran, A., Hess, M., Omelchenko, M. V., Kostandarithes, H. M., Makarova, K. S., Wackett, L. P., Fredrickson, J. K., & Ghosal, D. (2004). Accumulation of Mn(II) in *Deinicoccus radiodurans* facilitates gamma-radiation resistance. *Science (New York, N.Y.), 306*(5698), 1025–1028. Available from https://doi.org/10.1126/science.1103185.

De Luca, G., de Philip, P., Dermoun, Z., Rousset, M., & Verméglio, A. (2001). Reduction of technetium(VII) by *Desulfovibrio fructosovorans* is mediated by the nickel-iron hydrogenase. *Applied and Environmental Microbiology*, 4583–4587. Available from https://doi.org/10.1128/AEM.67.10.4583-4587.2001.

Dekker, L., Osborne, T. H., & Santini, J. M. (2014). Isolation and identification of cobalt- and caesium-resistant bacteria from a nuclear fuel storage pond. *FEMS Microbiology Letters, 359*(1), 81–84. Available from https://doi.org/10.1111/1574-6968.12562.

Dexter-Dyer, B., Kretzschmar, M., & Krumbein, W. E. (1984). Possible microbial pathways in the formation of precambrian ore deposits. *Journal - Geological Society London, 141*(2), 251–262. Available from https://doi.org/10.1144/gsjgs.141.2.0251.

Dhami, P. S., Gopalakrishnan, V., Kannan, R., Ramanujam, A., Salvi, N., & Udupa, S. R. (1998). Biosorption of radionuclides by *Rhizopus arrhizus*. *Biotechnology Letters, 20*(3), 225–228. Available from https://doi.org/10.1023/A:1005313532334.

Dua, M., Singh, A., Sethunathan, N., & Johri, A. (2002). Biotechnology and bioremediation: successes and limitations. *Applied Microbiology and Biotechnology, 59*(2-3), 143–152.

Elias, D. A., Suflita, J. M., McInerney, M. J., Krumholz, L. R. (2004). Periplasmic cytochrome c3 of Desulfovibrio vulgaris is directly involved in H2-mediated metal but not sulfate reduction. *Applied and Environmental Microbiology, 70*(1), 413–420.

Francis, A. J. (2007). Microbial mobilization and immobilization of plutonium. *Journal of Alloys and Compounds, 444–445*, 500–505. Available from https://doi.org/10.1016/j.jallcom.2007.01.132.

Francis, A., Dodge, C., & Gillow, J. (1992). Biodegradation of Metal Citrate Complexes and Implications for Toxic-metal Mobility. *Nature, 356*, 140–142.

Francis, A. J., Dodge, C. J., Lu, F., Halada, G. P., & Clayton, C. R. (1994). XPS and XANES studies of uranium reduction by *Clostridium* sp. *Environmental Science and Technology, 28* (4), 636–639. Available from https://doi.org/10.1021/es00053a016.

Francis, A. J., Gillow, J. B., Dodge, C. J., Harris, R., Beveridge, T. J., & Papenguth, H. W. (2004). Uranium association with halophilic and non-halophilic bacteria and archaea. *Radiochimica Acta, 92*(8), 481–488. Available from https://doi.org/10.1524/ract.92.8.481.39281.

Francis, A. J., & Nancharaiah, Y. V. (2015). *In situ and ex situ bioremediation of radionuclide-contaminated soils at nuclear and norm sites. Environmental remediation and restoration of contaminated nuclear and norm sites* (pp. 185–236). Elsevier Inc. Available from https://doi.org/10.1016/B978-1-78242-231-0.00009-0.

Fredrickson, J. K., Zachara, J. M., Balkwill, D. L., Kennedy, D., Li, S. M. W., Kostandarithes, H. M., Daly, M. J., Romine, M. F., & Brockman, F. J. (2004). Geomicrobiology of high-level nuclear waste-contaminated vadose sediments at the Hanford Site, Washington State. *Applied and Environmental Microbiology, 70*(7), 4230–4241. Available from https://doi.org/10.1128/AEM.70.7.4230-4241.2004.

Fredrikson, J. K., Kostantarithes, H. M., Li, S. W., Plymale, A. E., & Daly, M. J. (2000). Reduction of Fe(III), Cr(VI), U(VI), and Tc(VII) by *Deinococcus radiodurans* R1. *Appl. Environ. Microbiol.* 66 (5), 2006.

Gadd, G. M., & Fomina, M. (2011). Uranium and fungi. *Geomicrobiology Journal, 28*(5–6), 471–482. Available from https://doi.org/10.1080/01490451.2010.508019.

Gans, J., Wolinsky, M., & Dunbar, J. (2005). Computational improvements reveal great bacterial diversity and high metal toxicity in soil. *Science, 309*, 1387–1390.

Garcia-Leon, M. (1990). Determination and levels of 99Tc in environmental and biological samples. *Journal of Radioanalytical and Nuclear Chemistry, 138*(1), 171–179.

Gerbino, E., Carasi, P., Mobili, P., Serradell, M. A., & Gomez-Zavaglia, A. (2015). Role of S-layer proteins in bacteria. *World Journal of Microbiology and Biotechnology, 31*, 1877–1887.

Ghauri, M. A., & Johnson, D. B. (1991). Physiological diversity amongst some moderately thermophilic iron-oxidising bacteria. *FEMS Microbiology Ecology, 8*(4), 327–333. Available from https://doi.org/10.1111/j.1574-6941.1991.tb01778.x.

Green, S. J., Prakash, O., Jasrotia, P., Overholt, W. A., Cardenas, E., Hubbard, D., Tiedje, J. M., Watson, D. B., Schadt, C. W., Brooks, S. C., & Kostka, J. E. (2012). Denitrifying bacteria from the genus *Rhodanobacter* dominate bacterial communities in the highly contaminated subsurface of a nuclear legacy waste site. *Applied and Environmental Microbiology, 78*(4), 1039–1047. Available from https://doi.org/10.1128/AEM.06435-11.

Hamamura, N., Olson, S. H., Ward, D. M., & Inskeep, W. P. (2005). Diversity and functional analysis of bacterial communities associated with natural hydrocarbon seeps in acidic soils at Rainbow Springs, Yellowstone National Park. *Applied and Environmental Microbiology, 71* (10), 5943–5950. Available from https://doi.org/10.1128/AEM.71.10.5943-5950.2005.

Handley-Sidhu, S., Renshaw, J. C., Moriyama, S., Stolpe, B., Mennan, C., Bagheriasl, S., Yong, P., Stamboulis, A., Paterson-Beedle, M., Sasaki, K., Pattrick, R. A. D., Lead, J. R., & MacAskie, L. E. (2011). Uptake of Sr^{2+} and Co^{2+} into biogenic hydroxyapatite: Implications for biomineral ion exchange synthesis. *Environmental Science and Technology, 45*(16), 6985–6990. Available from https://doi.org/10.1021/es2015132.

Hansel, C. M., Fendorf, S., Jardine, P. M., & Francis, C. A. (2008). Changes in bacterial and archaeal community structure and functional diversity along a geochemically variable soil profile. *Applied and Environmental Microbiology, 74*(5), 1620–1633. Available from https://doi.org/10.1128/AEM.01787-07.

Head, I. M., Jones, D. M., & Röling, W. F. M. (2006). Marine microorganisms make a meal of oil. *Nature Reviews. Microbiology, 4*(3), 173–182. Available from https://doi.org/10.1038/nrmicro1348.

Icopini, G. A., Boukhalfa, H., & Neu, M. P. (2007). Biological reduction of Np(V) and Np (V) citrate by metal-reducing bacteria. *Environmental Science and Technology, 41*(8), 2764–2769. Available from https://doi.org/10.1021/es0618550.

Joner, J. E., Anders, J., P., L. A., Ann, dC. M., J., S. O. H., Jean-Marie, P., & Corinne, L. (2001). Rhizosphere effects on microbial community structure and dissipation and toxicity of polycyclic aromatic hydrocarbons (PAHs) in spiked soil. *Environmental Science & Technology, 35*, 2773–2777. Available from https://doi.org/10.1021/es000288s.

Jabbar, T., & Wallner, G. (2015). *Biotransformation of radionuclides: Trends and challenges. Radionuclides in the environment: Influence of chemical speciation and plant uptake on radionuclide migration* (pp. 169–184). Springer International Publishing. Available from https://doi.org/10.1007/978-3-319-22171-7_10.

Jiang, W., Saxena, A., Song, B., Ward, B. B., Beveridge, T. J., & Myneni, S. C. B. (2004). Elucidation of functional groups on gram-positive and gram-negative bacterial surfaces using infrared spectroscopy. *Langmuir: the ACS Journal of Surfaces and Colloids, 20*(26), 11433–11442, https://doi.org/10.1021/la049043 + .

Jorgensen, S.E. (2010). Ecotoxicology. A derivative from encyclopedia of ecology. Elsevier.

Keasling, J. D., & Hupf, G. A. (1996). Genetic manipulation of polyphosphate metabolism affects cadmium tolerance in *Escherichia coli. Applied and Environmental Microbiology, 62* (2), 743–746. Available from https://doi.org/10.1128/aem.62.2.743-746.1996.

Kim, S. J., Koh, D. C., Park, S. J., Cha, I. T., Park, J. W., Na, J. H., Roh, Y., Ko, K. S., Kim, K., & Rhee, S. K. (2012). Molecular analysis of spatial variation of iron-reducing bacteria in riverine alluvial aquifers of the Mankyeong River. *Journal of Microbiology, 50*(2), 207–217. Available from https://doi.org/10.1007/s12275-012-1342-z.

Kurnaz, A., Kucukomeroglu, B., Keser, R., Okumusoglu, N. T., Korkmaz, F., Karahan, G., & Cevik, U. (2007). Determination of radioactivity levels and hazards of soil and sediment samples in Firtina Valley (Rize, Turkey). *Applied Radiation and Isotopes, 65*, 1281–1289.

Landrigan, P. J., Fuller, R., Hu, H., Caravanos, J., Cropper, M. L., Hanrahan, D., Sandilya, K., Chiles, T. C., Kumar, P., & Suk, W. A. (2018). Pollution and global health—An agenda for prevention. *Environmental Health Perspectives, 126*(8). Available from https://doi.org/10.1289/EHP3141.

Liu, C., Zachara, J. M., Zhong, L., Kukkadupa, R., Szecsody, J. E., & Kennedy, D. W. (2005). Influence of sediment bioreduction and reoxidation on uranium sorption. *Environmental Science and Technology, 39*(11), 4125–4133. Available from https://doi.org/10.1021/es048501y.

Lloyd, J. R., & Gadd, G. M. (2011). The geomicrobiology of radionuclides. *Geomicrobiology Journal, 28*.

Lloyd, J. R., & Macaskie, L. E. (1996). A novel phosphorImager-based technique for monitoring the microbial reduction of technetium. *Applied and Environmental Microbiology, 62*(2), 578–582. Available from https://doi.org/10.1128/aem.62.2.578-582.1996.

Lloyd, J. R., & Renshaw, J. C. (2005a). Bioremediation of radioactive waste: Radionuclide-microbe interactions in laboratory and field-scale studies. *Current Opinion in Biotechnology, 16*(3), 254–260. Available from https://doi.org/10.1016/j.copbio.2005.04.012.

Lloyd, J. R., & Renshaw, J. C. (2005b). Microbial transformations of radionuclides: Fundamental mechanisms and biogeochemical implications. *Metal Ions in Biological Systems, 44*, 205–240. Available from https://doi.org/10.1201/9780849346071-8.

Lloyd, J. R., Ridley, J., Khizniak, T., Lyalikova, N. N., & Macaskie, L. E. (1999). Reduction of technetium by *Desulfovibrio desulfuricans*: Biocatalyst characterization and use in a flowthrough bioreactor. *Applied and Environmental Microbiology, 65*(6), 2691–2696. Available from https://doi.org/10.1128/aem.65.6.2691-2696.1999.

Lloyd, J. R., Yong, P., & Macaskie, L. E. (2000). Biological reduction and removal of Np(V) by two microorganisms. *Environmental Science and Technology, 34*(7), 1297–1301. Available from https://doi.org/10.1021/es990394y.

Lovley, D. R. (1991). Dissimilatory Fe(III) and Mn(IV) reduction. *Microbiological Reviews, 55*(2), 259–287. Available from https://doi.org/10.1128/mmbr.55.2.259-287.1991.

Lovley, D. R., Coates, J. D., Blunt-Harris, E. L., Phillips, E. J. P., & Woodward, J. C. (1996). Humic substances as electron acceptors for microbial respiration. *Nature, 382*(6590), 445–448. Available from https://doi.org/10.1038/382445a0.

Lovley, D. R., Giovannoni, S. J., White, D. C., Champine, J. E., Phillips, E. J. P., Gorby, Y. A., & Goodwin, S. (1993). *Geobacter metallireducens* gen. nov. sp. nov., a microorganism capable of coupling the complete oxidation of organic compounds to the reduction of iron and other metals. *Archives of Microbiology, 159*(4), 336–344. Available from https://doi.org/10.1007/BF00290916.

Lovley, D. R., & Phillips, E. J. P. (1994). Reduction of chromate by *Desulfovibrio vulgaris* and its c3 cytochrome. *Applied and Environmental Microbiology, 60*(2), 726–728. Available from https://doi.org/10.1128/aem.60.2.726-728.1994.

Lovley, D. R., Phillips, E. J. P., & Lonergan, D. J. (1989). Hydrogen and formate oxidation coupled to dissimilatory reduction of iron or manganese by *Alteromonas putrefaciens*. *Applied and Environmental Microbiology, 55*(3), 700–706. Available from https://doi.org/10.1128/aem.55.3.700-706.1989.

Lovley, D. R., Widman, P. K., Woodward, J. C., & Phillips, E. J. (1993). Reduction of uranium by cytochrome c3 of *Desulfovibrio vulgaris*. *Applied and Environmental Microbiology, 59*, 3572–3576. Available from https://doi.org/10.1128/aem.59.11.3572-3576.1993.

Lukyanova, E., Zakharova, E., Konstantinova, L., & Nazina, T. (2011). Sorption of radionuclides by microorganisms from a deep repository of liquid low-level waste. *Radiochemistry, 50*, 85–90.

Macaskie, L. E., Jeong, B. C., & Tolley, M. R. (1994). Enzymatically accelerated biomineralization of heavy metals: Application to the removal of americium and plutonium from aqueous flows. *FEMS Microbiology Reviews, 14*(4), 351–367. Available from https://doi.org/10.1016/0168-6445(94)90053-1.

Maila, M. P., Randima, P., Drønen, K., & Cloete, T. E. (2006). Soil microbial communities: Influence of geographic location and hydrocarbon pollutants. *Soil Biology and Biochemistry, 38*(2), 303–310. Available from https://doi.org/10.1016/j.soilbio.2005.05.006.

Merroun, M. L., & Selenska-Pobell, S. (2008). Bacterial interactions with uranium: An environmental perspective. *Journal of Contaminant Hydrology, 102*(3–4), 285–295. Available from https://doi.org/10.1016/j.jconhyd.2008.09.019.

Misra, H. S., Khairnar, N. P., Kota, S., Shrivastava, S., Joshi, V. P., & Apte, S. K. (2006). An exonuclease I-sensitive DNA repair pathway in *Deinococcus radiodurans*: A major determinant of radiation resistance. *Molecular Microbiology, 59*(4), 1308–1316. Available from https://doi.org/10.1111/j.1365-2958.2005.05005.x.

Naso, B., Perrone, D., Ferrante, M. C., Zaccaroni, A., & Lucisano, A. (2003). Persistent organo chlorine pollutants in liver of birds of different trophic levels from coastal areas of Campania, Italy. *Archives of Environmental Contamination and Toxicology, 45*(3), 407–414.

Paterson-Beedle, M., Macaskie, L. E., Lee, C. H., Hriljac, J. A., Jee, K. Y., & Kim, W. H. (2006). Utilisation of a hydrogen uranyl phosphate-based ion exchanger supported on a biofilm for the removal of cobalt, strontium and caesium from aqueous solutions. *Hydrometallurgy, 83*(1–4), 141–145. Available from https://doi.org/10.1016/j.hydromet.2006.03.020.

Phillips, R. W., Wiegel, J., Berry, C. J., Fliermans, C., Peacock, A. D., White, D. C., & Shimkets, L. J. (2002). *Kineococcus radiotolerans* sp. nov., a radiation-resistant, Gram-positive bacterium. *International Journal of Systematic and Evolutionary Microbiology, 52*(3), 933–938. Available from https://doi.org/10.1099/ijs.0.02029-0.

Pignolet, L., Auvray, F., Fonsny, K., Capot, F., & Moureau, Z. (1989). Role of various microorganisms on tc behavior in sediments. *Health Physics, 57*(5), 791–800. Available from https://doi.org/10.1097/00004032-198911000-00013.

Postgate J.R. (Ed.), 1979. The Sulphate Reducing Bacteria, Cambridge University Press, Cambridge, UK.

Rodrigue, A., Effantin, G., & Mandrand-Berthelot, M. A. (2005). Identification of rcnA (yohM), a nickel and cobalt resistance gene in *Escherichia coli*. *Journal of Bacteriology, 187*(8), 2912–2916. Available from https://doi.org/10.1128/JB.187.8.2912-2916.2005.

Rui, X., Kwon, M. J., O'Loughlin, E. J., Dunham-Cheatham, S., Fein, J. B., Bunker, B., Kemner, K. M., & Boyanov, M. I. (2013). Bioreduction of hydrogen uranyl phosphate: Mechanisms and U(IV) products. *Environmental Science and Technology, 47*(11), 5668–5678. Available from https://doi.org/10.1021/es305258p.

Schidlowski, M. (1988). A 3,800-million-year isotopic record of life from carbon in sedimentary rocks. *Nature, 333*.

Seilacher, A. (1992). Patterns of macroevolution: How to be prepared for extinction. *Journal of the Geological Society, 149*.

Shaw, G., & Bell, J. N. B. (1994). Plants and radionuclides. In M. E. Farago (Ed.), *Plants and chemical elements, biochemistry, uptake, tolerance and toxicity* (pp. 179–220). Weinheim (Federal Republic of Germany): VCH Publishers.

Shelobolina, E. S., Sullivan, S. A., O'Neill, K. R., Nevin, K. P., & Lovley, D. R. (2004). Isolation, characterization, and U(VI)-reducing potential of a facultatively anaerobic, acid-resistant bacterium from low-pH, nitrate- and U(VI)-contaminated subsurface sediment and description of *Salmonella subterranea* sp. nov. *Applied and Environmental Microbiology, 70*(5), 2959–2965. Available from https://doi.org/10.1128/AEM.70.5.2959-2965.2004.

Shuryak, I., Matrosova, V. Y., Gaidamakova, E. K., Tkavc, R., Grichenko, O., Klimenkova, P., Volpe, R. P., & Daly, M. J. (2017). Microbial cells can cooperate to resist high-level chronic ionizing radiation. *PLoS One, 12*(12). Available from https://doi.org/10.1371/journal.pone.0189261.

Sugiman-Marangos, S. N., Weiss, Y. M., & Junop, M. S. (2016). Mechanism for accurate, protein-assisted DNA annealing by *Deinococcus radiodurans* DdrB. *Proceedings of the National Academy of Sciences of the United States of America, 113*(16), 4308–4313. Available from https://doi.org/10.1073/pnas.1520847113.

Suzuki, Y., & Banfield, J. F. (1999). In F. C. Burns, & R. Finch (Eds.), *Uranium: Mineralogy, Geochemistry and the Environment* (vol. 38, p. 393). Washington DC, USA: Mineralogic Society of America.

Tabak, H. H., Lens, P., Van Hullebusch, E. D., & Dejonghe, W. (2005). Developments in bioremediation of soils and sediments polluted with metals and radionuclides - 1. Microbial processes and mechanisms affecting bioremediation of metal contamination and influencing metal toxicity and transport. *Reviews in Environmental Science and Biotechnology, 4* (3), 115–156. Available from https://doi.org/10.1007/s11157-005-2169-4.

Tamponnet, C., & Declerck, S. (2008). Editorial. *Journal of Environmental Radioactivity, 99*(5), 773–774. Available from https://doi.org/10.1016/j.jenvrad.2007.10.006.

Thomas, R. A. P., Beswick, A. J., Basnakova, G., Moller, R., & Macaskie, L. E. (2000). Growth of naturally occurring microbial isolates in metal-citrate medium and bioremediation of metal-citrate wastes. *Journal of Chemical Technology and Biotechnology, 75*(3), 187–195. Available from https://doi.org/10.1002/(SICI)1097-4660(200003)75:3 < 187::AID-JCTB206 > 3.0.CO;2-I.

Turner, J. S., & Robinson, N. J. (1995). Cyanobacterial metallothioneins: Biochemistry and molecular genetics. *Journal of Industrial Microbiology, 14*(2), 119–125. Available from https://doi.org/10.1007/BF01569893.

Venkateswaran, A., McFarlan, S. C., Ghosal, D., Minton, K. W., Vasilenko, A., Makarova, K., Wackett, L. P., & Daly, M. J. (2000). Physiologic determinants of radiation resistance in *Deinococcus radiodurans*. *Applied and Environmental Microbiology, 66*(6), 2620–2626. Available from https://doi.org/10.1128/AEM.66.6.2620-2626.2000.

Wildung, R. E., Gorby, Y. A., Krupka, K. M., Hess, N. J., Li, S. W., Plymale, A. E., McKinley, J. P., & Fredrickson, J. K. (2000). Effect of electron donor and solution chemistry on products of dissimilatory reduction of technetium by *Shewanella putrefaciens*. *Applied and Environmental Microbiology, 66*(6), 2451–2460. Available from https://doi.org/10.1128/AEM.66.6.2451-2460.2000.

Wronkiewicz, D., & Buck, E. (1999). In P. C. Burns, & R. Finch (Eds.), *Uranium, Geochemistry and the Environment* (vol. 38, pp. 475–494). Washington, DC: The Mineralogical Society of America.

Xie, S., Yang, J., Chen, C., Zhang, X., Wang, Q., & Zhang, C. (2008). Study on biosorption kinetics and thermodynamics of uranium by *Citrobacter freudii*. *Journal of Environmental Radioactivity, 99*(1), 126–133. Available from https://doi.org/10.1016/j.jenvrad.2007.07.003.

Yablokov, A. V., Nesterenko, V. B., & Nesterenko, A. V. (2009). Consequences of the chernobyl catastrophe for the environment. *Annals of the New York Academy of Sciences, 1181*, 221–286. Available from https://doi.org/10.1111/j.1749-6632.2009.04830.x.

Yong, P., & Macaskie, L. E. (1998). Bioaccumulation of lanthanum, uranium and thorium, and use of a model system to develop a method for the biologically-mediated removal of plutonium from solution. *Journal of Chemical Technology and Biotechnology, 71*(1), 15–26. Available from https://doi.org/10.1002/(SICI)1097-4660(199801)71:1 < 15::AID-JCTB773 > 3.0.CO;2-8.

Zhou, J., Xia, B., Treves, D. S., Wu, L. Y., Marsh, T. L., O'Neill, R. V., Palumbo, A. V., & Tiedje, J. M. (2002). Spatial and resource factors influencing high microbial diversity in soil. *Applied and Environmental Microbiology, 68*(1), 326–334. Available from https://doi.org/10.1128/AEM.68.1.326-334.2002.

CHAPTER

9

Microbial diversity in termite gut ecosystem and their role in lignocellulosic degradation

Gincy Marina Mathew[1], *Raveendran Sindhu*[1], *Chieh Chen Huang*[2], *Ashok Pandey*[3] and *Parameswaran Binod*[1]

[1]Microbial Processes and Technology Division, CSIR-National Institute for Interdisciplinary Science and Technology (CSIR-NIIST), Trivandrum, India, [2]Department of Life Sciences, National Chung Hsing University, Taichung, Taiwan, [3]Center for Innovation and Translational Research, CSIR-Indian Institute of Toxicology Research (CSIR-IITR), Lucknow, India

9.1 Introduction

Termites are mainly found in tropical areas and aid in carbon mineralization and nitrogen cycling (Kudo, 2009; Yamada et al., 2005). Termites are composed of 281 genera belonging to eight families. Among the termites, lower termites comprise seven families namely; Rhinotermitidae, Mastotermitidae, Kalotermitidae, Stolotermitidae, Archotermopsidae, Hodotermitidae, Serritermitidae, and higher termites (family: Termitidae). The Termitidae family forms 75% of the 2000 termite species comprising seven subfamilies belonging to Macrotermitinae, Termitinae, Sphaerotermitinae, Apicotermitinae, Foraminitermitinae, Nasutitermitinae, and Syntermitinae (Eggleton, 2011). Termites feed on different substrates like wood, grass, fungi, soil, dung, humus, and are classified into four main groups (Donovan et al., 2001; König et al., 2013) (Table 9.1). Termites are dependent on

Microbial Diversity and Ecology in Hotspots
DOI: https://doi.org/10.1016/B978-0-323-90148-2.00002-X

TABLE 9.1 Feeding groups of termites.

Group	Substrate	Termites
Group I (lower termites)	Wood, litter, and grass	*Mastotermes, Cryptotermes, Zootermopsis, Hodotermes, Coptotermes, Schedorhinotermes, Serritermes, Archotermopsis, Porotermes*
Group II (higher termites)	Wood, grass, fungus-feeders, and litter	*Macrotermes, Microtermes, Nasutitermes, Syntermes, Amitermes, Orientotermes, Protohamitermes, Cornitermes, Rhynchotermes*
Group III	Highly decomposed wood along rich in organic content	*Speculitermes, Amalotermes, Termes, Dicuspiditermes, Labritermes, Angularitermes, Armitermes, Foraminitermes*
Group IV	Soil feeders	*Coxotermes, Ophiotermes, Cubitermes, Orthotermes, Eburnitermes, Thoracotermes, Postsubulitermes, Labidotermes, Grigiotermes, Ateuchotermes*

cellulolytic enzymes along with cellulolytic bacteria like Spirochetes, Bacteriodetes, Firmicutes, Fibrobacteres (Anand et al., 2010; Brune, 2014). Unlike higher termites, lower termites, being xylophagous in nature are dependent on symbiotic protists (Cleveland, 1925) for lignocellulosic degradation, for example, *Coptotermes formosanus* (Xie et al., 2012). These protists are transmitted vertically to the offsprings by parents vertically by proctodeal feeding. This transmission involves cospeciation involving the termite host and the protist (Desai et al., 2010; Noda et al., 2007). Protists have a symbiotic relationship with symbiotic bacteria (ecto and endo) in the termite gut (Bauer et al., 2000). Unlike the lower termites, the higher termites do not have protists symbionts (flagellated) (Lo et al., 2011).

9.2 Termite gut

The alimentary tract of termite has three distinct areas forming the fore, mid, and hindgut (Engel & Moran, 2013) and requires 24 hours for the passage of the food material through the alimentary canal (Breznak, 2002). The foregut region is involved in the mastication of the woody matter. The midgut region secretes the digestive enzymes and absorbs the nutrients produced in the gut. The hindgut region is segmented into P1, P2, P3–P4, and P5 corresponding to ileum, enteric valve, colon, and rectum, respectively (Noirot, 2001; Watanabe & Tokuda, 2001). The wood foraging termites contain gut microbes that are compartmentalized in the hindgut and act as a fermentation chamber. The gut of the termite is analogous to the alimentary canal of ruminants. The P3 (Proctodeal segment) section of the gut has the highest microbial count containing microbes from the three domains: archae, bacteria, and eukaryotes which acts as a fermentation chamber and is relatively complex in the higher than in lower termites (Köhler et al., 2012). Termites recycle lignocellulose acting as a bioreactor to polymerize lignocellulose to ferment and provide nutrition to the host termite (He et al., 2013).

9.2.1 Composition of termite guts' microbiome

Each termite gut is a microenvironment having a complex ecosystem with unique species of microbes found in microtiter fluid and hence they function as hotspots. The framework of the termite gut microbiome is dependent on the diet of the termite (Mikaelyan et al., 2015). Mikaelyan et al. (2017) has reported the bacterial microflora of the termite gut based on their lignocellulosic diet during various stages of humification. Different gut compartments comprise microbes of special lineages that act as hotspots. The alkaline P1 comprised *Clostridiales,* the hindgut P3, P4 regions mainly contained phyla related to Spirochetes and Fibrobacteres and the TG3 phylum regions mainly contained Clostridiales and Bacteroidetes. The microbes of the lower termite gut of *Reticulitermes lucifugus* belonged to five main bacterial phyla consisting of Firmicutes, Proteobacteria, Spirochetes, Candidatus TG1, and Bacteroidetes, which were detected by 16S rRNA clonal library selection (Butera et al., 2016). Culturable cellulolytic and hemicellulolytic bacteria belonging to Firmicutes (*Bacillus* and *Paenibacillus* sp.) were isolated from *R. lucifugus. Microbacterium* sp. (belonging to Actinobacteria) were isolated from the higher termite *Zootermopsis angusticollis* capable of degrading cellulose and xylan (Wenzel et al., 2002). Through 454 pyrosequencing methods of 16S V5-V6 region of the microbes from the hindgut lumen of *Reticulitermes flavipes,* bacteria were affiliated to Phyla Spirochetes, Elusimicrobia, Firmicutes, Bacteroidetes, Proteobacteria, Fibrobacteres, Verrucomicrobia, Actinobacteria, and Tenericutes were identified (Boucias et al., 2013). The molecular diversity of wood feeder lower termite *Reticulitermes speratus* was assessed by 16S rRNA sequencing and showed Spirochetes followed by Bacteroides, Clostridia, Proteobacteria, Actinobacteria, Mycoplasma, and others (Hongoh et al., 2003). Termite gut microbes contain several microbes that are uncultivable but can be studied by metagenomics approaches, especially in complex environments. Warnecke et al. (2007) found that Bacteroidetes, Spirochetes, and Firmicutes phyla helped in the metabolism of carbohydrates and nitrogen. Various gut microbes can be cultivable using a specific culture medium. Various gut microbes involved in lignocellulosic degradation are mentioned in Table 9.2.

TABLE 9.2 Termite gut microflora and digestive enzymes involved in lignocellulosic degradation.

Chewing and mechanical breakdown of woody lignocellulosic biomass	Stage I polysaccharide degrading microbes	Stage II fermentative microorganisms (oligosaccharide degradation)	Stage III acetogenesis, methanogenesis	End products derived from fermentation
Cellulose (20%–50%),	Cellulolytic microbes **Protozoa** (in lower termites): *Trichonympha sphaerica* and *Trichomitopsis termopsidis* **Bacteria**: *Ochrobactrum* sp., *Escherichia coli*, *Shigella* and γ-*Proteobacteria*, *Rhizobia*, *Microbacterium*, *Cellulomonas*, *Brevibacillus*, *Paenibacillus*, *Bacillus*, *Klebsiella*, *Corynebacterium*, *Sphingomonas*, *Micromonospora* **Yeast**: *Debaryomyces hansenii*, *Pichia guilliermondii*, *Candida inconspicua*, *Candida orthopsilosis*, *Candida parasilosis* **Fungi**: *Termitomyces*, *Trichosporon*, *Cryptococcus*, *Sporothrix*	β-Glucosidase: *Rhizobia*, *Rhodococcus*, *Nocardia*, *Serratia*, *Enterococcus*, *Spirochetes*, *Klebsiella*	Acetinogenic bacteria: *Clostridium*, *Sporomusa*, and *Acetonema*Methanogenic bacteria: *Methanobrevibacter cuticularis*, *Methanobrevibacter curvatus*	Carbon dioxide, hydrogen, methane, partially digested feces forming the fungus comb (in Macrotermitine termites) and microbial cells
Hemicellulose (15%–35%)	Hemicellulose (xylan) degrading microbes: *Termitomyces*	*Arthobacter*, *Citrobacter*, *Pseudomonas*, *Bacillus*, *Aureobacterium*, *Actinobacteria*, *Escherichia coli*, *Streptomyces*, *Staphylococcus*		
Lignin (18%–35%)	Enzymes involved: laccases, lignin peroxidases	*Termitomyces* (detected by uncultivable methods)		*Termitomyces* that grow on the fungus comb help in degrading the lignin compounds

9.2.2 Symbiotic flagellates in lower termites

Most of the flagellated protozoans of the lower gut termite belonged to Oxymonads, Trichomonods, and *Hypermastigotes*. The microbial diversity (bacteria and flagellates) were higher in the feeding termites like *Reticulitermes* in comparison to the wood-inhabiting species *Cryptococcus domesticus, Crytotermes secundus,* and *Prorhinotermes simplex* (Waidele et al., 2017). Culture-independent methods studied the lignocellulosic degradation of protists due to their unculturability. According to Xie et al. (2012), the *C. formosanus* harbored three protists species that were identified as *Holomastigotoides mirabile, Spirotrichonympha leidyi,* and *Pseudotrichonympha grassii* using culture-independent methods with the help of phylogenetic markers functioned as hotspots for symbiotic breakdown of lignocellulosic plant matter. The mixed population in hindgut of *R. speratus* were analyzed by 16S rRNA sequences by the polymerase chain reaction (Ohkuma et al., 2000). Todaka et al. (2007) identified the lignocellulosic degradation genes from protists by biotinylated CAP trapper method and one-pass sequencing from the termite gut of *R. speratus.*

9.2.3 Yeast and fungi in termites

Some of the yeast and fungal species of the termite gut that was able to degrade cellulose and hemicellulose (xylan) were *Debaromyces, Sporothrix, Candida, Termitomyces, Trichosporon, Pichia, Cryptococcus,* etc. (Mathew et al., 2012; Wenzel et al., 2002). Rajagopal et al. (1979, 1981) isolated filamentous fungi from the termite's intestinal fluid which represented *Aspergillus* sp. (*A. clavatus, A. flavus, A. nidulans,* and *A. awamuri*), *Alternaria alternata, Rhizopus stolonifera,* and *Paecilomyces fusiporus.* The higher termite gut of *Macrotermes mulleri* developed a symbiotic relationship with the fungi *Termitomyces.* The xylanase produced from the termite gut showed resemblance to the enzyme secreted from the fungi, indicating them to be of gut origin (Rouland-Lefèvre et al., 2000).

9.2.4 Bacteria in the termite gut

The predominant category of bacteria in the termite gut belong to Spirochetes (König et al., 2013). They are found in the intestinal hindgut of all termites and range from size between 3 and 100 μm (length) and 0.2–1.0 μm (width) (Breznak, 2002). Spirochetes are dominant in wood-feeding termites in comparison to humus and fungus-growing termites (Macrotermitine) (Breznak, 2002). Spirochetes isolated from the lower termites were *Treponema primitia* and *Treponema azotonutricium* from *Z. angusticollis* (Breznak, 2002), *Spirochaeta coccoides* from *Neotermes castaneus* (Dröge et al., 2006) and *Treponema isoptericolens* sp. nov from *Incisitermes tabogae* (Dröge et al., 2006). Bacteroidetes are mainly found in fungus-growing termites (Macrotermitine). *Alistipes, Dysgonomonas termitidis,* and *Bacteroides reticulotermitis* (*R. speratus*), *Paludibacter,* and *Parabacteroides* were obtained from the guts (Pramono et al., 2015; Sakamoto & Ohkuma, 2013). Firmicutes from higher wood-eating termites belonged to Lachnospiraceae and Ruminococcaceae, *Clostridium termitidus,* and *Bacillus* sp. (Auer et al., 2017). *Nasutitermes* sp. detected the gut bacteria belonging to Spirochetes, candidate phylum TG3, Firmicutes, Fibrobacteres, Bacteroidetes, Proteobacteria, and Actinobacteria (Köhler et al., 2012). Schäfer et al. (1996) identified the presence of hemicellulose degrading bacteria of four termites (*Nasutitermes nigriceps, N. castaneus, Neotermes jouteli, Z. angusticollis, Zootermopsis nevadensis,* and *Macrotermes darwiniensis*) through enrichment and isolation by cultivable methods. Xylan degrading bacteria like *Bacillus* and *Clostridium* sp. were isolated from *Odontotermes formosanus,* the fungus-growing termite (Mathew et al., 2012). *Clostridium oroticum* was isolated from the gut homogenate of the termite *Pericapritermes latignathus* by alkaline enrichment methods (Thongaram et al., 2005). Cho et al. (2010) isolated cellulolytic bacteria like Dyella sp., *Bacillus* sp. and *Chryseobacterium* sp. from *R. speratus* gut. *Lactococcus* sp. isolated from carbohydrate utilizing microbes from the hindgut of lower termites like *M. darwiniensis* and *Cryptotermes primus* in higher termites such as *Nasutitermesarborum, Nasutitermes hainanensis, Thoracotermes macrothorax,* and *Anoplotermes pacificus* produced lactate as an end product by anaerobic action utilizing glucose (Brune & Ohkuma, 2011; Brune, 2014; Yang et al., 2016). By cultivable methods, *N. hainanensis,* a wood-feeding higher termite, *Lactococcus nasutitermitis* was isolated from the alimentary canal of termites (Yang et al., 2016). Proteobacteria are found abundantly in the Macrotermitine termites. Enterobacteriales bacterium was identified in the higher termite *O. formosanus* by denaturing gradient gel electrophoresis (DGGE) profiling using cellulolytic bacterial primers by PCR amplification (Mathew et al., 2012). Through culture-dependent methods, Proteobacteria like *Ochrobactrum* sp., *Escherichia coli, Shigella,* and *γ-Proteobacteria* are identified in the termite gut, whereas *Rhizobium* and *E. coli* were observed in the *O. formosanus* fungus combs (Mathew et al., 2012). An archaebacterium is found in termite guts (Boucias et al., 2013). *Methylocystis* (methanotrophic bacteria)were identified in the guts of *C. formosanus, Incisitermes marginipennis, M. darwiniensis* (Mathew et al., 2012), *Hodotermes mossambicus* by 16S rDNA analysis (Reuss et al., 2013). Methanogens isolated from *R. flavipes* belonged to *Methanobrevibacter* sp. viz., *Methanobrevibacter cuticularis* sp. nov., and *Methanobrevibacter curvatus* which utilized H_2 and CO_2 to form CH_4 (Leadbetter & Breznak, 1996).

$$4H_2 + CO_2 \rightarrow CH_4 + 2H_2O \tag{9.1}$$

Sulfate-reducing bacteria belonging to *Desulfovibrio* sp. from guts of termites elevated the rates of oxygen reduction and presence of hydrogen (Fröhlich et al., 1999; Trinkerl et al., 1990).

9.3 Lignocellulose and their degradation in termites

The lignocellulosic wood is composed of cellulose, hemicellulose, and lignin (Ni & Tokuda, 2013). The major constituent of lignocellulose, cellulose is the predominant component linked by β-1, 4-linkage attaching glucose monomers. Three main enzymes are responsible for the breakdown of cellulose to glucose; namely endoglucanases, cellobiohydrolases, and β-glucosidases. Endoglucanases (EC 3.2.1.4) hydrolyze the β-1, 4-linkage of cellulose; Cellobiohydrolases (EC 3.2.1.91) is an exocellulase that cleaves the 1, 4-β bonds at the ends of cellulose. β-Glucosidases (EC 3.2.1.21) cleaves the cellobiose units to form glucose units. Laccase (EC 1.10.3.2) enzymes are helpful for lignin breakdown. These lignocellulolytic enzymes are derived from the termites or by the symbiotic action of the gut symbionts. Lignocellulosic plant biomass is a major food source for higher and lower organisms (Pauly & Keegstra, 2008). The plant cell wall (PCW) is complex in nature preventing easy nutrient accessibility to various microorganisms. The end products of lignocellulosic polymer degradation are reabsorbed, aiding in the termite's metabolism, whereas microbial biomass provides nutrition for the host termite (Breznak & Brune, 1994). The cellulolytic enzymes produced by lower and higher termites are mentioned in Table 9.3.

TABLE 9.3 Cellulolytic enzymes in the lower and higher termites.

Lower termites	**GH family**	**Function**
Endo-1,4-β-D-glucanase	GH9	Hydrolyze amorphous cellulose
β-Glucosidase	GH1	Breakdown cellobiose units to glucose
Protistian cellulases β-1,4-glucosidase cellulase exoglucanase endoglucanase xylanase (protists)	GH3 GH5 GH7 GH45 GH11	Breakdown crystalline cellulose
Higher termites		
Endo-1,4-β-D-glucanase	GH9	Hydrolyze amorphous cellulose
β-Glucosidase	GH1	Breakdown cellobiose units to glucose
Bacterial cellulases β-glucosidase	GH1 and GH3	Hydrolyze cellobiose units to glucose
Endoglucanse	GH5, GH8, GH9, GH44, GH45, GH51, GH74	Hydrolyze amorphous cellulose
Bacterial xylanase	GH11, GH43	Breakdown of the xylan backbone
Fungal xylanases (*Termitomyces*)	GH10	Breakdown of the xylan backbone

9.3.1 Lignocellulosic degradation mechanism in lower termites

In lower termites, symbiotic flagellates along with the gut microbes assist in cellulosic digestion. The cellulosic and hemicellulosic materials are degraded within 24 hours in the termite's gut, whereas lignin is gradually degraded. Termites and their microbes also secrete glycoside hydrolases (GHs) that interplay in lignocellulosic degradation. Lower termites are composed of bacteria belonging to Spirochetes, Firmicutes, Proteobacteria, Bacteriodetes, and Elusibacteria, Archae and wood-degrading flagellates that help in the bioconversion of ingested plant biomass to methane, acetate, and hydrogen (Brune, 2014). The lignin is oxidized by the soil microbes belonging to Actinobacteria, γ-Proteobacteria, and α-Proteobacteria (Bugg et al., 2011). In the lower termite, strong endoglucanase activities are seen in the mandibular region (Tokuda et al., 2004). Strong cellulolytic activity is detected in the hindgut region, whereas β-glucosidase activity is detected both in the salivary gland and hindgut region (Tokuda et al., 2002). In the lower termites, wood is masticated (chewed) mechanically and mixed with the enzymes present in the salivary glands. The lignocellulosic wood particles are further broken down mechanically and mixed in the gizzard. The glucose released is reabsorbed in the midgut and the undigested wood is passed to the hindgut. The cellulolytic flagellates phagocytize the partially digested materials and degrade the polysaccharides with the help of cellulases and hemicellulases. The microbes present in the hindgut act as a fermentation chamber producing short-chain fatty acids are absorbed releasing lignin along with the feces, for example, *Trichomitopsis termopsidis* isolated from *Zootermopsis* (Kuwahara et al., 2017; Yamin & Trager, 1979). According to Yuki et al. (2015), lower termites were reported to host flagellates where around 19 species belonging to *Parabasilian* and *Oxymona* did, residing in the paunch of the termite gut. These protists contain digestive vacuoles harboring enzymes belonging to endoglucanases, β-glucosidases, xylanases, cellobiohydrolases (GH7), arabinosidases, and mannanosidases. These protists are attached to ectosymbionts or endosymbionts. Uncultivable symbionts belonging to Bacteroidales, *Candidatus Symbiothrix dinenymphae* colonize on the protist *Dinenympha* sp. (hemicellulolytic and cellulolytic) (Yuki et al., 2015). The cellulolytic parabasilid protist *Trichonympha agilis* in the termite gut of *R. speratus* hosts *Candidatus Endomicrobium trichonymphae* and *Candidatus Desulfovibrio trichonymphae,* respectively. The ectosymbiotic bacteria Ca. *Desulfovibrio trichonymphae* utilizes the hydrogen produced which could affect the fermentation carried by *T. agilis* and Ca. *Endomicrobium trichonymphae.*

9.3.2 Lignocellulosic degradation in higher termites

Higher termites (Macrotermitinae) belonged to fungus-growing termites helps in active lignocellulosic degradation and nutrient recycling.

The symbiotic association of termites, fungi, and gut microbes initiated lignocellulosic degradation (da Costa et al., 2019). In the higher termites, the workers forage lignocellulosic materials. The material is dependent on the availability of plant biomass according to seasonal changes. For example, in the macrotermitine termite *Macrotermes michaelseni*, the feces (fungus comb) comprised 70% grass material, followed by 30% woody material, which geographically varied to other regions. It may vary from grass to woody material. The partially digested plant material is excreted out to form a comb-like structure known as "fungus comb," which harbors *Termitomyces* spores. The host termites are dependent on the *Termitomyces* fungi for their sustenance by consuming their fungal nodules and mature parts of the fungus comb. Thus a tripartite symbiotic relationship is established between the termite, termite gut microbes, and *Termitomyces*. These *Termitomyces* spores were also detected in the termite gut of higher termites like *O. formosanus* (Mathew et al., 2012). These *Termitomyces* sp. produced glycosyl hydrolases (GH) which help in plant biomass degradation (Poulsen et al., 2014). In case of higher termites *Macrotermes natalensis* through metagenomics analysis, around 111 glycosyl hydrolase families (GHFs) showed that *Termitomyces* sp. helped in the breakdown of complex polysaccharides and the gut microbes of the worker termites contributed in the final digestion process of the oligosaccharides. *M. natalensis* contained GH9 cellulases which are responsible for cellulolytic degradation. The mound-building higher termite *Cornitermes cumulans* (subfamily Syntermitinae) feeds on grass and litter and forms the nest core with predigested plant materials, soil, and fecal matter in the form of nodules. These nests serve as food storage for *C. cumulans* and attract the colonization of other lignocellulosic microbes (Menezes et al., 2018). Through Illumina sequencing of 16S rRNA and ITS2 regions, it was observed that Actinobacteria and Sordariales populated the nest structure and food nodules in comparison to Spirochetes and Pleosporales that dominated the guts microflora of *C. cumulans*. The results proved the presence of cellulase and β-glucosidases of the accumulated food and termite gut, which showed a correlation with the symbiotic enzymes.

9.4 Metagenomic approaches in termites for detecting glycosyl hydrolase genes for lignocellulosic degradation

Metagenomic approaches have been used to determine the functional genes expressed by the gut microbes in lower and higher termites that are responsible for lignocellulosic degradation. The diversity of these flagellates can be detected by amplifying the small subunit ribosomal

RNA in the hindgut of lower termites. Ohkuma et al. (2000) identified 33 species of parabasalids from eight different termite species with the help of fluorescent-labeled oligonucleotide probes. From the hindgut of *C. formosanus,* three paraphyletic lineages of hypermastigotes encoding for *Pseudotrichonympha, Spirotrichonympha,* and *Trichonympha* were observed (Fujita et al., 2010; Ohkuma et al., 2000). The presence of GH genes was identified in the salivary glands, foregut, midgut, and hindgut of *C. formosanus* by comparative transcriptomics with RNA sequencing (high throughput). From their studies, 71,117 unigenes were organized and the transcriptomics studies revealed the presence of GH and redox enzyme genes in various gut compartments. GH9, GH22, and GH1 genes carry out the hydrolysis of plant biomass and separation of lignin and carbohydrate from the biomass. GH7 and GH5 gene encoded for the protists present in the hindgut. It was observed that 12 redox enzymes were initiated when fed on lignin-rich material. These studies show that GH and redox enzymes contribute to the lignin deconstruction of the host termite. Xie et al. (2012) carried out the metagenomics profiling of symbiotic protists that aids in the lignocellulosic degradation of *C. formosanus.* From their profiling, 118 GH from 18 different glycosyl hydrolase families were detected. Among them, GH10 coding for endo-1, 4-β-xylanase was characterized to understand the better functioning of protist hemicellulases. Ohtoko et al. (2000) identified the presence of diverse genes belong to GH45 coding for cellulolytic genes by PCR and cDNA screening in *R. speratus.* The whole-cell in situ hybridization with specific oligonucleotide probes helped to identify the genes to belong to the protists *Teranympha mirabilis* and *T. agilis.* Tokuda et al. (2018) through metatranscriptomic analysis in the hindgut of *Nasutitermes takasagoensis* identified the presence of glycoside hydrolase (GH) transcripts that were linked to the bacterial community of *Treponema* (Spirochetes phyla) that produced abundant GH11 transcripts coding for xylan hydrolysis, a constituent of hemicellulose. The *Treponema* sp. of the wood-feeding termite *N. takasagoensis* played an important role in hemicellulosic degradation. Metagenomic analysis of six higher termites viz., *Nasutitermes corniger* (wood feeders), *Microcerotermes parvus* (wood feeders), *Cornitermes* sp. (litter feeder), *Termes hospes* (humus feeder), *Neocapritermes taracua* (humus), and *Cubitermes ugandensis* (soil) with respect to the utilization of lignocellulosic degradation in different stages of humification were studied to assess their structural and functional community analysis. From the cluster analysis studies, it was observed that there were strong similarities between the orthologous gene clusters (COGs) in wood and litter feeding termites and drastically different with humus feeding termites. Through metagenomics and amplicon sequencing, similarities in the community and function were observed in various hindgut compartments (P1, P3, and P4). Liu et al. (2011) analyzed the functional screening for lignocellulose degrading genes from the metagenomics library of fungus-growing termites. Among the genes identified was one coding for xylanase clone and the other for 12 β-glucosidase. Mathew et al. (2012) identified the presence of cellulolytic gut microbes using DGGE using cellulolytic specific primers. Metatranscriptomic analysis (high throughput mRNA sequencing) has revealed the presence of symbiotic gut bacteria in higher termites (whole gut and luminal fluid) and the presence of carbohydrate-active enzymes (CAZy) (Marynowska et al., 2017). Through metagenomics and metatranscriptomic analysis, Spirochetes and Fibrobacteres dominated the gut community in *N. corniger*, whereas Firmicutes and Spirochetes dominated the hindgut paunch of *Amitermes wheeleri* (herbivore dung feeder) (He et al., 2013). Other *Nasutitermes* sp. from *Costa rican* rain forest in carbohydrate-active enzymes detected in the termite symbionts (lacking flagellates) (Warnecke et al., 2007), but lacked genes responsible for lignin degradation. Through their analysis, above 700 GH catalytic domains representing 45 CAZy enzymes coding for hemicellulases and cellulases were identified, responsible for xylan and cellulose hydrolysis. Poulsen et al. (2014) identified the Glycosyl hydrolase (GH) profile by cluster analysis in the higher termites, which is fungus-growing and nonfungus growing. The similarity was observed between microbes in the workers and soldiers of the *Macrotermitine* termite guts (e.g., *M. natalensis*), which included organisms belonging to *Burkholderia, Desulfovibrio, Alistipes, Clostridium*, and *Bacteroides* (Karen et al., 2015).

9.4.1 Glycoside hydrolases for lignocellulosic degradation

Intrinisic enzymes from the termites play an important role in lignocellulosic degradation (Breznak, 2002; Tokuda & Watanabe, 2007). The enzymes coding for hemicellulose and cellulose degradation are designated as more than 100 families that are endogenously produced by the termites. Lignocellulose degradation is initiated mainly with the help of carbohydrate-active enzymes (CAZymes). Among them, GH primarily breaks down the glycoside bonds. Other enzymes are carbohydrate esterases (CEs), polysaccharide lyases (PLs), and auxiliary enzymes (AAs). The GH genes expressed in the symbiotic microorganisms of host termites are mentioned in Tokuda and Watanabe (2007) (Table 9.4).

Among the GH family's GHF9 codes for cellulases of termite origin and GHF7 or GHF45 belonged to protist origin. Mechanical grinding by termites improves their digestion and increase access to the cellulolytic enzymes (Watanabe & Tokuda, 2001). The cellulolytic enzymes of *Coptotermes gestroi* produced CAZymes and pro-oxidant, antioxidant, and detoxification enzymes (PAD) that helped in the metabolism of

TABLE 9.4 GH genes expressed in the symbiotic microorganisms of host termites.

GH genes	Function	Microorganism	Host termite	Reference
GH1	β-Glucosidase, β -galactosidase, β -mannosidase,	*Treponema*	*Nasutitermes* sp.	Warnecke et al. (2007)
GH3	β -1,4-Glucosidase, β -1,4-xylosidase, β -1,3-glucosidase, a-L-arabinofuranosidase,	*Treponema*	*Nasutitermes* sp.	Warnecke et al. (2007)
GH5	Cellulase	*Treponema*	*Nasutitermes* sp.	Warnecke et al. (2007)
GH5	β -glucosidase	*Pedobacter sp.*	*Coptotermes getroi*	
GH7	Exoglucanase	*Holomastigotoides mirabile* and *Pseudotrichonympha grassi*	*Coptotermes getroi*	Xie et al. (2012)
GH8	Endo-1, 4 – β-D-glucanase	*Bacteria*	*Coptotermes formosanus*	Xie et al. (2012)
GH9	Endoglucanase, cellobiohydrolase, β -glucosidase	*Fibrobacter*	*Nasutitermes* sp.	Warnecke et al. (2007)
GH10	Xylanase, β -1,3-endoxylanase	*Treponema*	*Nasutitermes* sp.	Warnecke et al. (2007)
GH10	Xylanase	*Termitomyces*	*Macrotermes natalensis*	Poulsen et al. (2014)
GH11	Xylanase	*Treponema*	*M. natalensis*	Poulsen et al. (2014)
GH11	Xylanase	*Treponema*	*Nasutitermes takasagoensis*	Tokuda et al. (2018)
GH11	Xylanase	*Holomastigotoides mirabilis*	*Coptotermes gestroi*	Franco Cairo João et al. (2016)
GH26	β -1,3-Xylanase, mannanase	*Treponema*	*Nasutitermes* sp.	Warnecke et al. (2007)
GH43	Xylanase, β -xylosidase, a-L-arabinofuranosidase, arabinanase, others 1	*Treponema*	*Nasutitermes* sp.	Warnecke et al. (2007)

(Continued)

TABLE 9.4 (Continued)

GH genes	Function	Microorganism	Host termite	Reference
GH44	Endoglucanase, xyloglucanase	*Treponema*	*Nasutitermes* sp.	Warnecke et al. (2007)
GH47	mannosidase	*Termitomyces*	*M.natalensis*	Poulsen et al. (2014)
GH49	dextranases	*Termitomyces*	*M. natalensis*	Poulsen et al. (2014)
GH53	Endo-1, 4 − β -D-galactosidase	Bacteria	*Coptotermes formosanus*	Xie et al. (2012)
GH67	Alpha glucouronidase	Bacteria	*Coptotermes formosanus*	Xie et al. (2012)
GH79	Glucouronidase, heparanase	*Termitomyces*	*M. natalensis*	Poulsen et al. (2014)

lignin and carbohydrates. Among the dominant GH families related to lignocellulosic degradation in worker and soldier termites were GH9 coding for endoglucanases and GH1 for β-glucosidases. Other GH families taxonomically related to protists belonged to GH7 for exoglucanases, GH5 and GH45 representing endoglucanases and GH10 and GH11 for xylanases, highlighting the symbiotic role of protists in cellulolytic degradation along with *C. gestroi* (Do et al., 2014). GH families involved in hemicellulose degradation of the worker and soldier castes were GH3, GH8, GH10, GH11, GH16, GH26, GH29, and GH31. The host and gut symbionts of *R. flavipes* were identified using metatranscriptomic and metagenomic approaches in order to understand their role in lignocellulose degradation. The method consisted of high throughput sequencing of the host and symbiont cDNA library (Rajarapu et al., 2015; Tartar et al., 2009). In higher termites, the P3 region of *Nasutitermes* sp. contained catalytic domains of GH5 cellulases, GH94 cellobiose or cellodextrin phosphorylases, GH51 endoglucanase/arabinofuranosidases, and lower concentrations of GH8, GH9, GH44, GH45, and GH74 endoglucanases (Warnecke et al., 2007). Cellulase enzyme was detected to be secreted by the symbiotic bacteria dwelling in the termite hindgut of higher termites lacking symbiotic flagellates viz., *N. takasagoensis*, and *Nasutitermes walker*. Zymogram activity studies revealed that 59% of crystalline cellulose was degraded with the cellulase originating from hindgut microbes. Cellulolytic genes were analyzed in the transcriptome of dry-wood termite *C. domesticus* by Illumina paired-end sequencing technology. Eleven unigenes were identified to belong to endoglucanase of GH7, 9, 45 families (Wu et al., 2016). The five genes of GH7 showed similarity to the endoglucanase of the symbiotic protist from the termite *Neotermes koshunensis*. The four genes of GH9 showed homology to enzymes in termites of *O. formosanus*, *N. koshunensis*, *Hodotermopsis sjostedti*, and *Z. nevadensis*. Two genes of GH45 showed homology to the endoglucanase from the gut symbiont from the higher termite *M. darwiniensis*. The differential expression of the genes was analyzed when *C. domesticus* consumed three different wood species viz., *Casuarina equisetifolia* (CE), *Koompassia excelsa* (KE), and *Myristica* sp. (Ms). Expression analysis indicated that consuming various wood types activated or downregulated genes coding for carbohydrate metabolism, hydrolase, and catalytic activity.

9.4.2 Why consider termite gut ecosystem as hotspots for lignocellulosic degradation?

Termite gut acts a bioreactor as it hosts gut microbes along with in-house cellulolytic genes that help in the degradation of woody plant biomass. They function as a hotspot due to diverse and novel microbes residing in the gut, which varies depending on the termite's feeding habit. For example, in the case of higher termites, Termitomyces species are specifically found only Macrotermitine termites. The Termitomyces spores are transferred via the termite gut to the fungus comb. The fungus combs and termite guts of these Macrotermitine termites act as hotspots as they harbor Termitomyces sp. that are capable of lignocellulosic degradation (Mathew et al., 2012). The mechanism of degradation in termites can be used to establish natural biomass utilization systems (NBUS) for the development of biorefineries producing biofuels (Mathew et al., 2013; Xie et al., 2014). These strategies are cost-effective and sustainable as it mimics the natural environment for the degradation of lignocellulose. In the case of lower termites, lignocellulosic degradation is achieved effectively in few hours in comparison to using fungal systems that may require weeks and months for complete degradation (Ke et al., 2012). It is observed that the process of initially pretreatment by chewing is necessary in breaking down the complex plant cell wall by grinding action of mandibles to increase the surface area. Ke et al. (2012) analyzed the impact of chewing of softwood by *C. formosanus* and their chemical and structural modification of lignin and carbohydrates by pyrolysis gas chromatography-mass spectrometry, attenuated total reflectance Fourier transform infrared spectroscopy and thermogravimetry. On chewing, the cellulase enzyme released by mechanical grinding can initiate in the catalysis of the woody material (Ke et al., 2012). In *Rhinotermitidae* sp., the worker termites forage woody materials to form nest, which is also redigested (Emerson, 1938). On chewing, the woody material combined by the salivary and labial gland secretion of the termite help to cement the nesting material together (Emerson, 1938; Oshima, 1919). The worker termites are responsible for digesting the wood matter and feeding the community. The salivary glands of termites are rich in enzymes like lysozyme, cellobiase, cellulase etc. Fujita et al. (2001) determined the presence of lyzozymal activity and protease activity in the digestive guts and degutted guts of Japanese damp-wood termite, that is, *R. speratus*. Total cellulase activity of 80.8% was analyzed in the salivary glands of *C. formosanus* (Nakashima et al., 2002). Cellobiase found in the glands of *M. darwiniensis* and *N. koshunensis* was found to degrade cellobiose, a disaccharide (Tokuda et al., 2004; Veivers et al., 1982).

9.5 Future perspectives of termites/their gut microbes in lignocellulosic degradation

Termites can be used as an effective bioreactor model for biofuel production as they aid in the bioconversion of the lignocellulosic biomass at room temperature with any requirement of high temperature or pressure, thereby acting as an energy-efficient model. The xylophagous termites can effectively degrade 90% cellulosic material, 60% hemicellulosic substances and decompose 25% lignin, making them an ideal model for plant biomass utilization for biofuel production. Termite gut microbes are also used in artificial bioreactors for lignocellulosic biomass conversion (Auer et al., 2017). Different types of lignocellulosic biomass produce the expression of different gut microbes. Gut microbes of four termite species (*M. parvus, T. hospes, Nasutitermes ephratae,* and *N. lujae*) were cultured in controlled conditions with wheat straw and it was observed that 45% of the wheat straw was degraded with gut microbes from *N. ephratae*. Through the 16S rRNA analysis, it was observed that there was a shift in the initial gut community, which is Spirochetes and Fibrobacteres to Proteobacteria and Firmicutes at the end of wheat hydrolysis. Calusinska et al. (2020) analyzed the alimentary canal of the higher termite *Cortaritermes* sp. (Subfamily Nasutitermitinae). They studied the genetic profile of the microbial gut of two termite colonies consuming Miscanthus diet by 16S rRNA gene amplification. Through the integrative omics approaches of metatranscriptomics and metagenomics, it was observed that the major contributors of cellulose and arabinoxylan digestion in Miscanthus lignocellulose belonged to Spirochetes and Fibrobacteres by characterizing their carbohydrate-active enzymes. They identified 170 CAZy gene transcripts of which GH contained 66 genes. GH13 was the most abundant glucoside hydrolase coding for α-glucanases. From the studies, the Spirochetes were able to utilize the Miscanthus sugars compared to Fibrobacteres. Thus the termite hindgut microbes can be used for future bioprospecting.

9.6 Conclusion

Termites can degrade approximately 74%–99% lignocellulosic material by the symbiotic mechanism of the host enzymes and gut microbes. Their termite gut act as hotspots carrying unique microorganisms that are very specific and not found in the alimentary canals of other wood/lignocellulose consuming organisms like ruminants. For example, Termitomyces in Macrotermitine termites and protists like Oxymonads and Hypermastogotes found only in lower termites. Through different high put sequencing methods, that is metagenomics and metatranscriptomic approaches, various cellulase and hemicellulosic genes representing CAZymes for biomass breakdown from the host termites are identified. Through these methods, the bioprospecting applications of termite gut microbes as a sustainable method for the production of second-generation biofuel using lignocellulosic biomass are considered.

Acknowledgments

Gincy Marina Mathew would like to acknowledge the Kerala State Council for Science, Technology and Environment (KSCSTE) for the financial aid under the "Back-To-Lab" Post Doctoral Fellowship Program. Raveendran Sindhu is thankful to the Department of Science and Technology (DST), India, for sanctioning a project under DST.

References

Anand, A., Vennison, S., Sankar, S., Prabhu, D., Vasan, P., Raghuraman, T., Geoffrey, C. J., & Vendan, S. E. (2010). Isolation and characterization of bacteria from the gut of *Bombyx mori* that degrade cellulose, xylan, pectin and starch and their impact on digestion. *Journal of Insect Science (Online), 10*.

Auer, L., Lazuka, A., Sillam-Dussès, D., Miambi, E., O'Donohue, M., & Hernandez-Raquet, G. (2017). Uncovering the potential of termite gut microbiome for lignocellulose bioconversion in anaerobic batch bioreactors. *Frontiers in Microbiology, 8*. Available from https://doi.org/10.3389/fmicb.2017.02623.

Bauer, S., Tholen, A., Overmann, J., & Brune, A. (2000). Characterization of abundance and diversity of lactic acid bacteria in the hindgut of wood- and soil-feeding termites by molecular and culture- dependent techniques. *Archives of Microbiology, 173*(2), 126–137. Available from https://doi.org/10.1007/s002039900120.

Boucias, D. G., Cai, Y., Sun, Y., Lietze, V. U., Sen, R., Raychoudhury, R., & Scharf, M. E. (2013). The hindgut lumen prokaryotic microbiota of the termite *Reticulitermes flavipes* and its responses to dietary lignocellulose composition. *Molecular Ecology, 22*(7), 1836–1853. Available from https://doi.org/10.1111/mec.12230.

Breznak, J. A. (2002). Phylogenetic diversity and physiology of termite gut spirochetes. *Integrative and Comparative Biology, 42*(Issue 2), 313–318. Available from https://doi.org/10.1093/icb/42.2.313.

Breznak, J. A., & Brune, A. (1994). Role of microorganisms in the digestion of lignocellulose by termites. *Annual Review of Entomology, 39*, 453–487. Available from https://doi.org/10.1146/annurev.en.39.010194.002321.

Brune, A. (2014). Symbiotic digestion of lignocellulose in termite guts. *Nature Reviews. Microbiology*, *12*(3), 168–180. Available from https://doi.org/10.1038/nrmicro3182.

Brune, A., & Ohkuma, M. (2011). *Role of the termite gut microbiota in symbiotic digestion. Biology of termites: A modern synthesis* (pp. 439–475). Netherlands: Springer. Available from https://doi.org/10.1007/978-90-481-3977-4_16.

Bugg, T. D. H., Ahmad, M., Hardiman, E. M., & Rahmanpour, R. (2011). Pathways for degradation of lignin in bacteria and fungi. *Natural Product Reports*, *28*(12), 1883–1896. Available from https://doi.org/10.1039/c1np00042j.

Butera, G., Ferraro, C., Alonzo, G., Colazza, S., & Quatrini, P. (2016). The gut microbiota of the wood-feeding termite *Reticulitermes lucifugus* (Isoptera; Rhinotermitidae). *Annals of Microbiology*, *66*(1), 253–260. Available from https://doi.org/10.1007/s13213-015-1101-6.

Calusinska, M., Marynowska, M., Bertucci, M., Untereiner, B., Klimek, D., Goux, X., Sillam-Dussès, D., Gawron, P., Halder, R., Wilmes, P., Ferrer, P., Gerin, T., Roisin, Y., & Delfosse, P. (2020). Integrative omics analysis of the termite gut system adaptation to *Miscanthus* diet identifies lignocellulose degradation enzymes. *Communications Biology*, *3*, 275.

Cho, M.-J., Kim, Y.-H., Shin, K., Kim, Y.-K., Kim, Y.-S., & Kim, T.-J. (2010). Symbiotic adaptation of bacteria in the gut of *Reticulitermes speratus*: low endo-beta-1,4-glucanase activity. *Biochemical and Biophysical Research Communications*, *395*, 432–435.

Cleveland, L. R. (1925). The effects of oxygenation and starvation on the symbiosis between the termite, Termopsis, and its intestinal flagellates. *The Biological Bulletin*, *48*, 309–327.

da Costa, R. R., Hu, H., Li, H., & Poulsen, M. (2019). Symbiotic plant biomass decomposition in Fungus-Growing termites. *Insects*, *10*(4). Available from https://doi.org/10.3390/insects10040087.

Desai, M. S., Strassert, J. F. H., Meuser, K., Hertel, H., Ikeda-Ohtsubo, W., Radek, R., & Brune, A. (2010). Strict cospeciation of devescovinid flagellates and *Bacteroidales ectosymbionts* in the gut of dry-wood termites (Kalotermitidae). *Environmental Microbiology*, *12*(8), 2120–2132. Available from https://doi.org/10.1111/j.1462-2920.2009.02080.x.

Do, T. H., Nguyen, T. T., Nguyen, T. N., Le, Q. G., Nguyen, C., Kimura, K., & Truong, N. H. (2014). Mining biomass-degrading genes through Illumina-based de novo sequencing and metagenomic analysis of free-living bacteria in the gut of the lower termite *Coptotermes gestroi* harvested in Vietnam. *Journal of Bioscience and Bioengineering*, *118*(6), 665–671. Available from https://doi.org/10.1016/j.jbiosc.2014.05.010.

Donovan, S. E., Eggleton, P., & Bignell, D. E. (2001). Gut content analysis and a new feeding group classification of termites. *Ecological Entomology*, *26*(4), 356–366. Available from https://doi.org/10.1046/j.1365-2311.2001.00342.x.

Dröge, S., Fröhlich, J., Radek, R., & König, H. (2006). *Spirochaeta coccoides* sp. nov., a novel coccoid spirochete from the hindgut of the termite *Neotermes castaneus*. *Applied and Environmental Microbiology*, *72*(1), 392–397. Available from https://doi.org/10.1128/AEM.72.1.392-397.2006.

Eggleton, P. (2011). *An introduction to termites: Biology, taxonomy and functional morphology. Biology of termites: A modern synthesis* (pp. 1–26). Netherlands: Springer. Available from https://doi.org/10.1007/978-90-481-3977-4_1.

Emerson, A. E. (1938). Termite nests—A study of the phylogeny of behavior. *Ecological Monographs*, *8*, 247–284. Available from https://doi.org/10.2307/1943251.

Engel, P., & Moran, N. A. (2013). The gut microbiota of insects-diversity in structure and function. *FEMS Microbiology Reviews*, *37*(5), 699–735. Available from https://doi.org/10.1111/1574-6976.12025.

Fröhlich, J., Sass, H., Babenzien, H. D., Kuhnigk, T., Varma, A., Saxena, S., Nalepa, C., Pfeiffer, P., & König, H. (1999). Isolation of *Desulfovibrio intestinalis* sp. nov. from the hindgut of the lower termite *Mastotermes darwiniensis*. *Canadian Journal of Microbiology, 45*(2), 145–152. Available from https://doi.org/10.1139/cjm-45-2-145.

Fujita, A., Hojo, M., Aoyagi, T., Hayashi, Y., Arakawa, G., Tokuda, G., & Watanabe, H. (2010). Details of the digestive system in the midgut of *Coptotermes formosanus* Shiraki. *Journal of Wood Science, 56*(3), 222–226. Available from https://doi.org/10.1007/s10086-009-1088-3.

Fujita, A., Shimizu, I., & Abe, T. (2001). Distribution of lysozyme and protease, and amino acid concentration in the guts of a wood-feeding termite, *Reticulitermes speratus* (Kolbe): Possible digestion of symbiont bacteria transferred by trophallaxis. *Physiological Entomology, 26*, 116–123.

He, S., Ivanova, N., Kirton, E., Allgaier, M., Bergin, C., Scheffrahn, R. H., Kyrpides, N. C., Warnecke, F., Tringe, S. G., & Hugenholtz, P. (2013). Comparative metagenomic and metatranscriptomic analysis of hindgut paunch microbiota in wood- and dung-feeding higher termites. *PLOS ONE, 8*(4). Available from https://doi.org/10.1371/journal.pone.0061126.

Hongoh, Y., Ohkuma, M., & Kudo, T. (2003). Molecular analysis of bacterial microbiota in the gut of the termite *Reticulitermes speratus* (Isoptera; Rhinotermitidae). *FEMS Microbiology Ecology, 44*(2), 231–242. Available from https://doi.org/10.1016/S0168-6496(03)00026-6.

Karen, R., Carsten, D., Claire, T., Aram, M., Nonoh, J. O., Scheffrahn, R. H., David, S.-D., & Andreas, B. (2015). Metagenomic analysis of the microbiota in the highly compartmented hindguts of six wood- or soil-feeding higher termites. *Microbiome, 3*. Available from https://doi.org/10.1186/s40168-015-0118-1.

Ke, J., Laskar, D. D., Gao, D., & Chen, S. (2012). Advanced biorefinery in lower termite-effect of combined pretreatment during the chewing process. *Biotechnology for Biofuels, 5*. Available from https://doi.org/10.1186/1754-6834-5-11.

Köhler, T., Dietrich, C., Scheffrahn, R. H., & Brune, A. (2012). High-resolution analysis of gut environment and bacterial microbiota reveals functional compartmentation of the gut in wood-feeding higher termites (*Nasutitermes* spp.). *Applied and Environmental Microbiology, 78*(13), 4691–4701. Available from https://doi.org/10.1128/AEM.00683-12.

König, H., Li, L., & Fröhlich, J. (2013). The cellulolytic system of the termite gut. *Applied Microbiology and Biotechnology, 97*(18), 7943–7962. Available from https://doi.org/10.1007/s00253-013-5119-z.

Kudo, T. (2009). Termite-microbe symbiotic system and its efficient degradation of lignocellulose. *Bioscience, Biotechnology, and Biochemistry, 73*(12), 2561–2567. Available from https://doi.org/10.1271/bbb.90304.

Kuwahara, H., Yuki, M., Izawa, K., Ohkuma, M., & Hongoh, Y. (2017). Genome of "Ca. *Desulfovibrio trichonymphae*," an H 2 -oxidizing bacterium in a tripartite symbiotic system within a protist cell in the termite gut. *ISME Journal, 11*(3), 766–776. Available from https://doi.org/10.1038/ismej.2016.143.

Leadbetter, J. R., & Breznak, J. A. (1996). Physiological ecology of *Methanobrevibacter cuticularis* sp. nov. and *Methanobrevibacter curvatus* sp. nov., isolated from the hindgut of the termite *Reticulitermes flavipes*. *Applied and Environmental Microbiology, 62*(10), 3620–3631. Available from https://doi.org/10.1128/aem.62.10.3620-3631.1996.

Liu, N., Yan, X., Zhang, M., Xie, L., Wang, Q., Huang, Y., Zhou, X., Wang, S., & Zhou, Z. (2011). Microbiome of fungus-growing termites: a new reservoir for lignocellulase genes. *Applied and Environmental Microbiology, 77*(1), 48–56.

Lo, N., Tokuda, G., & Watanabe, H. (2011). *Evolution and function of endogenous termite cellulases. Biology of termites: A modern synthesis* (pp. 51–67). Netherlands: Springer. Available from https://doi.org/10.1007/978-90-481-3977-4_3.

Marynowska, M., Goux, X., Sillam-Dussès, D., Rouland-Lefèvre, C., Roisin, Y., Delfosse, P., & Calusinska, M. (2017). Optimization of a metatranscriptomic approach to study the lignocellulolytic potential of the higher termite gut microbiome. *BMC Genomics, 18*(1). Available from https://doi.org/10.1186/s12864-017-4076-9.

Mathew, G. M., Ju, Y. M., Lai, C. Y., Mathew, D. C., & Huang, C. C. (2012). Microbial community analysis in the termite gut and fungus comb of *Odontotermes formosanus*: The implication of Bacillus as mutualists. *FEMS Microbiology Ecology, 79*(2), 504–517. Available from https://doi.org/10.1111/j.1574-6941.2011.01232.x.

Mathew, G. M., Mathew, D. C., Lo, S. C., Alexios, G. M., Yang, J. C., Sashikumar, J. M., Shaikh, T. M., & Huang, C. C. (2013). Synergistic collaboration of gut symbionts in *Odontotermes formosanus* for lignocellulosic degradation and bio-hydrogen production. *Bioresource Technology, 145*, 337–344. Available from https://doi.org/10.1016/j.biortech.2012.12.055.

Menezes, L., Alvarez, T. M., Persinoti, G. F., Franco, J. P., Squina, F., Moreira, E. A., Alvaredo Paixão, D. A., Costa-Leonardo, A. M., da Silva, V. X., Clerici, M. T. P. S., & Arab, A. (2018). Food storage by the savanna termite *Cornitermes cumulans* (Syntermitinae): A strategy to improve hemicellulose digestibility? *Microbial Ecology, 76* (2), 492–505. Available from https://doi.org/10.1007/s00248-017-1128-2.

Mikaelyan, A., Meuser, K., & Brune, A. (2017). Microenvironmental heterogeneity of gut compartments drives bacterial community structure in wood- and humus-feeding higher termites. *FEMS Microbiology Ecology, 93*(1), fiw210.

Mikaelyan, A., Dietrich, C., Köhler, T., Poulsen, M., Sillam-Dussès, D., & Brune, A. (2015). Diet is the primary determinant of bacterial community structure in the guts of higher termites. *Molecular Ecology, 24*(20), 5284–5295. Available from https://doi.org/10.1111/mec.13376.

Nakashima, K., Watanabe, H., Saitoh, H., Tokuda, G., & Azuma, J. I. (2002). Dual cellulose-digesting system of the wood-feeding termite, *Coptotermes formosanus* Shiraki. *Insect Biochemistry and Molecular Biology, 32*(7), 777–784. Available from https://doi.org/10.1016/S0965-1748(01)00160-6.

Ni, J., & Tokuda, G. (2013). Lignocellulose-degrading enzymes from termites and their symbiotic microbiota. *Biotechnology Advances, 31*(6), 838–850. Available from https://doi.org/10.1016/j.biotechadv.2013.04.005.

Noda, S., Kitade, O., Inoue, T., Kawai, M., Kanuka, M., Hiroshima, K., Hongoh, Y., Constantino, R., Uys, V., Zhong, J., Kudo, T., & Ohkuma, M. (2007). Cospeciation in the triplex symbiosis of termite gut protists (*Pseudotrichonympha* spp.), their hosts, and their bacterial endosymbionts. *Molecular Ecology, 16*(6), 1257–1266. Available from https://doi.org/10.1111/j.1365-294X.2006.03219.x.

Noirot, C. (2001). The gut of termites (Isoptera) comparative anatomy, systematics, phylogeny. II. Higher termites (Termitidae). *Annales de La Societe Entomologique de France, 37* (4), 431–471.

Ohkuma, M., Ohtoko, K., Iida, T., Tokura, M., Moriya, S., Usami, R., Horikoshi, K., & Kudo, T. (2000). Phylogenetic identification of hypermastigotes, Pseudotrichonympha, Spirotrichonympha, Holomastigotoides, and parabasalian symbionts in the hindgut of termites. *Journal of Eukaryotic Microbiology, 47*(3), 249–259. Available from https://doi.org/10.1111/j.1550-7408.2000.tb00044.x.

Ohtoko, K., Ohkuma, M., Moriya, S., Inoue, T., Usami, R., & Kudo, X. (2000). Diverse genes of cellulase homologues of glycosyl hydrolase family 45 from the symbiotic protists in the hindgut of the termite Reticulitermes speratus. *Extremophiles, 4*(3), 343–349.

Oshima, M. (1919). Formosan termites and methods of preventing their damage. *Philippine Journal of Science, 15*, 319–383.

Pauly, M., & Keegstra, K. (2008). Cell-wall carbohydrates and their modification as a resource for biofuels. *Plant Journal, 54*(4), 559–568. Available from https://doi.org/10.1111/j.1365-313X.2008.03463.x.

Poulsen, M., Hu, H., Li, C., Chen, Z., Xu, L., Otani, S., Nygaard, S., Nobre, T., Klaubauf, S., M. Schindler, P., Hauser, F., Pan, H., Yang, Z., Sonnenberg, A. S. M., Wilhelm De Beer, Z.,

Zhang, Y., Wingfield, M. J., Grimmelikhuijzen, C. J. P., De Vries, R. P., & Zhang, G. (2014). Complementary symbiont contributions to plant decomposition in a fungus-farming termite. *Proceedings of the National Academy of Sciences of the United States of America, 111*(40), 14500–14505. Available from https://doi.org/10.1073/pnas.1319718111.

Pramono, A. K., Sakamoto, M., Iino, T., Hongoh, Y., & Ohkuma, M. (2015). *Dysgonomonas termitidis* sp. Nov., isolated from the gut of the subterranean termite *Reticulitermes speratus. International Journal of Systematic and Evolutionary Microbiology, 65*(2), 681–685. Available from https://doi.org/10.1099/ijs.0.070391-0.

Rajarapu, S. P., Shreve, J. T., Bhide, K. P., Thimmapuram, J., & Scharf, M. E. (2015). Metatranscriptomic profiles of Eastern subterranean termites, *Reticulitermes flavipes* (Kollar) fed on second generation feedstocks. *BMC Genomics, 16*(1). Available from https://doi.org/10.1186/s12864-015-1502-8.

Rajagopal, S., Rao, D. R., & Varma, A. K. (1979). Association of fungi in the termite gut. *Current Science, 48*, 998–999.

Rajagopal, S., Rao, D. R., & Varma, A. K. (1981). Fungi from worker termite gut, *Odontotermes obesus* (Rambur) from northern India. *Nova Hedwig, 34*, 97–100.

Reuss, J., König, H., & Dröge, S. (2013). *Biospektrum annual conference of the association for general and applied microbiology.*

Rouland-Lefèvre, C. Symbiosis With Fungi. In: Abe, T., Bignell, D. E., Higashi, M., editors. Termites: Evolution, Sociality, Symbioses, Ecology. Dordrecht: Springer; 2000.

Sakamoto, M., & Ohkuma, M. (2013). *Bacteroides reticulotermitis* sp. nov., isolated from the gut of a subterranean termite (*Reticulitermes speratus*). *International Journal of Systematic and Evolutionary Microbiology, 63*(2), 691–695. Available from https://doi.org/10.1099/ijs.0.040931-0.

Schäfer, A., Konrad, R., Kuhnigk, T., Kämpfer, P., Hertel, H., & König, H. (1996). Hemicellulose-degrading bacteria and yeasts from the termite gut. *Journal of Applied Microbiology, 80*, 471–478.

Tartar, A., Wheeler, M. M., Zhou, X., Coy, M. R., Boucias, D. G., & Scharf, M. E. (2009). Parallel metatranscriptome analyses of host and symbiont gene expression in the gut of the termite *Reticulitermes flavipes. Biotechnology for Biofuels, 2*(1). Available from https://doi.org/10.1186/1754-6834-2-25.

Thongaram, T., Hongoh, Y., Kosono, S., Ohkuma, M., Trakulnaleamsai, S., Noparatnaraporn, N., & Kudo, T. (2005). Comparison of bacterial communities in the alkaline gut segment among various species of higher termites. *Extremophiles: Life Under Extreme Conditions, 9*(3), 229–238. Available from https://doi.org/10.1007/s00792-005-0440-9.

Tokuda, G., Mikaelyan, A., Fukuia, C., Matsuuraa, Y., Watanabe, H., Fujishima, M., & Brune, A. (2018). Fiber-associated spirochetes are major agents of hemicellulose degradation in the hindgut of wood-feeding higher termites. *Proceedings of the National Academy of Sciences of the United States of America, 115*, E11996–E12004.

Tokuda, G., & Watanabe, H. (2007). Hidden cellulases in termites: revision of an old hypothesis. *Biology Letters, 3*(3), 336–339. Available from https://doi.org/10.1098/rsbl.2007.0073.

Tokuda, G., Lo, N., Watanabe, H., Arakawa, G., Matsumoto, T., & Noda, H. (2004). Major alteration of the expression site of endogenous cellulases in members of an apical termite lineage. *Molecular Ecology, 13*(10), 3219–3228. Available from https://doi.org/10.1111/j.1365-294X.2004.02276.x.

Tokuda, G., Saito, H., & Watanabe, H. (2002). A digestive β-glucosidase from the salivary glands of the termite, *Neotermes koshunensis* (Shiraki): Distribution, characterization and isolation of its precursor cDNA by 5′- and 3′-RACE amplifications with degenerate primers. *Insect Biochemistry and Molecular Biology, 32*(12), 1681–1689. Available from https://doi.org/10.1016/S0965-1748(02)00108-X.

Tokuda, G., & Watanabe, H. (2007). Hidden cellulases in termites: Revision of an old hypothesis. *Biology Letters, 3*(3), 336–339. Available from https://doi.org/10.1098/rsbl.2007.0073.

Trinkerl, M., Breunig, A., Schauder, R., & König, H. (1990). Desulfovibrio termitidis sp. nov., a carbohydrate-degrading sulfate-reducing bacterium from the hindgut of a termite. *Systematic and Applied Microbiology, 13*(4), 372–377. Available from https://doi.org/10.1016/S0723-2020(11)80235-3.

Veivers, P. C., Musca, A. M., O'Brien, R. W., & Slaytor, M. (1982). Digestive enzymes of the salivary glands and gut of *Mastotermes darwiniensis*. *Insect Biochemistry, 12*(1), 35–40. Available from https://doi.org/10.1016/0020-1790(82)90068-3.

Waidele, L., Korb, J., Voolstra, C. R., Künzel, S., Dedeine, F., & Staubach, F. (2017). Differential ecological specificity of protist and bacterial microbiomes across a set of termite species. *Frontiers in Microbiology, 8*. Available from https://doi.org/10.3389/fmicb.2017.02518.

Warnecke, F., Luginbühl, P., Ivanova, N., Ghassemian, M., Richardson, T. H., Stege, J. T., Cayouette, M., McHardy, A. C., Djordjevic, G., Aboushadi, N., Sorek, R., Tringe, S. G., Podar, M., Martin, H. G., Kunin, V., Dalevi, D., Madejska, J., Kirton, E., Platt, D., & Leadbetter, J. R. (2007). Metagenomic and functional analysis of hindgut microbiota of a wood-feeding higher termite. *Nature, 450*(7169), 560–565. Available from https://doi.org/10.1038/nature06269.

Watanabe, H., & Tokuda, G. (2001). Animal cellulases. *Cellular and Molecular Life Sciences, 58*(9), 1167–1178. Available from https://doi.org/10.1007/PL00000931.

Wenzel, M., Schönig, I., Berchtold, M., & Kämpfer, P. (2002). Aerobic and facultatively anaerobic cellulolytic bacteria from the gut of the termite *Zootermopsis angusticollis*. *Journal of Applied Microbiology, 92*(1), 32–40. Available from https://doi.org/10.1046/j.1365-2672.2002.01502.x.

Wu, W., Huang, Z., Li, Z., Zhang, S., Liu, X., & Gu, D. (2016). *De novo* transcriptome sequencing of *Cryptotermes domesticus* and comparative analysis of gene expression in response to different wood species. *Gene, 575*(2), 655–666. Available from https://doi.org/10.1016/j.gene.2015.09.055.

Xie, S., Syrenne, R., Sun, S., & Yuan, J. S. (2014). Exploration of Natural Biomass Utilization Systems (NBUS) for advanced biofuel—From systems biology to synthetic design. *Current Opinion in Biotechnology, 27*, 195–203. Available from https://doi.org/10.1016/j.copbio.2014.02.007.

Xie, L., Zhang, L., Zhong, Y., Liu, N., Long, Y., & Wang, S. (2012). Profiling the metatranscriptome of the protistan community in *Coptotermes formosanus* with emphasis on the lignocellulolytic system. *Genomics, 99*, 246–255.

Yamada, A., Inoue, T., Wiwatwitaya, D., Ohkuma, M., Kudo, T., Abe, T., & Sugimoto, A. (2005). Carbon mineralization by termites in tropical forests, with emphasis on fungus combs. *Ecological Research, 20*(4), 453–460. Available from https://doi.org/10.1007/s11284-005-0062-9.

Yamin, M. A., & Trager, W. (1979). Cellulolytic activity of an axenically-cultivated termite flagellate, *Trichomitopsis termopsidis*. *Journal of General Microbiology, 113*(2), 417–420. Available from https://doi.org/10.1099/00221287-113-2-417.

Yang, S. Y., Zheng, Y., Huang, Z., Wang, X. M., & Yang, H. (2016). *Lactococcus nasutitermitis* sp. nov. isolated from a termite gut. *International Journal of Systematic and Evolutionary Microbiology, 66*(1), 518–522. Available from https://doi.org/10.1099/ijsem.0.000743.

Yuki, M., Kuwahara, H., Shintani, M., Izawa, K., Sato, T., Starns, D., Hongoh, Y., & Ohkuma, M. (2015). Dominant ectosymbiotic bacteria of cellulolytic protists in the termite gut also have the potential to digest lignocellulose. *Environmental Microbiology, 17*(12), 4942–4953. Available from https://doi.org/10.1111/1462-2920.12945.

CHAPTER

10

Bacterial diversity from Garampani warm spring, Assam

Joyasree Das[1], *Pradipta Saha*[2] *and Srinivasan Krishnamurthi*[1]

[1]Microbial Type Culture Collection & Gene Bank (MTCC), CSIR-Institute of Microbial Technology, Chandigarh, India, [2]Department of Microbiology, The University of Burdwan, Golapbag Campus, Burdwan, India

10.1 Introduction

India houses four global biodiversity hotspots—the Himalaya, the Indo-Burma, the Western Ghats, and the Sundaland. It belongs to one of the 17 mega-diverse countries in the world. India, being a tropical country with a unique geographical location and the perfect blend of climates, offers a rich source of biological diversity. It is abundant in diverse endemic species of animals and plants and supports around 7.5% of total animal species of the world (Venkataraman & Sivaperuman, 2018). Besides a vast array of macroscopic life, it is a rich source in microbial biodiversity (Bhatnagar & Bhatnagar, 2005; Chaudhuri & Thakur, 2006; Ghosh et al., 2003; Thajuddin & Subramanian, 2005); but unfortunately, scanty information is available compared to macroscopic life. Therefore in order to exploit the prokaryotic diversity from these hotspots, its complex community structure and interplay with other organisms has to be unravelled. This will also shed light on their role in ecosystem shaping and bioprospecting potential (Bhattacharyya et al., 2014; Lyngwi et al., 2013; Nampoothiri et al., 2013; Ruckmani & Chakrabarti, 2011). The Himalayan biodiversity hotspot is the most interesting in terms of its diverse habitats including—world's highest mountain "Mount Everest," deepest rivers and a unique aquatic habitat; springs. The Himalayas is mainly divided into two main regions—Western Himalaya

Microbial Diversity and Ecology in Hotspots
DOI: https://doi.org/10.1016/B978-0-323-90148-2.00012-2

and Eastern Himalaya (Venkataraman & Sivaperuman, 2018). North-East India, a part of Eastern Himalaya and Indo-Burma hotspot, has a pristine natural ecosystem, presents an opportunity to discover several novel taxa and their biotechnological applications hitherto unrevealed to scientific world (Das et al., 2018; Panda et al., 2016; Ruckmani et al., 2011; Sharma & Thakur, 2020).

The diversity of microbial life of any habitat is difficult to ascertain due to the phenomenon of "Great plate count anomaly"—that is, the number of colonies able to grow on plate are well below than total number of cells observed microscopically from any environmental sample as majority of them could not be cultured with conventional axenic media (Staley & Konopka, 1985). The magnitude of this anomaly varies depending upon nature of the environmental sample, but mostly $\leq 1\%$ of bacteria are generally culturable (Amann et al., 1995). It is believed there exists about 100 bacterial phyla and only 30 phyla have culturable representatives and majority are poorly represented (Ley et al., 2006; Youssef et al., 2015). Thus contemporary approaches employ a combination of both culture-dependent and -independent techniques to unravel the diversity of an environmental habitat (Jackson et al., 2013; Zhang et al., 2019).

Among the natural environmental habitats, water springs are one of the most interesting and extreme ecosystems on earth (Reiss & Chifflard, 2017). These represent nutritionally depleted environments (Morita, 1997), but rich in prokaryotic diversity. The study of bacterial diversity and their basic adaptation in such oligophilic ecosystems could be a challenging and an interesting area of research to the scientific community. Bacterial strains that can thrive in such extreme low nutritional conditions are called "oligophiles" (Koch, 2001) and those growing in nutritionally rich are termed "copiophiles." A comparative analysis between the two populations could thus decipher the minimal nutritional conditions required for growth of bacteria and their basic adaptation mechanisms.

Keeping in perspective the above discussion, in this study we intended to analyze prokaryotic diversity from a warm spring water sample by both culture-dependent and culture-independent approaches with special focus on isolation of oligophiles. The site named as Garampani warm spring, was located on one of the major biodiversity hotspots "the Eastern Himalaya" in the state of Assam, India.

10.2 Methods

10.2.1 Sampling site description

The Barpung warm spring commonly known as "Garampani" because of the presence of the few more warm spring around this region, is situated in Namber reserve forest in the Karbi-Anglong district near Jorhat, Assam, India (located in "Eastern Ghats," a biodiversity hotspot). This region is part of the Mikir hill plateau (600–800 m) which lies between E–W trending Indo-Burman ranges in the East and Eastern Himalayas in the north–northeast. The entire plateau subjected by several faults and frauctures, stipulated active tectonic activity. The Garampani water was shallow (1.5–2 ft deep), clear close to origin points, but was brownish-purple alongside of the bank apart from the origin. The water temperature was 37°C–38°C and the pH was 7.2 at the source of the spring at the time of sampling. The chemical analysis of the Garampani water sample are as follows: BOD, 32 mg/L; COD, 58 mg/L; sulfide concentration, 0.50 mg/L; phosphate concentration, 0.130 mg/L; total nitrogen, 1 mg/L; total organic carbon, 0.15 mg/mL; and total organic matter was 0.26 mg/L.

10.2.2 Sample collection and isolation of bacteria

The warm spring water was collected in the month of December 2002. The sample was collected aseptically in a sterile glass bottle and kept at 4°C, while some aliquots of collected warm spring water were stored in glycerol (10% v/v) at −70°C for long-term storage. The sample was serially diluted and plated by standard spread plate technique on nutritionally rich tryptic soya broth agar (TSBA, Hi-Media) and one nutritionally poor medium, wherein tryptic soya broth was diluted 100 times with double distilled water and solidified with high purity agarose at a concentration of 1.5% (w/v) (labeled as TSBA100 in all further descriptions) media. The plates were subjected to incubation at 37°C for 7 days. The viable cell count (CFU) was read for 7 days and single isolated colonies were picked up based on their colony morphology and growth profile (time of appearance of colonies). The isolates were purified by restreaking on to their respective media plates. The pure cultures were preserved in 10% (v/v) glycerol stock at −70°C.

10.2.3 Culture-dependent study

10.2.3.1 Phenotypic characterization of isolates

The detailed characterization of isolates was carried out based on morphological, physiological, biochemical, chemotaxonomy, and molecular parameters.

Colony morphology (shape, size, opacity, margin, elevation, etc.) and cell morphology were recorded for isolate as detailed in Smibert and Krieg (1994). The characteristics were recorded after growth on respective media at 30°C for 1–3 days. The colony morphology was observed under low power stereomicroscope (Carl Zeiss, Germany) and cell morphology was observed under oil immersion lens (1000 ×) of microscope with simple staining or Gram staining. For some strains, cellular morphology was also recorded using electron photomicrography (SEM or TEM). Gram staining was done as per Murray et al. (1994) with slight modification in safranin counter staining for 1 minute. The Gram reaction also tested by nonstaining method (Powers, 1995). Briefly, a loop full of freshly grown cultures were taken in a drop of 3% KOH (w/v) and suspended, the Gram-negative cells formed a viscous suspension and strung out when the loop was lifted due to lipid-DNA complex, whereas Gram-positive cells did not lyse due to a thick peptidoglycan layer and gave a negative KOH reaction. For endospore staining, 3–4 days old culture from TSBA media plate was used and for some strains (strains that did not sporulate on TSBA media plate) sporulating agar media and/or TSBA100 media plate was used. Motility of individual isolates was checked by hanging drop method as per Murray et al. (1994) and also gliding motility was checked as described by Bowman (2000), where specific cultures were grown in nutritionally poor liquid broth media—MR2A (Modified Reasoner's 2A agar), MMR2A, or TSB100 solid and liquid media and observed under light microscope. The influence of temperature, pH, and NaCl on growth of individual culture was monitored. The cultures were grown on basal TSBA, TSBA100, or MR2A agar plates and growth was observed up to 7 days. For few strains liquid broth media were also used. The pH of the media was adjusted by 1 M NaOH or 1 M HCl. Growth under anaerobic condition was checked on basal R2A, MR2A, TSBA, TSBA100, and incubated at 30°C inside the Anoxy Mat anaerobic jar (Merck, Germany). The cross growth patterns on three different media were checked to group the isolates on the basis of nutritional requirement as copiophilic, oligophilic, and facultative oligophilic. The growth was observed up to 5 days. For confirming oligophily, growth was confirmed on various nutritionally rich media—LBA (Luria Bertani Agar), NA (Nutrient Agar), BHIA (Brain Heart Infusion), MHA (Mueller Hinton Agar), SBA (Sheep Blood Agar), YMA (Yeast Mannitol Agar), ZMA (Zobell Marine Agar), and their 100 times diluted media.

Biochemical tests such as catalase, oxidase, polymer hydrolysis, utilization of acetate, carbon sources and organic acids, acid production from different sugars, IMViC test, hemolytic reaction, H_2S and dihydroxy acetone production, gas production from glucose, growth on MacConkey agar and klinger's iron agar, decarboxylation of amino acids, activity of galactosidase, lecithinase, phenylalanine deaminase and lipase, nitrate reduction, antibiotic susceptibility test, oxidation-fermentation catabolism, tyrosin degradation, and lysozyme resistance were done as per standard methods (Claus et al., 1872; Cowan & Steel 1965; Smibert & Krieg, 1994).

10.2.3.2 Molecular characterization of isolates

Genomic DNA was isolated by Murmur's protocol (Johnson, 1994). Individual strains were grown overnight (fast growing strains) and/or 24–36 hours (slow grower bacterial strains) for the isolation of genomic DNA. PCR amplification, purification, and sequencing of 16S rRNA gene were done as described previously (Saha et al., 2005b). Isolates were identified by BLAST analysis of 16S rRNA gene sequences in EzTaxon server (Kim et al., 2012). Sequence alignment, editing, and phylogenetic tree reconstruction were carried out using MEGA7 software (Kumar et al., 2016) and tree visualization with iTOL software (Letunic & Bork, 2019). For some specific strains, DNA–DNA hybridization was carried out by the membrane filter methods (Tourova & Antonov, 1988) as described in Reddy et al. (2003) to determine their species affiliation.

10.2.4 Culture-independent study

The total community DNA was isolated from warm spring water sample by two different methods.

First, for the removal of coarse particulate matter, the water sample was filtered by cheesecloth and Whatman (no. 1) filter paper and then passed through 0.22-μm-pore size filters (25 mm in diameter, cellulose nitrate filter membrane, Millipore, United States). The filter paper was then aseptically cut into small pieces followed by placing into 1 mL TES (N-tris(hydroxymethyl) methyl-2-aminoethanesulfonic acid) buffer in a SS 34 polypropylene tube. In the second method, about 20 mL water sample was pelleted down and centrifuged for 1 hour at 12,000 rpm, followed by suspension in 1 mL TES buffer and 4 cycles of freezing (−70°C) and thawing (37°C). The suspension was homogenized for 10 minutes with 0.2 g of glass beads (0.1 mm, acid washed, dried, and autoclaved) by vigorous vortexing followed by incubation for 45 minutes at 37°C in lysozyme (3 mg). After addition of SDS (10%), proteinase K (20 mg/mL) incubation was continued at 37°C for 2 hours. Finally, 5 M NaCl and 0.4 mL of CTAB solution were incubated at 60°C for 30 minutes. The subsequent steps followed for both methods as per Marmur's protocol (Marmur, 1961). The isolated community DNA was further purified using Prep-A-Gene-DNA Purification kit (Bio-Rad) and stored at −20°C for further use. The concentration and purity of the purified DNA were determined at 260 nm with Lambda 35 spectrophotometer (Perkin Elmer, United States). The PCR amplification of the 16S rRNA gene and purification steps were carried out as mentioned in culturable diversity study (Section 10.2.3.2). The primer pairs utilized for amplification were 27f and 1500r, 27f and 1100r, 357f and 1500r, 357f and 1100r (Table 10.1). The amplified (~ 0.75–1.0 kb) fragment was cloned in chemically competent *Escherichia coli* DH5-α cells using

TABLE 10.1 List of primers.

Primer	Sequence
T7 Polymerase promoter	5′ TAA TAC GAC TCA CTA TAG GG 3′
SP6 Polymerase promoter	5′ TAT TTA GGT GAC ACT ATA G 3′
8–27f (rRNA gene primer)	5′ AGA GTT TGA TCC TGG CTC AG 3′
339F (rRNA gene primer)	5′ TCC TAC GGG AGG CAG CAG 3′
357f (rRNA gene primer)	5′ CTC CTA CGG GAG GCA GCA G 3′
342r (rRNA gene primer)	5′ CTG CTG CSY CCC GTA CG 3′
530f (rRNA gene primer)	5′ GTG CCA GCM GCC GCG GTA A 3′
685r (rRNA gene primer)	5′ TCT ACG CAT TTC ACC GCT AC 3′
704f (rRNA gene primer)	5′ GTA GCG GTG AAA TGC GTA GA 3′
926f (rRNA gene primer)	5′ AAA CTC AAA GGA ATT GAC GG 3′
1100r (rRNA gene primer)	5′ GGG TTG CGC TCG TTG 3′
1114f (rRNA gene primer)	5′ GCA ACG AGC GCA ACC C 3′
1382r (rRNA gene primer)	5′ GGC GGT GTG TAC AAG GCC C 3′
1492r (rRNA gene primer)	5′ TAC GGY TAC CTT GTT ACG ACT T 3′
1500r (rRNA gene primer)	5′ AGA AAG GAG GTG ATC CAG CCA 3′

PGEM-T Easy vector kit (Promega, United States) by TA cloning strategy and blue/white colony selection. Recombinant white colonies were picked, purified on an ampicillin-LB plate, and preserved in glycerol stock at −70°C. The isolation of plasmid DNA from recombinant clones was carried out by the alkaline lysis method (Sambrook et al., 1989) and QIAprep Spin miniprep kit (QIAGEN, Germany). The primers used for sequencing of the cloned 16S rRNA gene were T7, SP6 (from the vector), and 357f (from the insert) (Table 10.1). The protocol for sequencing and phylogenetic analyses were the same as detailed in the culturable study.

10.2.4.1 Amplified ribosomal DNA restriction analysis

The 16S rRNA gene fragment from individual clones were PCR amplified by T7 and the SP6 primers. Approximately, 1.0 kb amplified gene product was purified and digested using restriction enzyme AluI. Enzyme buffer (1 ×), 2 units of restriction enzyme, and 300 ng of DNA were aliquoted, mixed, and incubated for 4 hours at 37°C followed by 65°C for 20 minutes. The restriction profile was visualized in a 2% (w/v) agarose gel by staining with ethidium bromide (0.5 μg/ml). The digested banding profiles of different clones were used for the reconstruction of a dendrogram using Image Master 1D Elite (version 4.0).

10.3 Results

10.3.1 Diversity of culturable isolates

The viable cell count (CFU) of bacteria isolated from warm spring water was 4.65×10^4 CFU/mL on nutritionally poor media (TSBA100) and 8.8×10^3 CFU/mL on nutritionally rich media (TSBA). A total of 28 and 31 isolates were purified and preserved from TSBA (labeled as GPTSA) and TSBA100 media (labeled as GPTSA100), respectively, on the basis of detailed morphological characteristics (colony color, size, shape, texture, and slow grower and/or fast grower). This exercise was aimed at reducing the taxonomic redundancy of the isolates. Preliminarily characterization based on a combination of different phenotypic characteristics (Gram staining reaction, endospores, motility, catalase, oxidase fermentation of glucose/lactose, colony color, surface texture, growth pattern, O_2 requirement, slow growing, positive growth on nutritionally poor media) of 59 isolates facilitated classification into 32 groups (Tables 10.2–10.4; Figs. 10.1 and 10.2). The GPTSA isolates were categorized into three main groups, that is, Gram staining positive with endospore, Gram staining variable with endospore, and Gram staining negative, which were further subdivided into 6, 2, and 11 subgroups, respectively, based on flowchart 1 (Figs. 10.1, Table 10.4). The GPTSA100 isolates were categorized into two main groups, that is, Gram staining positive and Gram staining negative which were further subdivided into 6 and 7 subgroups, respectively, based on flowchart 2 (Figs. 10.2, Table 10.4). Subsequently from each subgroup, one or more representative members were selected for further detailed characterization (nitrate reduction, indole production, MR-VP, H_2S gas production, citrate utilization, acetate utilization, hydrolysis of polymers, acid production from several different carbohydrates, temperature, pH and NaCl range determination, whole cell fatty acid analysis, etc.) (data unpublished). Six GPTSA and nine GPTSA100 isolates from Gram-positive group, two GPTSA isolates from Gram-variable group and thirteen GPTSA and sixteen GPTSA100 isolates from Gram-negative group were finally selected (total 46 isolates; Table 10.4) for 16S rRNA gene sequence-based identification and phylogenetic analysis (Tables 10.5 and 10.6; Fig. 10.3, Fig. 10.4).

Based on 16S rRNA gene sequence analysis and low similarity to existing cultivated representatives, detailed polyphasic taxonomic characterization of eight strains was further pursued on priority (Fig. 10.5). On the basis of 16S rRNA gene sequence phylogeny, all the 46 strains were placed in four bacterial phyla (Fig. 10.6). In total 17 strains represented phylum *Firmicutes*, 26 belonged to Proteobacteria, 2 to *Bacteroidetes*, and 1 identified as a member of *Deinococcus-Thermus*. About 22 different genera were

TABLE 10.2 The preliminary characterization of GPTSA strains from TSBA plates.

GPTSA strain	1	2	3	4	5	6	7	8	9	10	11	12	13	14	15	16	17	18	19	20	21	22	23	24	25	26	27	28
Shape	R	R	R	R	R	R	R	R	R	R	R	R	R	R	R	CB	R	R	R	R	R	R	R	R	R	R	R	R
Gram reaction	+	+	−	+	−	−	−	+	+	−	v	−	−	−	−	−	+	−	−	−	v	+	+	+	−	+	+	−
Spores	+	+	−	+	−	−	−	+	+	−	+	−	−	−	−	−	+	−	+	−	+	+	+	+	−	+	+	-
Motility	−	−	+	−	+	+	+	+	−	−	+	+	+	+	+	+	−	+	+	+	+	+	−	+	+	+	+	+
Aerobic growth	+	+	+	+	+	+	+	+	+	+	+	+	+	+	+	+	+	+	+	+	+	+	+	+	+	+	+	+
Anaerobic growth	+	+	+	+	−	+	-	−	+	−	−	−	−	−	−	−	−	−	+	−	+	+	+	+	+	+	−	−
Catalase	+	+	+	+	+	+	+	+	+	+	+	+	+	+	+	+	+	+	+	+	+	+	+	+	+	+	+	+
Oxidase	+	+	+	+	+	+	+	−	+	+	+	+	+	+	+	+	+	+	+	+	+	+	+	+	+	+	+	+
Acid from dextrose	+	+	+	+	−	+	−	+	−	+	+	−	+	−	−	+	+	−	+	−	+	+	+	+	+	+	+	+
O/F/-	F	F	F	F	−	F	−	WO	−	F	−	−	−	−	−	−	WF.	−	WF.	-	WF.	F	F	F	F	F	F	−
Growth on MAC	−	−	+	−	+	+	+	−	−	+	−	+	−	−	−	−	−	−	−	−	−	−	−	−	+	−	−	+
Lactose fermentation	−	−	−	−	−	+	−	−	−	−	−	−	−	−	−	−	−	−	−	−	−	−	−	−	−	−	−	−

MAC, MacConkey agar; O/F, oxidation/fermentation.

TABLE 10.3 The preliminary characterization of GPTSA100 strain from TSBA100 media.

Strain GPTSA100	1	2	3	4	5	6	7	8	9	10	11	12	13	14	15	16	17	18	19	20	21	22	23	24	25	26	27	28	29	30	31
Shape	R	R	R	R	R	R	R	R	R	R	R	R	R	R	R	R	R	R	R	R	R	R	R	C	R	R	R	R	R	R	R
Gram reaction	+	+	–	–	–	+	+	+	–	–	+	–	+	+	–	+	–	–	–	+	–	–	–	+	–	–	–	–	–	–	+
Spores	+	+	–	–	–	+	+	+	–	–	+	–	+	+	–	+	–	–	–	+	–	–	–	–	–	–	–	–	–	–	+
Motility	+	–	+	–	+	–	+	–	–	+	–	+	+	–	–	–	–	–	+	–	+	+	–	–	+	+	–	–	+	–	–
Aerobic growth	+	+	+	+	+	+	+	+	+	+	+	+	+	+	+	+	+	+	+	+	+	+	+	+	+	+	+	+	+	+	+
Anaerobic growth	–	–	–	–	+	–	– – +	+	–	–	–	–	–	+	–	–	–	–	+	–	–	–	–	–	+	–	–	–	–	–	–
Catalase	+	+	+	+	+	+	+	+	+	+	+	+	+	+	+	+	+	+	+	+	+	+	+	+	+	+	+	+	+	+	+
Oxidase	+	+	+	+	+	+	+	+	+	+	+	+	+	+	+	+	+	+	+	+	+	+	+	+	+	+	+	+	+	+	+
Acid from dextrose	+	+	+	–	+	+	+	+	+	+	+	+	+	+	+	+	+	+	+	–	–	–	+	+	+	–	+	–	–	–	+
O/F/ –	ng	ng	–	ng	F	ng	ng	ng	ng	O	ng	ng	ng	ng	ng	ng	ng	ng	F	ng	ng	ng	–	–	F	ng	ng	–	–	O	O
Growth on MAC	–	–	–	–	–	–	–	–	–	–	–	–	–	–	–	–	–	–	–	–	–	–	–	–	+	–	–	–	–	–	–
Lactose fermentation	–	–	–	–	–	–	–	–	–	–	–	–	–	–	–	–	–	–	–	–	–	–	–	–	–	–	–	–	–	–	–
Growth on TSBA	+	+	–	–	–	+	+	+	–	–	+	–	+	+	–	+	–	+	+	+	–	–	–	–	+	–	–	–	+	+	+
Growth on MR2A or R2A	+	+	+	+	+	+	+	+	+	+	+	+	+	+	+	+	+	+	+	+	+	+	+	+	+	+	+	+	+	+	+

MAC, MacConkey Agar; *O/F*, oxidation/fermentation; *R2A*, Reasoner's 2A agar; *MR2A Modified*, Modified Reasoner's 2A agar; *TSBA*, trypticase soy broth agar; *ng*, no growth.

TABLE 10.4 Strains selected for polyphasic study after preliminary characterization.

														Total strain
Strains selected from (TSBA) plate	**Gram positive**	**Groups**	Group A1	Group A2	Group A3	Group A4	Group A5	Group A6						
		Strains ID	GPTSA24	GPTSA8	GPTSA27	GPTSA17	GPTSA1	GPTSA9						6
	Gram variable	**Groups**	Group B1			Group B2								2
		Strains ID	GPTSA11			GPTSA21								
	Gram negative	**Groups**	Group C1	Group C2	Group C3	Group C4	Group C5	Group C6	Group C7	Group C8	Group C9	Group C10	Group C11	13
		Strains ID	GPTSA5	GPTSA 12	GPTSA 20	GPTSA 14, 15, 18	GPTSA 10	GPTSA 28	GPTSA 13	GPTSA 16	GPTSA 19	GPTSA 6	GPTSA 3	
	Total number of selected strains from TSBA media													**21**
Strains selected from (TSBA100) plate	**Gram positive**	**Groups**	Group D1	Group D2	Group D3	Group D4	Group D5	Group D6						9
		Strains ID	GPTSA100 24	GPTSA100 1, 6	GPTSA100 16, 20	GPTSA100 31	GPTSA100 13	GPTSA100 7, 8						
	Gram variable	**Groups**	No representative											0
		Strains ID												
	Gram-Negataive	**Groups**	Group E1	Group E2	Group E3	Group E4	Group E5	Group E6	Group E7					16
		Strains ID	GPTSA100 18, 19, 30	GPTSA100 3, 21, 22, 26	GPTSA100 9	GPTSA100 10, 12, 17, 27	GPTSA100 15	GPTSA100 5	GPTSA100 19, 25					
	Total number of selected strains from TSBA100 media													**25**
Total number of selected strains for further polyphasic characterization														**46**

Group A1—Motile, facultatative anaerobic; **Group A2**—Motile, aerobic, red color colony; **Group A3**—Motile, aerobic, yellow color colony; **Group A4**—Nonmotile, aerobic; **Group A5**—Nonmotile, facultative anaerobic, glucose fermenter; **Group A6**—Nonmotile, facultative anaerobic, glucose nonfermenter.

Group B1—Motile; **Group B2**—Nonmotile.

Group C1—Aerobic, glucose nonfermenter, able grew on MAC media, yellow mucoid colony; **Group C2**—Aerobic, glucose nonfermenter, able grew on MAC media, yellow colony adhered to media; **Group C3**—Aerobic, glucose nonfermenter, unable grew on MAC media, yellow color colony, slow grower; **Group C4**—Aerobic, glucose nonfermenter, unable grew on MAC media, yellow color colony, fast grower; **Group C5**—Aerobic, glucose fermenter, pink color colony; **Group C6**—Aerobic, glucose fermenter, cream yellow color colony; **Group C7**—Aerobic, glucose fermenter, spreading growth colony; **Group C8**—Aerobic, glucose fermenter, colony adhered to medium; **Group C9**—Facultative anaerobic, endospore forming; **Group C10**—Facultative anaerobic, nonendospore forming, lactose fermenter; **Group C11**—Facultative anaerobic, nonendospore forming, lactose nonfermenter.

Group D1—cocci, nonendospore forming, unable to grew on TSBA media; **Group D2**—Rod shaped, endospores forming, able to grew on TSBA media, aerobic, rhizoid/spreading colony; **Group D3**—Rod shaped, endospores forming, able to grew on TSBA media, aerobic, creamish color colony; **Group D4**—Rod shaped, endospores forming, able to grew on TSBA media, aerobic, orange-yellow color colony; **Group D5**—Rod shaped, endospores forming, able to grew on TSBA media, aerobic, colonies concentric with outer raised and inner hyaline; **Group D6**—Rod shaped, endospores forming, facultative anaerobic, rhizoid/spreading colony.

Group E1—Aerobic grew on TSBA; **Group E2**—Aerobic, unable to grew on TSBA but grew on R2A and MR2A, mucoid colony; **Group E3**—Aerobic, unable to grew on TSBA but grew on R2A and MR2A, yellow color colony; **Group E4**—Aerobic, unable to grew on TSBA but grew on R2A and MR2A, small size colony, slow grower; **Group E5**—Aerobic, unable to grew on TSBA but grew on R2A and MR2A, light pinkish color colony; **Group E6**—Facultative anaerobic, unable to grew on TSBA but grew on R2A and MR2A; **Group E7**—Facultative anaerobic, grew on TSBA.

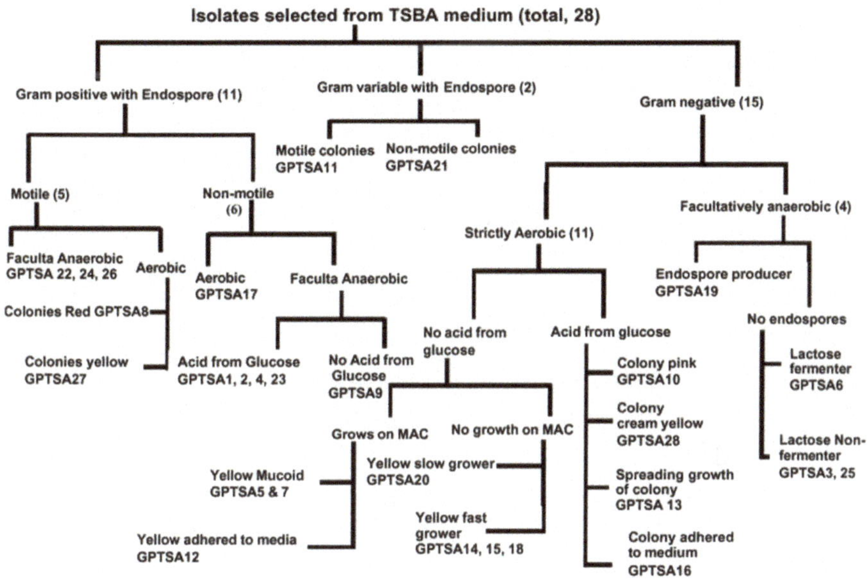

FIGURE 10.1 Preliminary grouping of GPTSA isolates. Preliminary grouping of GPTSA isolates based on phenotypic characteristics. Source: *Adapted from Pradapita Saha's thesis. Saha P. (2006).* Bacterial diversity of water of a warm spring located in Assam. [Doctoral dissertation], CSIR-IMTech/*Jawaharlal Nehru University. CSIR-IMTech Digital Repository. http://crdd.osdd.net/open/1422/.*

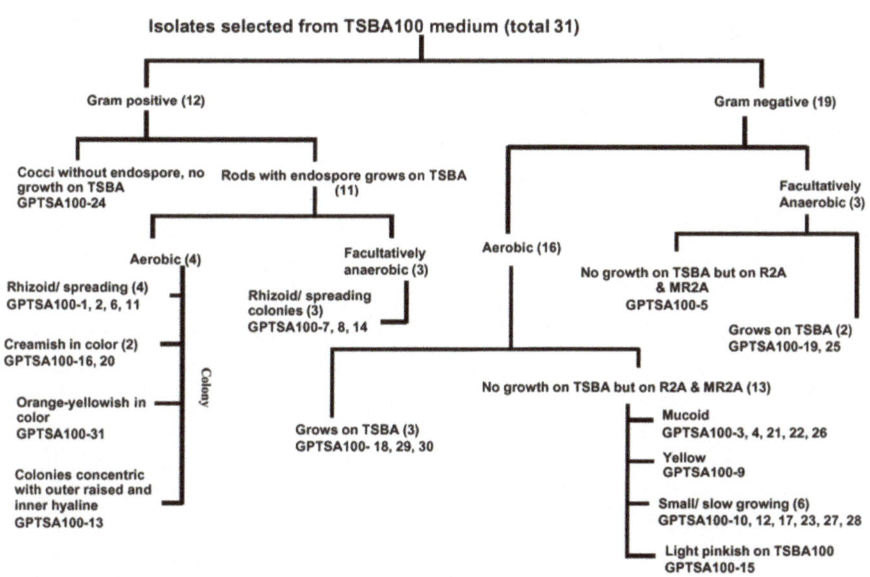

FIGURE 10.2 Preliminary grouping of GPTSA100 isolates. Preliminary grouping of GPTSA100 isolates based on phenotypic characteristics. Source: *Adapted from Pradapita Saha's thesis. Saha, P. (2006).* Bacterial diversity of water of a warm spring located in Assam. [Doctoral dissertation], CSIR-IMTech/*Jawaharlal Nehru University. CSIR-IMTech Digital Repository. http://crdd.osdd.net/open/1422/.*

TABLE 10.5 16S rRNA gene sequence similarity of GPTSA strains.

Strain isolated on rich TSBA medium 21 in number	Accession number (length of nucleotide sequence)	Closest relative(s) in terms of (%) 16S rRNA gene sequence similarity	Phylum level affiliation
GPTSA1	DQ854975 (1480)	*Bacillus proteolyticus* (99.7)	*Firmicutes*
GPTSA3	DQ859920 (1315)	*Aeromonas veronii* (100)	*Proteobacteria*
GPTSA5	DQ854969 (1392)	*Comamonas jiangduensis* (99.2)	*Proteobacteria*
GPTSA6	**DQ013306 (1430)**	***Pseudaeromonas pectinilytica* (96.8)**	*Proteobacteria*
GPTSA8	DQ288677 (1411)	*Falsibacillus pallidus* (99.8)	*Firmicutes*
GPTSA9	DQ854976 (1493)	*Paenibacillus popilliae* (99)	*Firmicutes*
GPTSA10	DQ851174 (1430)	*Rubrivivax benzoatilyticus* (99.7)	*Proteobacteria*
GPTSA11	**AY884046 (1466)**	*Paenibacillus taiwanensis* (95.6)	*Firmicutes*
GPTSA12	DQ851175 (1349)	*Azoarcus communis* (99.6)	*Proteobacteria*
GPTSA13	DQ851176 (1473)	*Pseudaeromonas flexibillis* (99.7)	*Proteobacteria*
GPTSA14	DQ854970 (1441)	*Hydrogenophaga atypica* (98.8)	*Proteobacteria*
GPTSA15	DQ851177 (1466)	*Thauera propionica* (99.8)	*Proteobacteria*
GPTSA16	AY544767 (1417)	*Aquabacterium commune* (97.4)	*Proteobacteria*
GPTSA17	DQ854977 (1322)	*Bacillus aryabhattai* (100)	*Firmicutes*
GPTSA18	DQ851178 (1467)	*Thauera propionica* (99.8)	*Proteobacteria*
GPTSA19	**DQ023221 (1482)**	***Fontibacillus solani* (97.5)**	*Firmicutes*
GPTSA20	**AY544768 (1433)**	***Chiayiivirga flava* (94.7)**	*Proteobacteria*
GPTSA21	DQ854978 (1488)	*Paenibacillus cucumis* (98.9)	*Firmicutes*
GPTSA24	DQ854979 (1468)	*Bacillus altitudinis* (99.8)	*Firmicutes*
GPTSA27	DQ288678 (1479)	*Bacillus koreensis* (100)	*Firmicutes*
GPTSA28	DQ851179 (1433)	*Comamonas jiangduensis* (99)	*Proteobacteria*

Note: Novel taxa based on 16S rRNA gene sequence similarity have been highlighted in bold.

identified—*Bacillus* (12), *Falsibacillus* (1), *Paenibacillus* (3), *Fontibacillus* (1), *Comamonas* (2), *Chitinimonas* (1), *Niveibacterium* (1), *Emticicia* (1), *Ideonella* (1), *Curvibacter* (1), *Rubrivivax* (1), *Azoarcus* (1), *Rheinheimera* (3), *Hydrogenophaga* (4), *Thauera* (2), *Aquabacterium* (2), *Aeromonas* (3), *Pseudoaeromonas* (2), *Pseudomonas* (1), *Aquimonas* (1), *Flavobacterium* (1), and *Deinococcus* (1) (Fig. 10.7).

The detailed phylogeny and comparative phenotypic characterization led to the proposal of four novel genera and two novel species (strains GPTSA20, GPTSA6, GPTSA11, GPTSA19, GPTSA100-15, GPTSA100-9). All the strains were Gram staining negative except strain GPTSA11 which

TABLE 10.6 16S rRNA gene sequence similarity of GPTSA100 strains.

Strain isolated on TSB A100 medium (25 in number)	Accession number (length of nucleotide sequence)	Closest relative(s) in terms of (%) 16S rRNA gene sequence similarity	Phylum level affiliation
GPTSA100-1	DQ854980 (1447)	*Bacillus acidiceler* (99.3)	*Firmicutes*
GPTSA100-3	DQ854971 (1468)	*Chitinimonas taiwanensis* (99.5)	*Proteobacteria*
GPTSA100-5	DQ288676 (1315)	*Niveibacterium umoris* (98.9)	*Proteobacteria*
GPTSA100-6	DQ854981 (1472)	*Bacillus wiedmannii* (100)	*Firmicutes*
GPTSA100-7	DQ854982 (1478)	*Bacillus cereus* (100)	*Firmicutes*
GPTSA100-8	DQ854983 (1464)	*Bacillus pseudomycoides* (98.9)	*Firmicutes*
GPTSA100-9	**NR_074422-1504**	***Flavobacterium haoranii* (94.9)**	*Bacteroidetes*
GPTSA100-10	DQ854972 (1448)	*Hydrogenophaga bisanensis* (100)	*Proteobacteria*
GPTSA100-12	DQ854968 (1437)	*Hydrogenophaga flava* (98.8)	*Proteobacteria*
GPTSA100-13	DQ859924 (1004)	*Bacillus acidiceler* (99.8)	*Firmicutes*
GPTSA100-15	**AY904352 (1385)**	***Emticicia sediminis* (97.7)**	*Bacteroidetes*
GPTSA100-16	DQ854984 (1027)	*Bacillus aryabhattai* (99.9)	*Firmicutes*
GPTSA100-17	DQ854973 (993)	*Ideonella dechloratanas* (99.1)	*Proteobacteria*
GPTSA100-18	DQ683360 (1458)	*Aquabacterium olei* (99.2)	*Proteobacteria*
GPTSA100-19	DQ859921 (1448)	*Aeromonas dhakensis* (100)	*Proteobacteria*
GPTSA100-20	DQ854985 (1464)	*Bacillus aryabhattai* (99.8)	*Firmicutes*
GPTSA100-21	DQ914844 (1431)	*Rheinheimera coerulae* (98.9)	*γ-proteobacteria*
GPTSA100-22	DQ914844 (1431)	*Rheinheimera texasensis* (98.9)	*Proteobacteria*
GPTSA100-24	(631, submitted)	*Deinococcus radiotolerans* (99.8)	*Deinococcus-Thermus*
GPTSA100-25	DQ859922 (1412)	*Aeromonas jandaei* (99.7)	*Proteobacteria*
GPTSA100-26	DQ914846 (1428)	*Rheinheimera coerulae* (99.00)	*Proteobacteria*
GPTSA100-27	**DQ854967 (1443)**	***Curvibacter gracilis* (97), *Acidovorax monticola* (97%)**	*Proteobacteria*
GPTSA100-29	DQ859923 (1392)	*Pseudomonas alcaligenes* (99.5)	*Proteobacteria*
GPTSA100-30	DQ854974 (1423)	*Hydrogenophaga bisanensis* (99.7)	*Proteobacteria*
GPTSA100-31	DQ854986 (1451)	*Bacillus vietnamensis* (99.7)	*Firmicutes*

Note: Novel taxa based on 16S rRNA gene sequence similarity have been highlighted in bold.

showed a Gram-variable reaction. The strain GPTSA20 was validly described as a novel genus *Aquimonas voraii* gen. nov., sp. nov. and strain GPTSA6 as a novel species *Aeromonas sharmana* sp. nov. under the class *γ-Proteobacteria* (Saha et al., 2005a; Saha & Chakrabarti, 2006a), strain

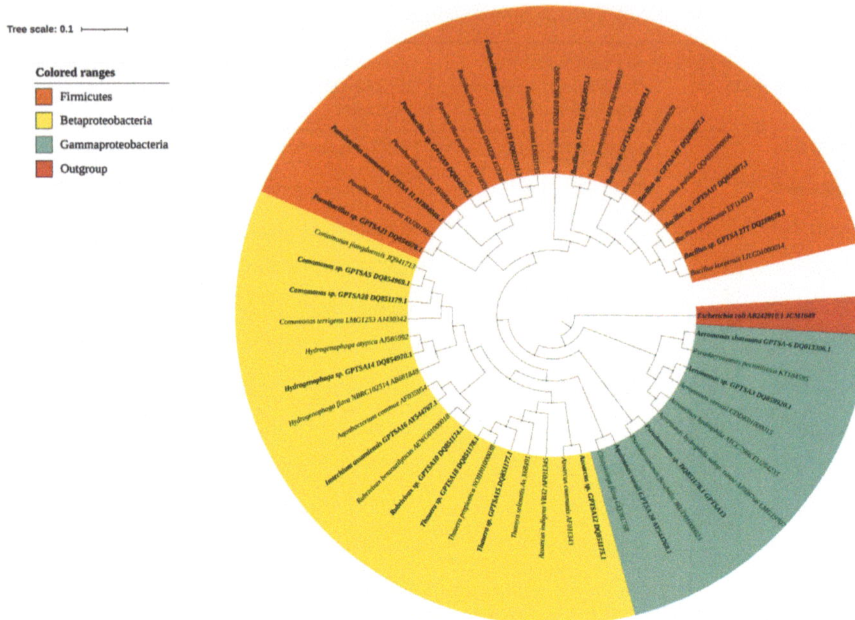

FIGURE 10.3 Evolutionary relationships of GPTSA strains with their closest hit. The evolutionary history was inferred using the Neighbor-Joining method (Saitou & Nei, 1987). The optimal tree with the sum of branch length = 1.33978026 is shown. The percentage of replicate trees in which the associated taxa clustered together in the bootstrap test (1000 replicates) are shown next to the branches 9 (Felsenstein, 1985). The tree is drawn to scale, with branch lengths in the same units as those of the evolutionary distances used to infer the phylogenetic tree. The evolutionary distances were computed using the Kimura 2-parameter method (Kimura, 1980) and are in the units of the number of base substitutions per site. The rate variation among sites was modeled with a gamma distribution (shape parameter = 5). The analysis involved 49 nucleotide sequences. All ambiguous positions were removed for each sequence pair. There were a total of 1459 positions in the final dataset. Evolutionary analyses were conducted in MEGA7 (Kumar et al., 2016). Tree visualization were conducted in iTOL (Letunic & Brok, 2019).

GPTSA100-15 described as novel genera *Emticicia oligotrophica* gen. nov., sp. nov. and GPTSA100-9 as novel species *Flavobacterium indicum* sp. nov., respectively, within the phylum *Bacteroidetes* (Saha & Chakrabarti, 2006b, 2006c), and strains GPTSA11 as a novel species *Paenibacillus assamensis* sp. nov. and GPTSA19 as novel genus *Fontibacillus aquaticus* gen. nov., sp. nov., within phylum *Firmicutes* (Saha et al., 2005b, 2010). It is interesting to note that at the time of proposal strain GPTSA6 fell outside the radiation of the genus *Aeromonas* based on phenotypic characteristics and phylogeny (Saha & Chakrabarti, 2006a). But due to lack of supportive evidence in terms of MLST (Multi locus sequence typing based on housing keeping genes), it was identified only as a novel species within the genus. However, this loose

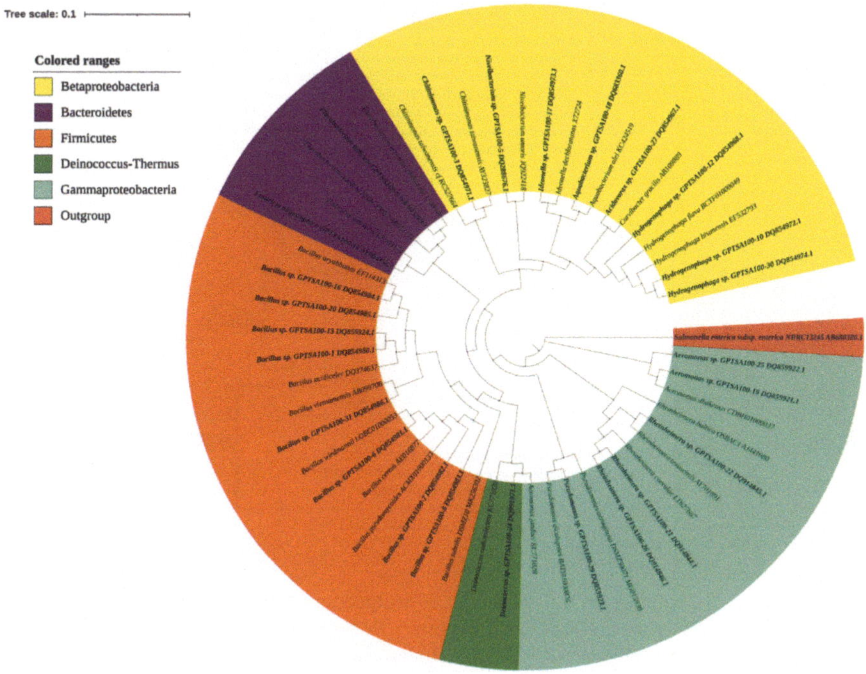

FIGURE 10.4 Evolutionary relationships of GPTSA100 strains with their closest hit. The evolutionary history was inferred using the Neighbor-Joining method (Saitou & Nei, 1987). The optimal tree with the sum of branch length = 1.84973861 is shown. The percentage of replicate trees in which the associated taxa clustered together in the bootstrap test (1000 replicates) are shown next to the branches (Felsenstein, 1985). The tree is drawn to scale, with branch lengths in the same units as those of the evolutionary distances used to infer the phylogenetic tree. The evolutionary distances were computed using the Kimura 2-parameter method (Kimura, 1980) and are in the units of the number of base substitutions per site. The rate variation among sites was modeled with a gamma distribution (shape parameter = 5). The analysis involved 52 nucleotide sequences. All ambiguous positions were removed for each sequence pair. There were a total of 1408 positions in the final dataset. Evolutionary analyses were conducted in MEGA7 (Kumar et al., 2016). Tree visualization were conducted in iTOL (Letunic & Brok, 2019).

association with the genus was subsequently reiterated by Martínez-Murcia et al. (2007) based on detailed phylogeny, signature nucleotides, and phenotypic characteristics described in the original publication and finally a decade later based on phylogeny of the housekeeping genes (gyrB, rpoD, and cpn60), investigators recommended transfer of the species to a novel genus—*Pseudaeromonas sharmana* gen. nov. comb. nov. (Padakandla & Chae, 2017).

Further, based on 16S rRNA gene sequence similarity threshold of 98.5% for species delineation and phylogeny (Barco et al., 2020; Kim et al., 2014), strains GPTSA16 and GPTSA100-27 were identified as

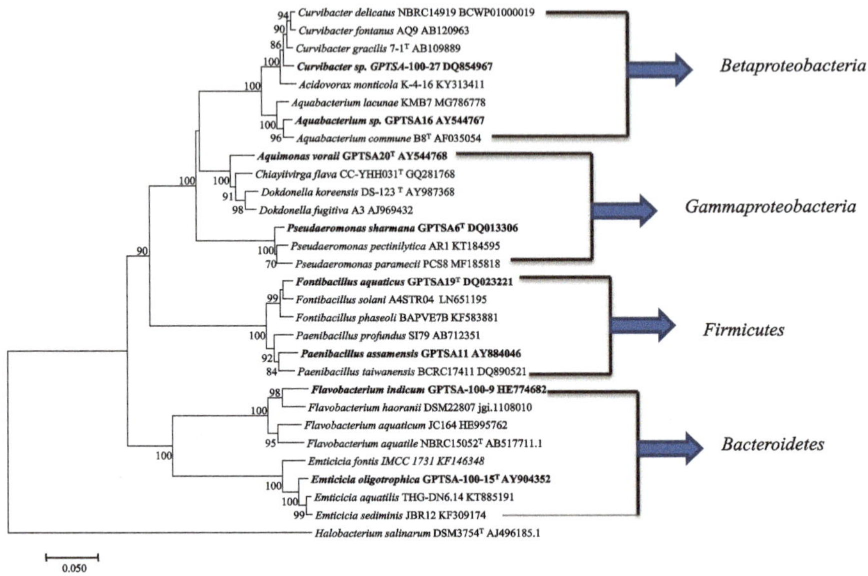

FIGURE 10.5 Evolutionary relationships of eight novel taxa with their closest hit, procured during the study of total cultarable diversity from "Garampani warm spring," Assam, India. The evolutionary history was inferred using the Neighbor-Joining method (Saitou & Nei, 1987). The optimal tree with the sum of branch length = 1.50006639 is shown. The percentage of replicate trees in which the associated taxa clustered together in the bootstrap test (1000 replicates) are shown next to the branches (Felsenstein, 1985). The tree is drawn to scale, with branch lengths in the same units as those of the evolutionary distances used to infer the phylogenetic tree. The evolutionary distances were computed using the Kimura 2-parameter method (Kimura, 1980) and are in the units of the number of base substitutions per site. The rate variation among sites was modeled with a gamma distribution (shape parameter = 5). The analysis involved 30 nucleotide sequences. All positions containing gaps and missing data were eliminated. There were a total of 1285 positions in the final dataset. Evolutionary analyses were conducted in MEGA7 (Kumar et al., 2016).

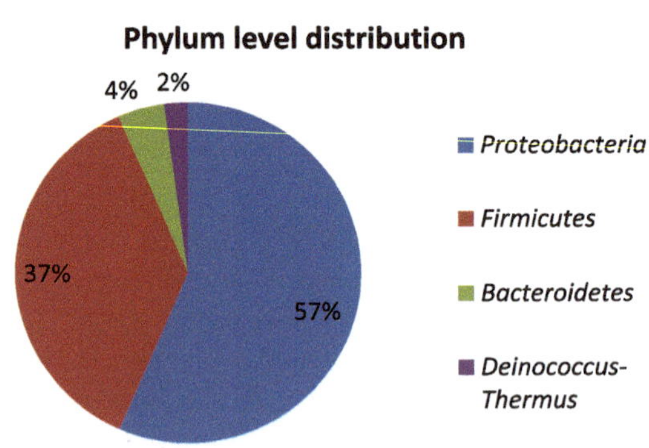

FIGURE 10.6 Diversity of bacterial taxa (phylum level) from Garampani warm spring, Assam, India, by culture-dependent study.

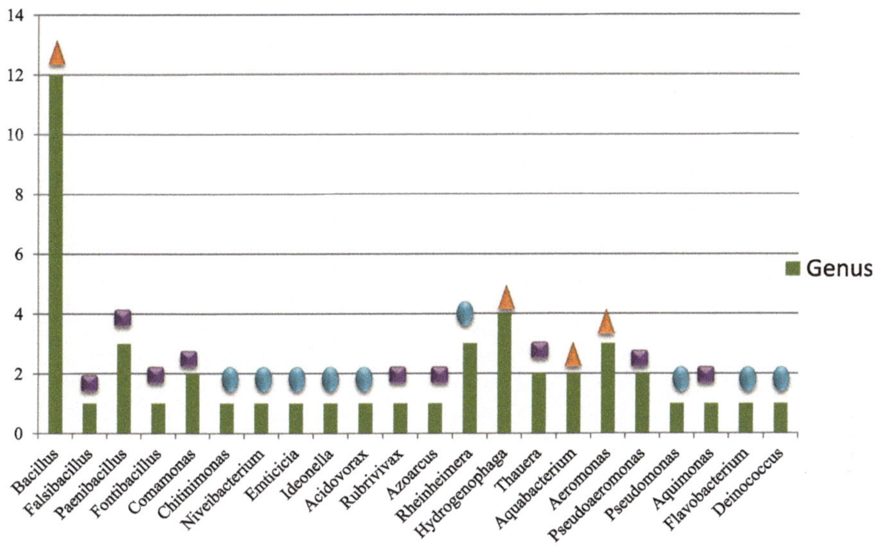

FIGURE 10.7 Diversity of bacterial taxa (genus level) from Garampani warm spring, Assam, India, by culture-dependent study. ■ Represent strains recovered from TSBA media, ● represent strains recovered from TSBA100 media, and ▲ represent strains recovered both from TSBA and TSBA100 media. Numerical value on the *y*-axis shows the number of strains isolated form each genus.

potential novel species belonging to the genera *Aquabacterium* and *Curvibacter*, respectively (Fig. 10.5). Strain GPTSA100-27 were unable to grow on most of the nutritionally rich media like LB and TSBA agar but grew well on nutritionally poor TSBA100, R2A, and MR2A media. It is pertinent to mention that in terms of 16S rRNA gene sequence similarity, the strain showed close identity (97.0%) to both *Acidovorax monticola* and *Curvibacter gracilis* (Table 10.6). Therefore a detailed polyphasic taxonomic characterization is awaited before a final phylogenetic placement can be agreed for both the strains.

10.3.2 Cross growth pattern analysis

In the course of the isolations, about 12 strains were observed to be oligophiles (able to grow only in nutritionally poor TSBA100 media, but not in nutritionally rich TSBA media). About 46 strains were identified as facultatives that could grow both on both TSBA and TSBA100, whereas interestingly only one strain (GPTSA100-27) was identified copiophilic due to its inability to grow on nutritionally poor TSBA100 media (Table 10.7). Further, growth pattern was checked on several

TABLE 10.7 Cross growth pattern analysis of oligophilic strains.

Garampani isolates	Isolate no	TSBA	TSBA-NaCl	TSBA100	LBA	LBA-NaCl	LBA 100	NA	NA-NaCl	NA 100	PCA	PCA 100	MHA	MHA 100	BHIA	BHIA 100	SBA	SBA 100	YMA	YMA 100	R2A	R2A 100	R3A	R3A 100	Mod. R2A	Mod. 2A100	RM	RM 100
Isolated on poor TSBA100 medium	5	−	−	+	−	−	+	−	w+	w+	+	+	−	+	−	+	−	+	−	+	+	+	+	+	+	+	−	+
	6	+	+	+	+	+	+	+	+	+	+	+	+	+	+	+	+	+	+	w+	+	w+	w+	+	+	w+	+	+
	9	−	−	+	−	+	+	+	+	+	+	w+	−	+	−	+	−	+	+	w+	+	+	+	+	+	+	w+	+
	15	−	−	+	−	−	+	−	−	+	−	+	−	+	−	+	−	+	−	w+	+	+	+	+	+	+	−	+
	17	−	−	+	−	+	+	−	w+	+	+	+	−	+	−	+	−	+	w+	+	+	+	+	+	+	+	−	+
	18	+	+	+	+	+	+	+	+	+	+	+	+	+	+	+	−	+	+	+	+	+	+	+	+	+	w+	+
	19	+	+	+	+	+	+	+	+	+	+	+	+	+	+	+	+	+	+	+	+	+	+	+	+	+	+	+
	21	−	+	+	−	+	+	−	+	+	+	+	−	+	−	+	−	+	+	w+	+	w+	+	w+	w+	+	−	+
	22	+	+	+	+	+	+	+	+	+	+	+	+	+	+	+	−	+	+	w+	+	w+	+	w+	w+	+	−	+
	23	−	+	+	−	+	+	−	+	+	+	+	−	+	−	+	−	+	+	w+	+	w+	+	w+	w+	+	−	+
	24	w+	+	+	−	+	+	+	+	+	+	+	+	+	−	+	−	+	+	w+	+	w+	+	+	+	w+	w+	w+
	26	−	+	+	−	+	+	−	+	+	+	+	−	+	−	+	−	+	+	w+	+	w+	+	w+	w+	+	−	+
	27	−	−	w+	−	−	w+	+	−	w+	−	w+	−	w+	−	+	−	+	−	−	+	w+	+	w+	+	w+	−	w+
Isolated on rich TSBA medium	3	+	+	+	+	+	+	+	+	+	+	w+	+	+	+	+	+	+	+	+	+	+	+	+	+	w+	w+	+
	11	+	+	+	+	-	+	+	+	+	+	+	+	+	+	+	−	+	+	+	+	+	+	+	+	+	−	+
	18	+	+	+	+	+	+	+	+	+	+	+	+	+	+	+	+	+	+	+	+	+	+	+	+	-	+	+
	21	+	+	+	+	+	+	+	+	+	+	+	+	+	+	+	+	+	+	+	+	+	+	+	+	+	+	+
	27	+	+	+	+	+	+	+	+	+	+	w+	+	+	+	+	+	+	+	+	+	+	+	+	+	w+	w+	+
Strain Kaza27		−	−	+	−	+	+	−	+	−	−	+	−	+	−	+	−	+	+	+	+	+	+	+	+	+	−	+
Escherichia coli[T]		+	+	+	+	+	+	+	+	+	+	+	+	+	+	+	+	+	+	+	+	+	+	+	+	+	+	+
Micrococcus luteus[T]		+	+	+	+	+	+	+	+	+	+	+	+	+	+	+	+	+	+	+	+	+	+	+	+	+	+	+

Note: BHIA, Brain heart infusion agar; *LBA*, Luria Bertani agar; *MHA*, Mueller Hinton agar; *Mod. R2A*, modified R2A; *NA*, nutrient agar; *PCA*, plate count agar; *R2A*, Reasoner's 2A agar; *R3A*, Reasoner's 2A agar; *SBA*, sheep blood agar; *TSBA*, trypticase soy broth agar; *YMA*, yeast mannitol agar.

other nutritionally rich and poor media formulations (TSBA, TSBA100, TSBA-NaCl, LBA, LBA100, LBA-NaCl, NA, NA100, NA-NaCl, PCA, PCA100, MHA, MHA100, BHIA, BHIA100, SBA, SBA100, YMA, YMA100, R2A, R2A100, R3A, R3A100, Mod.R2A, Mod.R2A100, RM, RM100) (Table 10.7). In this manner, nine strains each (GPTSA100-5, 9, 15, 17, 21, 23, 24, 26, 27), (GPTSA3, 11, 18, 21, 27 and GPTSA100-6, 18, 19, 22) were identified as oligophilic and facultative, respectively. For comparative purposes, two reference strains (strain Kaza27 an oligophilic strain isolated from our lab previously, *E. coli MTCC1610*^{T}, and *Micrococcus luteus*) were included in the analysis. The results indicated that majority of the nine oligophilic strains grew on only 100 times diluted media, but no growth was observed on full strength media. The strain GPTSA100-15 (identified as *Emticicia oligotrophica* gen. nov., sp. nov.) showed growth only on the diluted media formulations and full strength R2A and R3A media (poor nutrient media) and was unable to grow on all tested nutritionally rich media (Table 10.7). On the basis of this observation, the isolate GPTSA100-15 was considered as a true oligophilic representative.

10.3.3 Culture-independent study

Out of the total 154 recombinant clones preserved, 65 clones were digested with *EcoR I* restriction endonuclease and analyzed by agarose gel electrophoresis to ascertain correct site of insert amplicon. The results suggested that the library contained 92% recombinant clones and only 6 clones out of 65 did not show insert. Amplified ribosomal DNA restriction analysis (ARDRA) was performed in order to determine the heterogeneity, and the restriction profile revealed 38 unique patterns which could be potentially treated as different phylotypes (Fig. 10.8) and all the

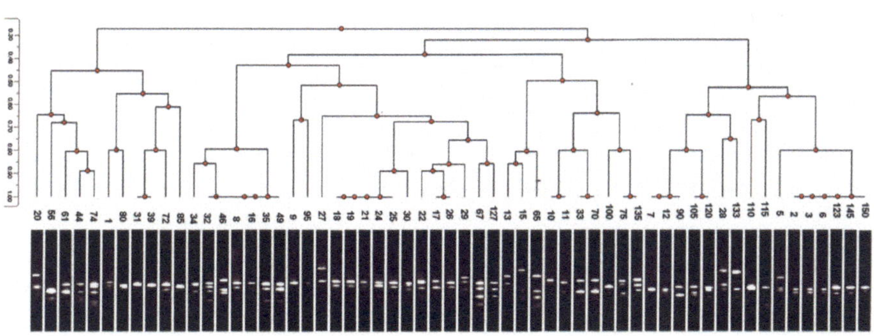

FIGURE 10.8 Dendrogram of 59 clonal sequences by ADRA. Source: *Adapted from Pradapita Saha's thesis Saha, P. (2006).* Bacterial diversity of water of a warm spring located in Assam. [Doctoral dissertation], CSIR-IMTech/*Jawaharlal Nehru University. CSIR-IMTech Digital Repository. http://crdd.osdd.net/open/1422/.*

TABLE 10.8 Result of 28 clonal library that belonged to Cyanobacterial 16S rRNA gene sequence analysis of warm spring.

Clone	Closest match	16S rRNA gene sequence similarity (in percentage)
GPENV2 (DQ512804)	*Cyanobacterium aponinum*	97.3
GPENV9 (DQ512805)	*Geminocystis hermanii*	93.8
GPENV10 (DQ512806)	*Geminocystis hermanii*	93.8
GPENV11 (DQ512807)	*Geminocystis hermanii*	96.3
GPENV18 (DQ512808)	*Geminocystis hermanii*	93.9
GPENV19 (DQ512809)	*Geminocystis hermanii*	88.5
GPENV20 (DQ512810)	*Geminocystis hermanii*	94.2
GPENV21 (DQ512811)	*Geminocystis hermanii*	97.3
GPENV25 (DQ512812)	*Cyanobacterium aponinum*	91.8
GPENV26 (DQ512813)	*Geminocystis hermanii*	95.6
GPENV27 (DQ512814)	*Geminocystis hermanii*	96.9
GPENV28 (DQ512815)	*Geminocystis hermanii*	96.9
GPENV33 (DQ512816)	*Cyanobacterium aponinum*	96.5
GPENV34 (DQ512817)	*Cyanobacterium aponinum*	93.7
GPENV39 (DQ512818)	*Cyanobacterium aponinum*	93.8
GPENV44 (DQ512819)	*Geminocystis hermanii*	93.5
GPENV46 (DQ512820)	*Geminocystis hermanii*	95.7
GPENV49 (DQ512821)	*Cyanobacterium aponinum*	96.7
GPEVV61 (DQ512822)	*Geminocystis hermanii*	95.5
GPENV72 (DQ512823)	*Woronichinia naegeliana*	93
GPENV75 (DQ512824)	*Geminocystis hermanii*	97.1
GPENV80 (DQ512825)	*Cyanobacterium aponinum*	96.6
GPENV90 (DQ512826)	*Geminocystis hermanii*	91.9
GPENV95 (DQ512827)	*Geminocystis hermanii*	94.5
GPENV110 (DQ512828)	*Geminocystis hermanii*	97.1
GPENV115 (DQ991973)	*Geminocystis hermanii*	93.7
GPENV120 (DQ512829)	*Geminocystis hermanii*	95.4
GPENV123 (DQ512830)	*Cyanobacterium aponinum*	90.7
GPENV127 (DQ512831)	*Cyanobacterium aponinum*	96.3
GPENV133 (DQ512831)	*Geminocystis herdmanii*	96.6
GPENV152 (DQ512833)	*Geminocystis herdmanii*	93.9
GPENV154 (DQ512834)	*Geminocystis herdmanii*	96.4

TABLE 10.9 Sequence analysis of the 16S rRNA gene from the clones of the spring water showing their affiliation, nearest relative with their source of isolation, and percent similarity.

Clone (accession no.)	Affiliation	Closest match	16S gene sequence similarity (in percentage)
GPENV1 (DQ512835)	α-Proteobacteria	*Acidimangrovimonas indica*	86.1
GPENV3 (DQ512836)	γ-Proteobacteria	*Aeromonas rivipollensis*	99.8
GPENV4 (DQ512837)	β-Proteobacteria	*Janthinobacterium lividum*	87.7
GPENV5 (DQ512838)	β-Proteobacteria	*Aeromonas rivipollensis*	99.4
GPENV6 (DQ512839)	β-Proteobacteria	*Azonexus fungiphilus*	94
GPENV7 (DQ512840)	γ-Proteobacteria	*Aeromonas schubertii*	91.1
GPENV8 (DQ512841)	β-Proteobacteria	*Polynucleobacter wuianus*	98.5
GPENV15 (DQ512842)	α-Proteobacteria	*Rhabdaerophilum calidifontis*	96.9
GPENV16 (DQ512843)	β-Proteobacteria	*Azonexus fungiphilus*	92.3
GPENV17 (DQ512844)	β-Proteobacteria	*Malikia granosa*	92.1
GPENV22 (DQ512845)	β-Proteobacteria	*Undibacterium macrobrachii*	95.2
GPENV29 (DQ512846)	α-Proteobacteria	*Acidimangrovimonas indica*	85.8
GPENV30 (DQ512847)	γ-Proteobacteria,	*Enterobacter asburiae*	82.5
GPENV31 (DQ512848)	β-Proteobacteria	*Janthinobacterium lividum*	98.4
GPENV35 (DQ512849)	β-Proteobacteria	*Hydrogenophaga caeni*	96.3
GPENV56 (DQ512850)	β-Proteobacteria	*Hydrogenophaga caeni*	97.4
GPENV65 (DQ512851)	β-Proteobacteria	*Inhella inkyongensis*	89.6
GPENV67 (DQ512852)	β-Proteobacteria	*Comamonas composti*	86.2
GPENV70 (DQ512853)	β-Proteobacteria	*Undibacterium danionis*	96.1
GPENV74 (DQ512854)	β-Proteobacteria	*Hydrogenophaga caeni*	96.4
GPENV85 (DQ512855)	β-Proteobacteria	*Noviherbaspirillum soli*	90.2
GPENV100 (DQ991972)	β-Proteobacteria	*Leptothrix discophora*	85
GPNV105 (DQ512856)	γ-Proteobacteria	*Aeromonas salmonicida*	94
GPENV135 (DQ512857)	γ-Proteobacteria	*Acinetobacter bereziniae*	95.1
GPENV145 (DQ991974)	γ-Proteobacteria	*Klebsiella huaxiensis*	90

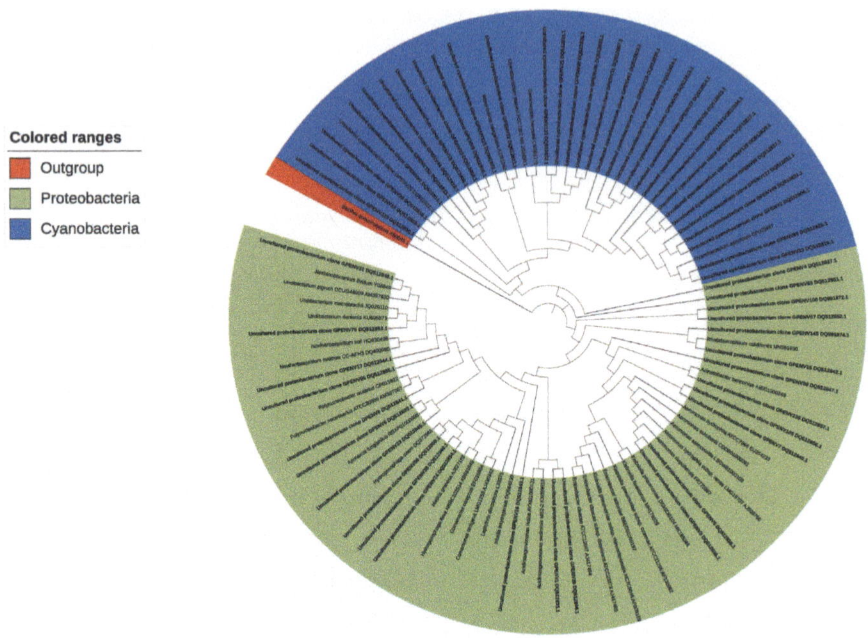

FIGURE 10.9 Evolutionary relationships of clonal sequences with their closest hit. The evolutionary history was inferred using the Neighbor-Joining method (Saitou & Nei, 1987). The optimal tree with the sum of branch length = 3.21685768 is shown. The percentage of replicate trees in which the associated taxa clustered together in the bootstrap test (1000 replicates) are shown next to the branches (Felsenstein, 1985). The tree is drawn to scale, with branch lengths in the same units as those of the evolutionary distances used to infer the phylogenetic tree. The evolutionary distances were computed using the Kimura 2-parameter method (Kimura, 1980) and are in the units of the number of base substitutions per site. The rate variation among sites was modeled with a gamma distribution (shape parameter = 5). The analysis involved 94 nucleotide sequences. All ambiguous positions were removed for each sequence pair. There were a total of 1051 positions in the final dataset. Evolutionary analyses were conducted in MEGA7 (Kumar et al., 2016). Tree visualization were conducted in iTOL (Letunic & Brok, 2019).

59 clones were sequenced. Out of 59 clonal sequences, 2 were chimeric and were not considered further. Phylogenetic analysis of the 57 clone sequences revealed that 32 belonged to *Cyanobacteria* and 25 showed affiliation to *Proteobacteria* (3 to -Proteobacteria, 16 to −*Proteobacteria*, and 6 to -*Proteobacteria*) (Tables 10.8 and 10.9). The tree depicting phylogenetic placements of these clones are shown in Fig. 10.9.

Most of the cyanobacterial clonal sequences showed closest similarity with *Geminocystis hermanii* (88%−97%) and *Cyanobacterium aponinum* (90%−97%). Many different genera were identified within Proteobacterial clone sequences, that is, *Acidimangrovimonas, Aeromonas, Janthinobacterium, Azonexus, Polynucleobacter, Rhabdaerophilum, Malikia, Undibacterium,*

Anterobacter, Hydrogenophaga, Inhella, Comamonas, Noviherbaspirillum, Leptothrix, Acinetobacter, Klebsiella. However, majority of the clones GPENV (1, 4, 6, 7, 9, 10, 16, 17, 18, 19, 25, 29, 30, 34, 39, 44, 65, 67, 72, 90, 95, 100, 105, 115, 123, 145, 152) showed very low 16S rRNA gene sequence similarity (in the range between 82% and 95%) and no closest cultivated representative could be identified, therefore their taxonomic affiliation is still inconclusive. Suggesting that majority of the sequences represent promising novel lineages.

10.4 Discussion

In the present study, our goal was to decipher bacterial diversity from "Garampani water" situated in Eastern Himalayan region. Since, it is a well-known fact, that majority of the bacterial representatives cannot be isolated by conventional culturing methods (Amann et al., 1995), we choose to study bacterial diversity both by culture-dependent and culture-independent approaches. Preferably, a combination of different media formulations should be used to target culturing a diverse group of bacteria including novel taxa. Among various traditionally used media, TSBA supports the growth of different kinds of bacteria including *Actinomycetes*. However, such nutritionally rich media are also known to kill cells through different phenomena, i.e., oxidative stress, substrate accelerated death, lack of cell-to-cell communication etc. (Bruns et al., 2002, 2003). Therefore in this study, the medium was diluted 100 times and solidified in high purity agarose (low EEO SRL to prevent inhibition of growth of oligotrophic bacteria through contamination by bacteriological grade agar) (Kato et al., 2018). Interestingly, the viable count of bacteria in nutritionally poor TSBA100 medium was five times more than in the nutritionally rich parent TSBA which was not surprising considering the fact that the warm spring was not a nutrient rich environment. A total of 59 culturable bacterial representatives (31 morphotypes from TSBA100 and 28 morphotypes from TSBA) were isolated, purified, and preserved. Based on preliminary phenotypic characterization, the number of these distinct morphotypes was reduced to 46 unique taxa. The detailed characterization by polyphasic approach and the taxonomic status of these strains are listed in Tables 10.4–10.6, Figs. 10.3 and 10.4. Some of the strains have been described as novel genera and species as described below.

1. Strain GPTSA20—*Aquimonas voraii* gen. nov., sp. nov.
2. Strain GPTSA100-15—*Emticicia oligotrophica* gen. nov., sp. nov.
3. Strain GPTSA6—*Pseudaeromonas sharmana* gen. nov., sp. nov.
4. Strain GPTSA11—*Paenibacillus assamensis* sp. nov.

5. Strain GPTSA19—*Fontibacillus aquaticus* gen.nov., sp. nov.
6. Strain GPTSA100-9—*Flavobacterium indicum* gen. nov., sp. nov.
7. Strain GPTSA16—*Aquibacterium* sp. nov.
8. Strain GPTSA100-27—*Curvibacter* sp. nov.

One other aim of the study was based on the hypothesis that a vast majority of bacterial cells adapt to low nutrient conditions and do not grow in regularly used laboratory media. Therefore TSB medium diluted 100 times was used to isolate oligophilic bacteria from the Garampani water sample. On the basis of cross growth patterns on various kinds of culture media, 12 strains were considered oligophiles (GPTSA 5, 6, 9, 15, 17, 18, 19, 21, 22, 23, 24, 26, Table 10.7). On the basis of detailed cross growth pattern analysis (results shared in Table 10.7), GPTSA100-15 (*Emiticicia oligotrophica* gen nov., sp. nov) was finally selected for further experiments. In order to understand the rationale for inability of the strain GPTSA100-15 to grow on nutritionally rich media, a comparative study of viable cell count in TSBA and TSBA100 media was conducted and expectedly there was a significant drop in viable cell count of three orders of magnitude when transferred to nutritionally rich TSBA media within an hour, following a typical death curve (killing curve, data not shown). In order to understand the cellular changes, whole cell fatty acids analysis was performed along with some other facultative oligophiles (isolated from this study) and reference organisms (data not shown). There was, however, no significant change in fatty acid profiles of the microbes to draw any conclusion

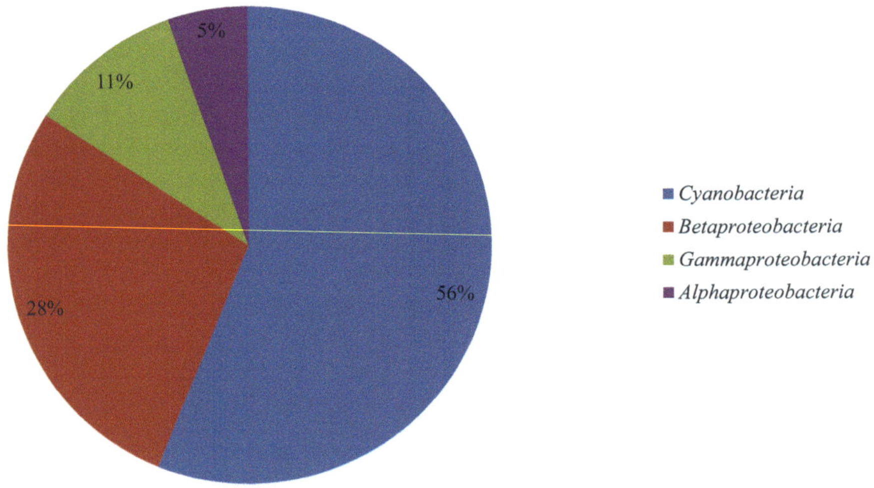

FIGURE 10.10 Diversity of bacterial taxa (phylum and class level) from Garampani warm spring, Assam, India by culture-independent study.

regarding their role (either as cause/effect) in nutrient tolerance of oligophilic bacteria.

The culture-independent study showed the presence of only phyla *Cyanobacteria* and *Proteobacteria* which could not be detected by culturing techniques. About 56.1% clones represented sequences belonging to the *Cyanobacteria* and 43.9% belonged to *Proteobacteria*. Within this, 28.1% was identified within subclass β-*Proteobacteria* followed by γ-*Proteobacteria* (10.5%) and α-*Proteobacteria* (5.3%) (Fig. 10.10). Majority of the sequences showed very low sequence similarity (82.50%–90%) at the level of family and genus, thus indicating the presence of novel bacterial lineages that yet remain to be cultured and described from this site.

10.5 Conclusion

The exploration of bacterial diversity from Garampani warm spring was conducted by both culture-dependent and -independent approaches. A special emphasis was placed on isolation of oligophilic bacteria and studying their properties by using a combination of nutritionally rich and poor media. In this process isolates belonging to different genera, that is, *Bacillus* (12), *Chitinimonas* (1), *Niveibacterium* (1), *Hydrogenophaga* (4), *Emticicia* (1), *Ideonella* (1), *Aquabacterium* (2), *Aeromonas* (3), *Rheinheimera* (3), *Deinococcus* (1), *Curvibacter* (1), *Comamonas* (2), *Pseudaeromonas* (2), *Falsibacillus* (1), *Paenibacillus* (3), *Rubrivivax* (1), *Azoarcus* (1), *Thauera* (2), *Fontibacillus* (1), *Aquimonas* (1), *Pseudomonas* (1), *Flavobacterium* (1) were successfully isolated and described. Eight novel taxa including four novel genera and four novel species were successfully cultivated and taxonomically described. About 12 oligophilic strains were isolated and one of them designated as GPTSA100-15 was identified as a member of a novel genus *E. oligotrophica* gen. nov., sp. nov. In culture-independent analysis, only two phyla, *Cyanobacteria* and *Proteobacteria*, were recognized with majority of clone sequences showing no affiliation to a culturable representative. Overall our research has indicated a very rich and complex microbial diversity in this unique habitat that needs further exploration with contemporary high-throughput sequencing techniques and novel cultivation approaches. With our best knowledge, this is the first and only report of a vast investigation of bacterial diversity from Garampani warm spring, Jorhat, Assam.

Acknowledgments

We are highly grateful to our mentor, teacher, and philosopher Dr. Tapan Chakrabarti, founder head of Microbial Type Culture Collection & Gene Bank (MTCC), CSIR-Institute of Microbial Technology (IMTECH), Chandigarh, a great luminary of microbial diversity and taxonomy research in India, who established a new school of young taxonomists and with his legacy of thoughts and vision his disciples are moving ahead. His pioneering thoughts on oligophily, novel methods for culturing microbes from various unexplored Indian ecological niches are noteworthy contributions in microbiology that gave great impetus to the future generations to focus on this important area of research. He with his research group substantially contributed towards description of several microbial taxa and biodiversity research. We are also grateful to CSIR-IMTECH for providing all the necessary infrastructure support and also to Dr. S. Mayilraj (another eminent taxonomist and former scientist, MTCC, CSIR-IMTECH) for his valuable scientific inputs, suggestions, and constant encouragement. The authors also thank DBT and CSIR for project funding and supporting fellowships. Joyasree Das is a recipient of UGC fellowship.

We are also thankful to Dr. T. C. Bora of RRL, Jorhat (Now CSIR-NEIST, Jorhat) for his help in sample collection and providing primary physicochemical data for the water sample. Financial support of CSIR and DBT, Government of India is thankfully acknowledged. The IMTECH communication number is 084/2020.

References

Amann, R. I., Ludwig, W., & Schleifer, K. H. (1995). Phylogenetic identification and in situ detection of individual microbial cells without cultivation. *Microbiological Reviews, 59*(1), 143–169. Available from https://doi.org/10.1128/mmbr.59.1.143-169.1995.

Barco, R. A., Garrity, G. M., Scott, J. J., Amend, J. P., Nealson, K. H., & Emerson, D. A. (2020). Genus definition for *Bacteria* and *Archaea* based on a standard genome relatedness index. *mBio, 11*, e02475–19. Available from https://doi.org/10.1128/mBio.02475-19.

Bhatnagar, A., & Bhatnagar, M. (2005). Microbial diversity in desert ecosystems. *Current Science, 89*(1), 91–100.

Bhattacharyya, P. N., Tanti, B., Barman, P., & Jha, D. K. (2014). Culture-independent metagenomic approach to characterize the surface and subsurface soil bacterial community in the Brahmaputra valley, Assam, North-East India, an Indo-Burma mega-biodiversity hotspot. *World Journal of Microbiology and Biotechnology, 30*(2), 519–528. Available from https://doi.org/10.1007/s11274-013-1467-1.

Bowman, J. P. (2000). Description of *Cellulophaga algicola* sp. nov., isolated from the surfaces of Antarctic algae, and reclassification of *Cytophaga uliginosa* (ZoBell and Upham 1944) Reichenbach 1989 as *Cellulophaga uliginosa* comb. nov. *International Journal of Systematic and Evolutionary Microbiology, 50*(5), 1861–1868. Available from https://doi.org/10.1099/00207713-50-5-1861.

Bruns, A., Cypionka, H., & Overmann, J. (2002). Cyclic AMP and acyl homoserine lactones increase the cultivation efficiency of heterotrophic bacteria from the central Baltic Sea. *Applied Environmental Microbiology, 68*, 3978–3987. Available from https://doi.org/10.1128/AEM.68.8.3978-3987.2002.

Bruns, A., Nübel, U., Cypionka, H., & Overmann, J. (2003). Effect of signal compounds and incubation conditions on the culturability of freshwater bacterioplankton. *Applied Environmental Microbiology, 69*, 1980–1989. Available from https://doi.org/10.1128/AEM.69.4.1980-1989.2003.

Chaudhuri, S. R., & Thakur, A. R. (2006). Microbial genetic resource mapping of East Calcutta wetlands. *Current Science, 91*(2), 212–217. Available from http://www.ias.ac.in/currsci/jul252006/217.pdf.

Claus, D., Berkeley, R. C. W., & Cohn, G. B. (1872). In P. H. A. Sneath, N. S. Mair, M. E. Sharpe, & J. G. Holt (Eds.), *Bergey's manual of systematic bacteriology* (2, pp. 1105–1139). Baltimore: The Williams & Wilkins Co., 1986.

Cowan, S. T., & Steel, K. J. (1965). *Manual for the identification of medical bacteria* (pp. 22–29). London: Cambridge University Press, ISBN 0521203996.

Das, R., Romi, W., Das, R., Sharma, H. K., & Thakur, D. (2018). Antimicrobial potentiality of actinobacteria isolated from two microbiologically unexplored forest ecosystems of Northeast India. *BMC Microbiology, 18*(1). Available from https://doi.org/10.1186/s12866-018-1215-7.

Felsenstein, J. (1985). Confidence limits on phylogenies: An approach using the bootstrap. *Evolution, 39*, 783–791. Available from https://doi.org/10.1111/j.1558-5646.1985.tb00420.x.

Ghosh, D., Bal, B., Kashyap, V. K., & Pall, S. (2003). Molecular phylogenetic exploration of bacterial diversity in a Bakreshwar (India) hot spring and culture of Shewanella-related thermophiles. *Applied and Environmental Microbiology, 69*(7), 4332–4336. Available from https://doi.org/10.1128/AEM.69.7.4332-4336.2003.

Jackson, C. R., Randolph, K. C., Osborn, S. L., & Tyler, H. L. (2013). Culture dependent and independent analysis of bacterial communities associated with commercial salad leaf vegetables. *BMC Microbiology, 13*(1). Available from https://doi.org/10.1186/1471-2180-13-274.

Johnson, J. L. (1994). Similarity analysis of DNAs. In P. Gerhardt, R. G. E. Murray, W. A. Wood, & N. R. Krieg (Eds.), *Methods for molecular and general bacteriology* (pp. 656–682). Washington, DC: American Society for Microbiology.

Kato, S., Yamagishi, A., Daimon, S., Kawasaki, K., Tamaki, H., Kitagawa, W., Abe, A., Tanaka, M., Sone, T., Asano, K., & Kamagata, Y. (2018). Isolation of previously uncultured slowgrowing bacteria by using a simple modification in the preparation of agar media. *Applied and Environmental Microbiology, 84*(19). Available from https://doi.org/10.1128/AEM.00807-18.

Kim, O. S., Cho, Y. J., Lee, K., Yoon, S. H., Kim, M., Na, H., Park, S. C., Jeon, Y. S., Lee, J. H., Yi, H., Won, S., & Chun, J. (2012). Introducing EzTaxon-e: A prokaryotic 16s rRNA gene sequence database with phylotypes that represent uncultured species. *International Journal of Systematic and Evolutionary Microbiology, 62*(3), 716–721. Available from https://doi.org/10.1099/ijs.0.038075-0.

Kim, M., Oh, H. S., Park, S. C., & Chun, J. (2014). Towards a taxonomic coherence between average nucleotide identity and 16S rRNA gene sequence similarity for species demarcation of prokaryotes. *International Journal of Systematic and Evolutionary Microbiology, 64*(2), 346–351. Available from https://doi.org/10.1099/ijs.0.059774-0.

Kimura, M. (1980). A simple method for estimating evolutionary rates of base substitutions through comparative studies of nucleotide sequences. *Journal of Molecular Evolution, 16*, 111–120. Available from https://doi.org/10.1007/BF01731581.

Koch, A. L. (2001). Oligotrophs vs copiotrophs. *BioEssays, 23*(7), 657–661. Available from https://doi.org/10.1002/bies.1091.

Kumar, S., Stecher, G., & Tamura, K. (2016). MEGA7: Molecular evolutionary genetics analysis version 7.0 for bigger datasets. *Molecular Biology and Evolution, 33*(7), 1870–1874. Available from https://doi.org/10.1093/molbev/msw054.

Letunic, I., & Bork, P. (2019). Interactive Tree Of Life (iTOL) v4: Recent updates and new developments. *Nucleic Acids Research, 47*, W256–W259. Available from https://doi.org/10.1093/nar/gkz239.

Ley, R. E., Harris, J. K., Wilcox, J., Spear, J. R., Miller, S. R., Bebout, B. M., Maresca, J. A., Bryant, D. A., Sogin, M. L., & Pace, N. R. (2006). Unexpected diversity and complexity of the Guerrero Negro hypersaline microbial mat. *Applied and Environmental Microbiology, 72*(5), 3685–3695. Available from https://doi.org/10.1128/AEM.72.5.3685-3695.2006.

Lyngwi, N. A., Koijam, K., Sharma, D., & Joshi, S. R. (2013). Cultivable bacterial diversity along the altitudinal zonation and vegetation range of tropical Eastern Himalaya. *Revista de Biologia Tropical, 61*(1), 467–490. Available from https://doi.org/10.15517/rbt.v61i1.11141.

Marmur, J. (1961). A procedure for the isolation of deoxyribonucleic acid from microorganisms. *Journal of Molecular Biology, 3*.

Martínez-Murcia, A. J., Figueras, M. J., Saavedra, M. J., & Stackebrandt, E. (2007). The recently proposed species *Aeromonas sharmana* sp. nov., isolate GPTSA-6T, is not a member of the genus *Aeromonas. International Microbiology, 10*(1), 61–64. Available from https://doi.org/10.2436/20.1501.01.9.

Morita, R. Y. (1997). *Bacteria in oligotrophic environments: Starvation-survival life style* (pp. 1–4). New York: Chapman & Hall, ISBN 9780412106613.

Murray, R. G. E., Doetsch, R. N., & Robinow, C. F. (1994). Determinative and cytological light microscopy. In P. Gerhard, R. G. E. Murray, W. A. Wood, & N. R. Krieg (Eds.), *Methods for general and molecular bacteriology* (pp. 21–41). Washington, DC: American Society for Microbiology.

Nampoothiri, K. M., Ramkumar, B., & Pandey, A. (2013). Western ghats of India: Rich source of microbial biodiversity. *Journal of Scientific and Industrial Research, 72*(9–10), 617–623. Available from http://nopr.niscair.res.in/bitstream/123456789/20956/1/JSIR%2072%289-10%29%20617-623.pdf.

Padakandla, S. R., & Chae, J. C. (2017). Reclassification of *Aeromonas sharmana* to a new genus as *Pseudaeromonas sharmana* gen. nov., comb. nov., and description of *Pseudaeromonas pectinilytica* sp. nov. isolated from a freshwater stream. *International Journal of Systematic and Evolutionary Microbiology, 67*(4), 1018–1023. Available from https://doi.org/10.1099/ijsem.0.001734.

Panda, A. K., Bisht, S. S., De Mandal, S., & Kumar, N. S. (2016). Bacterial and archeal community composition in hot springs from Indo-Burma region, North-east India. *AMB Express, 6*(1). Available from https://doi.org/10.1186/s13568-016-0284-y.

Powers, E. M. (1995). Efficacy of the Ryu nonstaining KOH technique for rapidly determining Gram reactions of food-borne and waterborne bacteria and yeasts. *Applied and Environmental Microbiology, 61*(10), 3756–3758. Available from https://doi.org/10.1128/aem.61.10.3756-3758.1995.

Reddy, G. S. N., Prakash, J. S. S., Srinivas, R., Matsumoto, G. I., & Shivaji, S. Y. (2003). *Leifsoniarubra* sp. nov. and *Leifsoniaaurea* sp. nov., psychrophiles from a pond in Antarctica. *International Journal of Systematic and Evolutionary Microbiology, 53*, 977–984. Available from https://doi.org/10.1099/ijs.0.02396-0.

Reiss, M., & Chifflard, P. (2017). An opinion on spring habitats within the earth's critical zone in headwater regions. *Water (Switzerland), 9*(9). Available from https://doi.org/10.3390/w9090645.

Ruckmani, A., & Chakrabarti, T. (2011). Analysis of bacterial community composition of a spring water from the Western Ghats, India using culture dependent and molecular approaches. *Current Microbiology, 62*(1), 7–15. Available from https://doi.org/10.1007/s00284-010-9663-9.

Ruckmani, A., Kaur, I., Schumann, P., Klenk, H. P., & Mayilraj, S. (2011). *Calidifontibacter indicus* gen. nov., sp. nov., a member of the family *Dermococcaceae* isolated from a hot spring, and emended description of the family *Dermacoccaceae. International Journal of Systematic and Evolutionary Microbiology, 61*(10), 2419–2424. Available from https://doi.org/10.1099/ijs.0.025593-0.

Saha, P., & Chakrabarti, T. (2006a). *Aeromonas sharmana* sp. nov., isolated from a warm spring. *International Journal of Systematic and Evolutionary Microbiology, 56*(8), 1905–1909. Available from https://doi.org/10.1099/ijs.0.63972-0.

Saha, P., & Chakrabarti, T. (2006b). *Emticicia oligotrophica* gen. nov., sp. nov., a new member of the family *"Flexibacteraceae,"* phylum *Bacteroidetes. International Journal of Systematic and Evolutionary Microbiology, 56*(5), 991–995. Available from https://doi.org/10.1099/ijs.0.64086-0.

Saha, P., & Chakrabarti, T. (2006c). *Flavobacterium indicum* sp. nov., isolated from warm spring water in Assam, India. *International Journal of Systematic and Evolutionary Microbiology, 56*(11), 2617–2621. Available from https://doi.org/10.1099/ijs.0.64309-0.

Saha, P., Krishnamurthi, S., Mayilraj, S., Prasad, G. S., Bora, T. C., & Chakrabarti, T. (2005a). *Aquimonas voraii* gen. nov., sp. nov., a novel gammaproteobacterium isolated from a warm spring of Assam, India. *International Journal of Systematic and Evolutionary Microbiology, 55*(4), 1491–1495. Available from https://doi.org/10.1099/ijs.0.63552-0.

Saha, P., Mondal, A. K., Mayilraj, S., Krishnamurthi, S., Bhattacharya, A., & Chakrabarti, T. (2005b). *Paenibacillus assamensis* sp. nov., a novel bacterium isolated from a warm spring in Assam, India. *International Journal of Systematic and Evolutionary Microbiology, 55*(6), 2577–2581. Available from https://doi.org/10.1099/ijs.0.63846-0.

Saha, P., Krishnamurthi, S., Bhattacharya, A., Sharma, R., & Chakrabarti, T. (2010). *Fontibacillus aquaticus* gen. nov., sp. nov., isolated from a warm spring. *International Journal of Systematic and Evolutionary Microbiology, 60*(2), 422–428. Available from https://doi.org/10.1099/ijs.0.012633-0.

Saitou, N., & Nei, M. (1987). The neighbor-joining method: A new method for reconstructing phylogenetic trees. *Molecular Biology and Evolution, 4*, 406–425. Available from https://doi.org/10.1093/oxfordjournals.molbev.a040454.

Sambrook, J., Fritsch, E. F., & Maniatis, T. (1989). *Molecular Cloning: A Laboratory Manual* (2nd ed.). Harbor, NY: Cold Spring Harbor Laboratory, Cold Spring.

Sharma, P., & Thakur, D. (2020). Antimicrobial biosynthetic potential and diversity of culturable soil actinobacteria from forest ecosystems of Northeast India. *Scientific Reports, 10*(1). Available from https://doi.org/10.1038/s41598-020-60968-6.

Smibert, R. M., & Krieg, N. R. (1994). Phenotypic characterization. In P. Gerhardt, R. G. E. Murray, W. A. Wood, & N. R. Krieg (Eds.), *Methods for general and molecular bacteriology* (pp. 607–654). Washington, DC: American Society for Microbiology.

Staley, J. T., & Konopka, A. (1985). Measurement of in situ activities of nonphotosynthetic microorganisms in aquatic and terrestrial habitats. *Annual Review of Microbiology, 39*, 321–346. Available from https://doi.org/10.1146/annurev.mi.39.100185.001541.

Thajuddin, N., & Subramanian, G. (2005). Cyanobacterial biodiversity and potential applications in biotechnology. *Current Science, 89*(1), 47–57.

Tourova, T. P., & Antonov, A. S. (1988). Identification of microorganisms by rapid DNA-DNA hybridization. *Methods in Microbiology, 19*(C), 333–355. Available from https://doi.org/10.1016/S0580-9517(08)70414-8.

Venkataraman, K., & Sivaperuman, C. (2018). *Biodiversity hotspots in India, . Indian hotspots: Vertebrate faunal diversity, conservation and management* (Volume 1, pp. 1–27). Singapore: Springer. Available from https://doi.org/10.1007/978-981-10-6605-4_1.

Youssef, N. H., Couger, M. B., McCully, A. L., Criado, A. E. G., & Elshahed, M. S. (2015). Assessing the global phylum level diversity within the bacterial domain: A review. *Journal of Advanced Research, 6*(3), 269–282. Available from https://doi.org/10.1016/j.jare.2014.10.005.

Zhang, X., Li, L., Butcher, J., Stintzi, A., & Figeys, D. (2019). Advancing functional and translational microbiome research using *meta*-omics approaches. *Microbiome, 7*(1). Available from https://doi.org/10.1186/s40168-019-0767-6.

CHAPTER

11

Diversity and biotechnological importance of cellulolytic microorganisms from biodiversity hotspots

Himanshu and Jitendra Kumar Saini

Department of Microbiology, Central University of Haryana, Mahendergarh, Haryana, India

11.1 Lignocellulose: composition and availability

Lignocellulosic biomass is the most abundant source of renewable energy and promising raw material due to its accessibility, cost-efficiency, and noncompetency with the food. Because of the issues of economy and environmental-related concerns, the worldwide exploration interest for manufacture of different chemicals, powers, energy, and different materials as an alternate to petroleum-based products (Hemansi et al., 2018). Lignocellulose comprise of three polymers including cellulose, hemicellulose and lignin (Fig. 11.1). These components after enzymatic degradation can be used for production of various value-added biochemical and bioproduct after their transformation to simple sugars. Another polymer is lignin with high economic significance, which can be used for cogeneration, production of phenols and its components and different other chemicals (De Bhowmick et al., 2018).

11.1.1 Cellulose

Cellulose is one of the major components of lignocellulosic biomass. It is an anhydroglucose homopolymer, with glucose residues bound in

Microbial Diversity and Ecology in Hotspots
DOI: https://doi.org/10.1016/B978-0-323-90148-2.00006-7

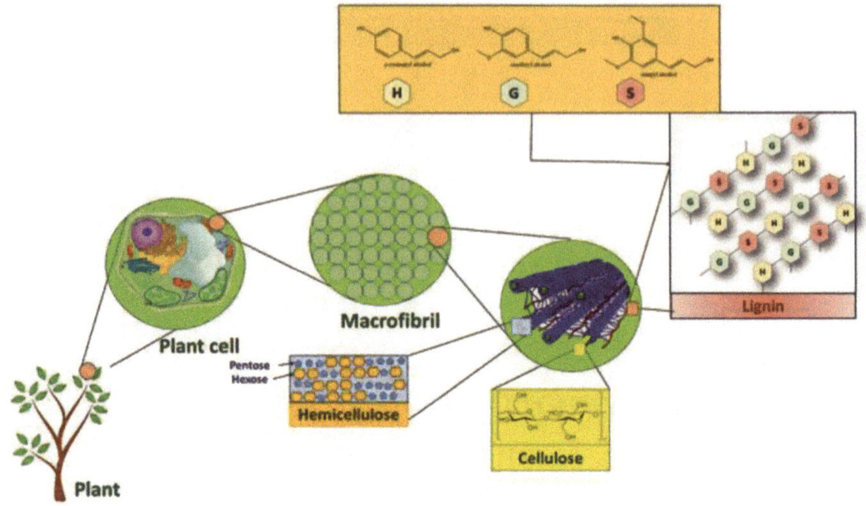

FIGURE 11.1 Structural components of lignocellulosic biomass. *G*, Coniferyl alcohol; *H*, *p*-coumaryl alcohol; *S*, sinapyl alcohol. Source: *Adapted from Saini, J.K., Gupta, R., Verma, A., Gaur, P., Saini, R., Shukla, R., & Kuhad, R.C. (2019). Integrated lignocellulosic biorefinery for sustainable bio-based economy. In* Sustainable approaches for biofuels production technologies *(pp. 25–46). Springer, Cham.*

β-1,4 manner at C2, C3, and C6 and is responsible for the mechanical strength of the plant cell walls. Being homopolymer of glucose, it is a good source of fermentable sugar and utilized for the production of bioethanol and other biochemicals. Cellulose is crystalline in nature (Kumbar et al., 2014). The crystalline fiber structure of the cellulose is the result of superior hydrogen bonds. Though cellulose is majorly present in the plants, few bacteria are also responsible for producing unusual type of cellulose which can also be exploited for various kind of applications. Bacteria belong to genera *Gluconacetobacter*, *Sarcina*, and *Agrobacterium* usually produce cellulose (Costa et al., 2017).

11.1.2 Hemicellulose

Hemicellulose is another component of lignocellulosic biomass and the second most abundant biopolymer in nature. Hemicellulose is heteropolysaccharides with complex structure. Contrasting to cellulose, hemicellulose comprises of different pentoses such as xylose, mannose, arabinose and hexoses like galactose, glucose, fructose as its structural units. Hemicellulose is graded as xylan, mannans, and glucans according to principal sugar residue in the backbone. Different subclasses of hemicellulose can found depending on the species of plant, developmental stages and tissue variance. On the basis of hydration of fibers, these various subtypes can be grouped into two general groups via hydrogen-bonding nitration with cellulose and covalent interaction with lignin, low hydration polysaccharides work to stabilization of the plant cell wall. Due to their branched structure, they are water-soluble. Another form is composed of hydrocolloids, which serves predominately as storage device for extracellular energy and raw material, and also act as mechanism for water-holding in seeds.

11.1.3 Lignin

Lignin is perhaps of the most generous aromatic biopolymers that represent up to 30% of the earth's organic carbon and can be viewed as a viable feasible source for the production of the aromatic chemicals as well as for the energy production. It has an amorphous and highly branched structure which provides structural support to the plant. Lignin is formed by three monomers, guaiacyl (G), syringyl (S), and *p*-hydroxyphenyl (H) units (Poovaiah et al., 2014). Lignin present in the gymnosperms contains almost entirely G unit; G and S units are present in dicotyledonous angiosperms, whereas the monocotyledonous lignin contains all the G, S, and H units (Poovaiah et al., 2014). Ferulates and coumarates units with less quantity are present in the lignin of LCBM (Molinari et al., 2013). The lignin content of softwood, hardwood and grasses is shown in Table 11.1.

11.2 Lignocellulolytic enzyme system

11.2.1 Cellulase

In the environment, lignocellulosic biomass converts into monomeric form via depolymerization process by number of enzymes produced by microorganisms. Hydrolytic enzymes like Cellulases represents three interdependent enzymes complex consisting exoglucanases (cellobiohydrolases, CBH), endoglucanases (EG), and β-glucosidases (cellobiase,

TABLE 11.1 Lignin content in softwood, hardwood, and grasses.

Typical lignin content	
Source	**Quantity of lignin (%)**
Softwood	24–33
Hardwood	19–28
Grasses	15–25

BG) are associated with depolymerization of biomass in collaboration (Behera & Ray, 2016). Upon acting on cellulosic chains, EG makes two kinds of reactive ends for CBHs. CBH I and CBH II act on reducing and nonreducing ends, respectively. Subsequently, cellulose degrades into cellobiose by stepwise catalysis process. BG uses cellobiose and changes over it into glucose (Kuhad et al., 2011). CBH gets restrained by cellobiose, hence; BG assumes a vital job in reduction of final product restraint and depolymerization of cellulose totally. There can be complex or noncomplex microbial cellulase system. Cellulolytic complexes can be produced by both bacteria and fungi. Cellulase is applicable in several industrial and biotechnological applications of cellulose in different industries as shown in Fig. 11.2.

11.2.2 Hemicellulase

To increase chemical delignification, hemicellulases is used to change the structure of xylan and glucomannan in pulp fibers. With different type of bleaching process and pulping processes, this technology can be combined.

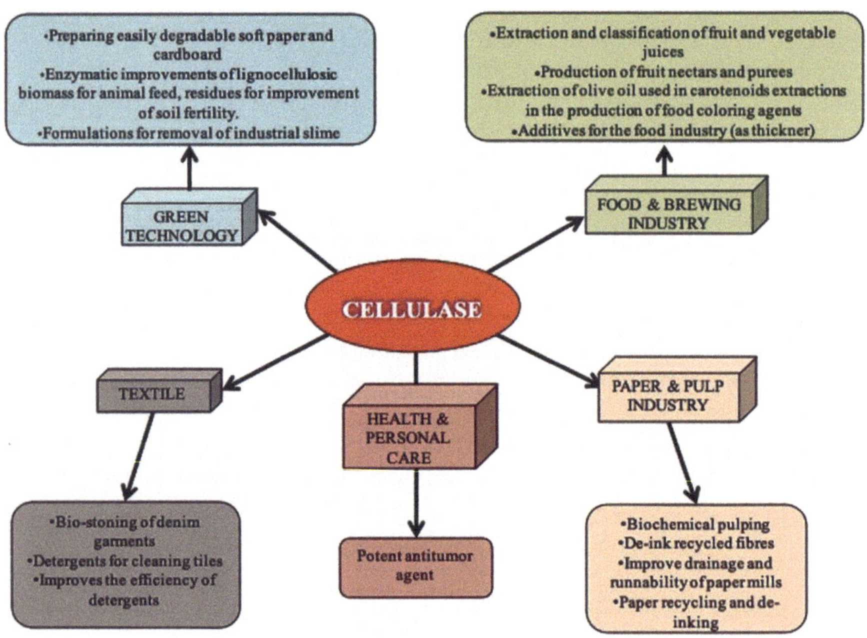

FIGURE 11.2 Industrial and biotechnological applications of cellulase in different industries. Source: *Adapted from Siqueira, J.G.W., Rodrigues, C., de Souza Vandenberghe, L.P., Woiciechowski, A.L., & Soccol, C.R. (2020). Current advances in on-site cellulase production and application on lignocellulosic biomass conversion to biofuels: A review.* Biomass and Bioenergy, ***132**, 105419.*

The goals of enzymatic prebleaching depends on the real condition of the mill and may or may not be linked to environment criteria, decrease in the cost of chemical bleaching and/or improvement in product quality.

The first-scale applications in pulp and paper industry are bleaching with hemicellulase enzymes. Hemicellulase is defined and identified as a group of enzymes. The hemicellulose is generally categorized as D-galactan, D-mannan, D-xylan according to sugar residues that are present, for example, L-arabinans are often associated with many plant sources of pectin polysaccharides, but are usually included in the hemicellulose group. However, most hemicellulose does not occur as homoglycans, but as heteroglycans, comprising various kind of sugar residues, mostly as short appendages connected to main backbone chain.

The definition of "hemicellulases" refer to those enzymes that explicity degrade hemicellulose, and do not contain glycosidases (A-L-arabinofuranosidase, a- and P-D-galactosidases and P-D-mannosidases, and P-D-xylosidases) which in addition to their action on low molecular weight glycosides are also able to hydrolyze the short-chain or monosaccharides appendages from the main hemicellulose backbone chain. Therefore typical hemicellulases are the L-arabinanases, D-galactanases, D-mannanase, and D-xylanases, respectively.

Clarke and coworker demonstrated the preparation of the cell-free enzymes from three *Butyrivibrio* fibrisolvens rumen. The strains were capable of releasing L-arabinose from sugar beet. L-arabinases have been found in terrestrial plants, but it has not been documented that germinating seed produce them. Most L-arabinases of fungal origin are usually secreted extracellularly into the medium in which the organism is grown, but intracellularly L-arabinanases have also been found to exist, for example within the sclerotinia fructigenu mycelial cells.

11.2.3 Accessory lignocellulolytic enzymes

11.2.3.1 Laccases

Laccases are the subsequent group of lignin-degrading and dye-decolorizing peroxidase (DyP). Lignin peroxidase (LiP), manganese peroxidase (MnP), and versatile peroxidase (VP) are part of the superfamily of plant and microbial peroxidases. These are extracellular fungal peroxidases, class II (Karigar & Rao, 2011). LiP are secreted extracellularly as a secondary metabolic product by white-rot fungi. When cosubstrates like hydrogen peroxide and mediator like veratryl alcohol are present, lignin and other phenolic compounds can be broken down by LiP. For the biodegradation of lignin, LiP is most important (Azadfar et al., 2015). LiPs are effective oxidants that have low optimum pH and high redox potential. LiP can oxidize lignin's nonphenolic structures (Fig. 11.3).

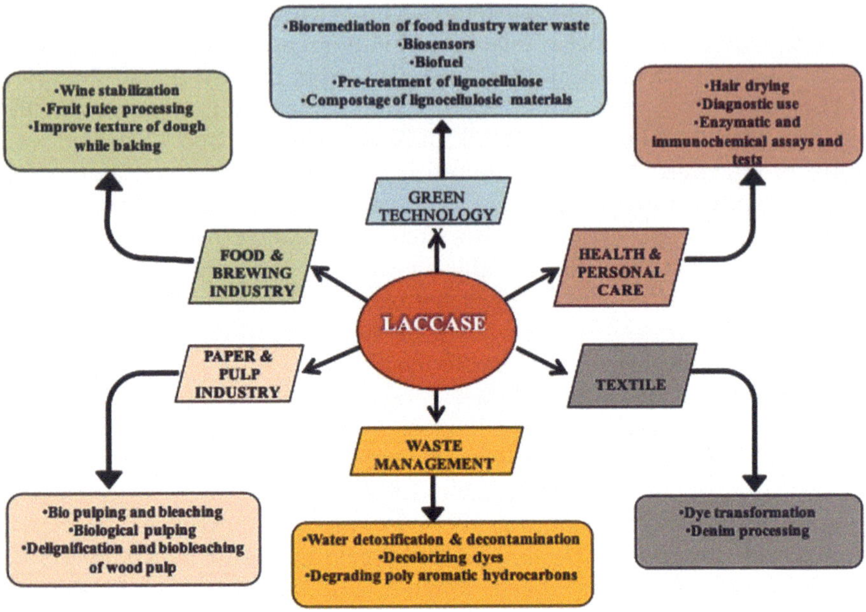

FIGURE 11.3 Industrial and biotechnological applications of laccase. Source: *Adapted from Nunes, C.S., and Kunamneni, A. (2018). Laccases—properties and applications. In* Enzymes in human and animal nutrition *(pp. 133–161). Academic Press.*

11.3 Applications of ligninolytic enzymes

Lignin-degrading enzymes play critical role in biofuel production by using lignocellulosic biomass. Delignification helps in the release of cellulose and hemicellulose, depolymerization of carbohydrate polymers for the production of monomers and their fermentation of sugar into valued-added products, laccase and LiP, MnP, VP, work to substitute chlorine-based delignification in the biobleaching of wood pulp. They can also be used to decolorize textile dye wastewater, and effluent treatment in distilleries (Asgher et al., 2016). They are also involved in the humification of different phenolic substances formed in the soil and are helpful in detoxifying xenobiotics (Fig. 11.4).

11.4 Applications of cellulolytic and xylanase enzymes

Hydrolytic enzymes are in demand as biodegradation of organic chemical compound and polymers that do not easily dissolve in the atmosphere (Kim, 2018). In various areas, such as food and chemical industries, feed additives and biomedical-sciences amylases, proteases, and xylanases are being used (Zhang et al., 2008).

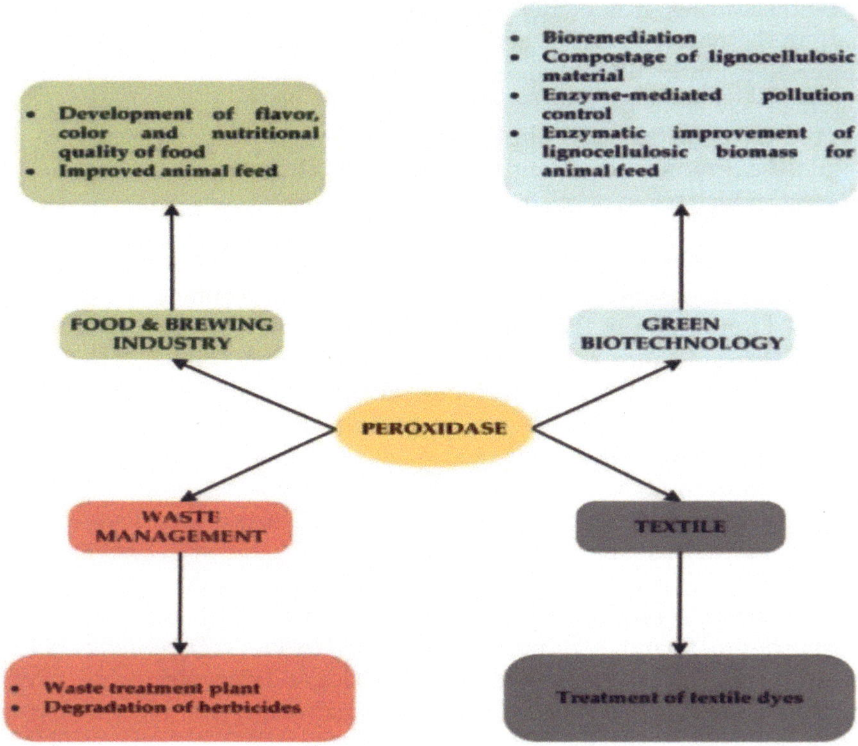

FIGURE 11.4 Industrial and biotechnological applications of peroxidase. Source: *Adapted from Nunes, C.S., and Kunamneni, A. (2018). Laccases—properties and applications. In* Enzymes in human and animal nutrition *(pp. 133–161). Academic Press.*

Xylanases are used in the improvement of bread quality in the treatment of hemicellulose waste; in the clarification of fruit and vegetable juices in the pretreatment of forage crops to increase forage production which increases the digestibility of ruminant feed; ethanol and xylitol manufacturing and degumming of bast fibers. In the bakery, amylases have major applications along with animal nutrition, health care, aquaculture, biofuel, dishwashing and laundry detergent and paper industries.

11.5 Diversity of culturable cellulolytic microbes

11.5.1 Cellulolytic microbes and their habitats

Cellulolytic microorganisms are enticing area of research because of their monstrous potential for production of hydrolytic enzymes.

Notwithstanding, it is evident that more proficient species are still out there in the nature unexplored by researchers. Further investigation and comprehension of the shrouded mechanism behind the action of these enzymes are considerably more significant. Microbial cellulases are favored for their expected applications in a broad scope of industries. In cellulose-degrading ecosystems, a variety of cellulolytic fungi and bacteria exist, which act synergistically.

11.5.1.1 Aerobic microbes (bacteria, fungi, others)

Numerous research were endorsed out for cellulose-degrading enzymes applicable in Biorefinery processes by several microbes counting fungi for instance *Trichoderma* sp., *Penicillium* sp., and *Aspergillus* sp. (Gautam et al., 2012). The oxygen-consuming microbes are typically found in soil, water, plant materials, humus, dung, and leaf litter. As of now a noteworthy assortment of in excess of 14,000 organisms dynamic against cellulose and insoluble strands were Collected (Mandels et al., 1976). *Thermocellum* sp., *Ruminococcus* spp., Clostridium, *Pseudomonas* spp., *Cellulomonas* spp., *Streptomyces* spp., *Serratia* sp., *Bacillus* spp., *Proteus*, *Staphylococcus* spp., and *Bacillus subtilis* are wide variety of bacteria reported to produce cellulase (Gautam et al., 2010). And one of the most examined and significantly high producer is *Trichoderma reesei*, and the sort *Trichoderma* contains 92 species, a large portion of which are unexplored, proposing extraordinary capability for recognizing as novel cellulase-producing strains.

Previously, several investigations were carried out for bacterial assorted variety dependent on metagenomics, culture and revealed different phyla of bacterial, for example, Actinobacteria, Proteobacteria, Firmicutes, Acidobacteria, and Bacteroidetes from the Himalayan mountain range (Gangwar et al., 2009; Gupta et al., 2015; Pradhan et al., 2010; Shivaji et al., 2011; Srinivas et al., 2011). A few examinations uncovered the existence of biocatalyst like lipase, phytase, amylase, cellulase, protease, and so forth from microorganisms of Himalayan areas (Bhat et al., 2013; Gangwar et al., 2009; Pradhan et al., 2010; Sahay et al., 2013; Salwan et al., 2010; Shivaji et al., 2011; Venkatachalam et al., 2015).

Some studies also detailed the presence of cellulose-degrading bacteria from extreme environments, for example, cool dry areas or high elevation (Soares et al., 2012; Venkatachalam et al., 2015). Interior portion of northeastern Brazil and Antarctica show dominancy in several genera of bacteria belonging to Actinobacteria, Firmicutes, Bacteroidetes (Soares et al., 2012) and Firmicutes, Actinobacteria from range of the Nepal and Indian Himalayas were accounted for potential cellulose producers (Venkatachalam et al., 2015). Comprehensively, over the previous years, numerous genera, for instance, *Rhodothermus*, *Micrococcus* sp., *Geobacillus* *Cellulomonas fimi*, *Bacillus* spp., *Burkholderia cepacia*, *Pandoraea pnomenusa*,

Arthrobacter sp., *Pseudomona* spp., *Citrobacter* spp., *Paenibacillus polymyxa*, have been accounted for cellulolytic potential (Behera & Ray, 2016; Hreggvidsson et al., 1996; Kuhad et al., 2011; Sadhu et al., 2011).

Pseudoaltermonas haloplanktis a marine psychrotolerant bacterium isolated Antarctic showed pectinolytic activity with maximum activity at 30°C with pH range of 9–10 on citrus agar plates (Truong et al., 2001, 2001). Not only the psychrophiles, several thermophiles are also known for cellulolytic activity. For instance, *Acidothermus cellulolyticus, Thermobifida fusca, Cellulomonas flavigena* were isolated from hot springs, compost and leaf litter (Abt et al., 2010, 2010; Barabote et al., 2009, 2009; Lykidis et al., 2007, 2007). Diverse range of mesophiles from compost, sludge, marshy area, and soil harbor several species of microbes includes *Bacillus pumilis, Brevibacterium linens, P. polymyxa, Saccharophagus degradans* essential for plant cell wall degradation (Doi, 2008). A gastropod native to East Africa reported with high cellulolytic activity was *Achatina fulica*, for the most part because of the plenitude of glycoside hydrolases delivered by both animal and its resident microbiota. Land snails composed of fascinating and versatile microbiota enhancing the quick bioconversion of lignocellulosic content of plant biomass, with their 60%–80% digestive efficiency (Pinheiro et al., 2015). The presence of individuals of Enterobacteriaceae in the gastrointestinal tract of *A. fulica* showing potential CMCase as well as avicelase activity has been reported by Pawar et al. (2014).

Not only gastropods, but insects are also well known for effective degradation of lignocellulosic biomass as source of energy (Willis et al., 2010). In this, the Isoptera has gained importance because the gut of termites is a rich wellspring of assorted microorganisms, similar to other cellulolytic microscopic organisms, parasites, and protozoans which can effectively debase cellulose and hemicellulose. There have been various reports on the following activity by a number of insects, for instance, *Reticulitermes flavipes* (Zhou et al., 2007) *Pachnoda marginata* (Cazemier et al., 1997), *Dendroctonus frontalis*, and *Saperda vestita* (Delalibera et al., 2005).

Insects living on the ground are not only potential degraders, but even water living phostosynthetic microalga *Chlamydomonas reinhardtii* utilize cellulose as a carbon source by producing endocellulases (Blifernez-Klassen et al., 2012). It has been discovered that the existence of catalytic domains of algae resembles with Glycoside Hydrolase 9 family (conserved sequences found in cellulolytic organisms) (Guerriero et al., 2018). The diversity of aerobic microorganisms is shown in Table 11.2.

11.5.1.2 Anaerobic bacteria, fungi, and others

Not only aerobic microorganisms, anaerobic microbes also proficiently debase agricultural residues by creating an extracellular protein complex called the cellulosomes. The anaerobic microscopic organisms are abundantly isolated from soil, rotting plant components, swamps digestive tracts of

TABLE 11.2 Diversity of aerobic microorganisms.

Group	Microorganism	Habitat	Enzyme activity	References
Bacteria	*Paenibacillus terrae*	Soil, starch, compost	2.08 (CMCase)	Liang et al. (2014)
	Pseudomonas fluorescens	Soil	0.48	Sethi et al. (2013)
	Cellulomonas sp. YJ5	Soil	60.0 (CMCase)	Yin et al. (2010)
	Bacillus sp.	Cowdung	68.0 (CMCase)	Sadhu et al. (2013)
	Paenibacillus sp.	Sugar molasses	2655.0 IU/mg protein (CMCase)	Islam and Roy (2018)
	Bacillus vallismortis	Soil	4105.0 (CMCase)	Gaur and Tiwari (2015)
	Bacillus licheniformis C55	Waste of carrageenan	0.13 (CMCase)	Listyaningrum et al. (2018)
Fungi	*Myceliophthora thermophila* JCP 1−4	Decomposing organic materials (corn cob, sugarcane bagasse silage, fruits, wood, animal feed)	357.51 IU/g (endoglucanase)	Cassia et al. (2015)
			45.42 IU/g (β-glucosidase)	
			931.11 IU/g (xylanase)	
			3.58 IU/g (avicelase)	
	Aspergillus fumigatus M 7.1	Soil, decaying organic matter, Sugarcane piles	7238.0 IU/g (xylanase)	Moretti et al. (2012)
	Myceliophthora thermophila M 7.7		40.40 IU/g (β-glucosidase)	
			1044.60 IU/g (xylanase)	
			54.0 IU/g (CMCase)	
	Trichoderma reesei ZU-02	Purchased from ATCC56764	5.48 (FPase)	Liming and Xueliang (2004)
	Aspergillus fumigatus Z5	Compost of a fertilizer factory	4.94 (β-glucosidase)	Liu et al. (2011)
Actinomycetes	*Streptomyces argenteolus* AE58P	Plant biomass (*Arundo donax, Eucalyptus camaldulensis*, and *Populus nigra*)	0.42 (CMCase)	Ventorino et al. (2016)
	Streptomyces sp.	Marine sediments	0.26 (CMCase)	Fatokun et al. (2016)
			18.54 (Xylanase)	

ruminants, muds, decaying vegetable matter, sludge effluents, gut of termites, wooden chips heaps, fertilizer heaps, and papers mills. To be precisely, well-known genera of anerobic fungi are Anaeromyces, Neocallimastix, Cyllamyces, Orpinomyces, and Caecomyces from Neocallimastigaceae family (Nicholson et al., 2005). A mesophilic bacterium *Clostridium cellulovorans* genome codes for 57 cellulosomal genes and also for proteinase inhibitors, lipases and peptidase notwithstanding two novel genes CbpB and CbpC. The genome development of *C. cellulovorans* incorporated with qualities more identified with deprivation of polysaccharides, for example, hemicellulose and gelatin, compared to celluloses. Rising metagenomic techniques ought to give experiences into the profundity and intricacy of microbial frameworks; such as the rumen of large herbivorous animals contain potential cellulosome producing anaerobic fungi and bacteria (Yang et al., 2020).

Some anaerobic cellulolytic thermophillic bacteria, for example, *Caldicellulosiruptor* sp. produce multidomain cellulases that contain CBMs. These microorganisms have undeniable plant cell wall components degradation scheme that hydrolyzes plant biomass without any pretreatment not at all like most other cellulolytic microorganisms (Blumer-Schuette et al., 2010). Several anaerobic cellulolytic eubacteria exhibit potential Endoglucanase, Exoglucanase, Cellobiohydrolase activity at very extreme temperature from 60°C to 80°C and pH range from 7.0 to 8.0. Microbes like *Clostridium thermocellum, Spirochaeta thermophila, C. thermocellum* (Aksenova et al., 1992), *Thermotoga neapolitana* (Sunna et al., 1997), *Thermoanaerobacter cellulolyticus* (Bredholt et al., 1995), *Dictyoglomus turgidus* (Svetlichnii, 1988), *Clostridium stercorarium* (Riedel et al., 1997), *Thermotoga maritime* (Liebl et al., 1996) are well known for their enzymatic activity.

The previous scarcely any decades have seen an expansion in the enthusiasm for the detachment of cellulolytic actinomycetes. With certain specific bacteria and fungi, actinomycetes are also as known to proficiently debase lignocellulosic biomass, as they produce a large group of extracellular hydrolases, as cellulases, hemicellulases, xylanases, proteases, pectinases and amylases. The cellulase creating actionomycetes have been accounted to belonging with the accompanying genera *Streptomyces, Micromonospora, Nocardia, Microbispora, Streptoverticillium, Actinoplanes, Thermoactinomycetes Saccharopolysopra, Pseudonocardia, Cellulomonas,* and *Thermobifida*. In the majority of the actinomycetes, cellulases are produced extracellularly, utilizing explicit emission pathways, that is, sec general secretion system and sec independent twin-arginine translocation (TAT) frameworks, found in *Thermobifida* (Lykidis et al., 2007).

Mesophilic bacteria such as *Acetivibrio cellulolyticus, B. fibrisolvens, Clostridium cellulolyticum, Eubacterium cellulosolvens, Ruminococcus albus* isolated from rumen of herbivorous, sludge, rot grass are potential producers in category of anaerobic producers (Krause et al., 2003). *C. thermocellum, S. thermophila, T. neapolitana* are thermophilic producers isolated from hot springs (Akinosho et al., 2014). The diversity of anaerobic microorganisms is represented in Table 11.3.

11.5.1.3 Hotspots of cellulolytic microbial diversity

India positions fourth in Asia, and tenth on the planet in plant biodiversity. Being a nation with various geological and environmental highlights, India represents world's microbial diversity 10.6%, 17.9%, 20.1%,

TABLE 11.3 Diversity of anaerobic microorganisms.

Group	Microorganism	Habitat	Cellulolytic activity	References
Bacteria	*Acetivibrio cellulolyticus*	Sewage sludge	NA	Patel et al. (1980)
	Caldicellulosiruptor owensensis DSM 13100	Purchased from the DSMZ	532.4 mU/mg (β-D-glucosidase)	Peng et al. (2015)
	Fervidobacterium islandicum	Icelandic hot spring	NA	
	Thermoanaerobacter cellulolyticus	Icelandic hot spring	NA	Bredholt et al. (1995)
	Dictyoglomus turgidus sp.	Volcano caldera	NA	Svetlichnii et al. (1988)
Fungi	*Anaeromyces mucronatus* KF8	Cow	NA	Novotna et al. (2010)
	Piromyces rhizinflatus	Buffalo	2475.1 U/mg	Chu et al. (2011)
	Neocallirnastix sp.	Sheep	NA	
	Piromyces E2 sp.	Elephant		
	Piromyces R1 sp.	Rhinoceros		

Adapted from German Collection of Microorganisms and Cell Cultures, Braunschweig, Germany.

16.4% of virus, bacteria, algae, fungi, lichen, respectively. India host four biodiversity hotspots viz., the Himalayas, the Western Ghats, the Indo-Burma region, and the Sundaland. Out of which Sundaland is the richest hotspot, but unexplored region in terms of microbes. Screening of undiscovered soils of Western Ghats shows the presence of some significant hydrolytic proteins such as cellulase, lipase, amylase, and protease. And cellulolytic includes several genera, for instance, *Aspergillus*, *Bacillus*, *Streptomyces* well known for lignocellulosic biomass degradation and microbes includes fungal isolates, *Chaetomium* sp., *Penicillium* sp., *Aspergillus* sp., *Cladosporium*, *Trichoderma* sp., predominated. Exploration of Western Ghats hotspot shows a diverse range of hydrolytic enzyme producers as shown in Table 11.3. A huge diverse nature microbial community present in Indo-Burma hotspot include aerobic and cellulolytic microorganisms. A comparative study shows that fungal and actinomycetes species are less than bacterial species from eight different forests in Sikkim, Manipur, Assam, Meghalaya, Arunachal Pradesh, where the annual temperature ranges from 2°C to 37°C. This study shows that many isolates indicated action on the CMC plate and few showed FPAase activity (Goyari et al., 2014). Not only from forest soil, even terrestrial invertebrates have potential cellulolytic activity. Gut microflora of earthworms found in following region which harbors *Streptomyces* sp., *Paenibacillus barcinonensis* CB10, *Aspergillus terreus* CF15, and *Clostridium* sp. FAB6, showed cellulase activity in the range of 0.013–0.036 IU/mL (https://shodhganga.inflibnet.ac.in/bitstream/10603/184032/14/14_summary%20and%20conclusion.pdf). From temperate to extreme environment, a wide range of ecosystem is favored by extremely cold Himalaya.

The Himalayan ranges are the most dynamic mountains, one of the cold environment regions includes Proteobacteria, *Firmicutes*, Bacteroidetes, and Actinobacteria. Phylogenetic investigation includes uncovered plenty of Firmicutes with 50 isolates belonging to following genera in particular *Geobacillus* sp., *Bacillus* sp., *Saccharococcus* sp., *Staphylococcus* sp., *Paenibacillus* sp., *Exiguobacterium* sp. Proteobacteria was second most bountiful phylum with β-proteobacteria (5%), α-proteobacteria (2%), and γ-proteobacteria (27%). This phylum is likewise discovered to be the most different with 12 genera, for example, *Sphingomonas* sp., *Janthinobacterium* sp., *Burkholderia* sp., *Citrobacter* sp., *Enterobacter* sp., *Acinetobacter* sp., *Pseudomonas* sp., *Caulobacter*, *Duganella*, *Stenotrophomonas*, and *Lelliottia*. Also, four genera resembling to phylum Actinobacteria to be specific *Agromyces* sp., *Streptomyces* sp., *Arthrobacter* sp., *Microbacterium* sp., and three from Bacteroidetes, for example, Flavobacterium, Pedobacter, Chryseobacterium. Among the genus, *Pseudomonas* sp. was generally prevailing (22 isolates) trailed by Bacillus (21 isolates), Geobacillus (16 isolates), Arthrobacter (7 isolates), Flavobacterium sp., Paenibacillus sp., and Chryseobacterium sp. each with five isolates (Thakur et al., 2018). The great Himalayan also facilitate the hot temperature, Geothermal valley Manikaran and Yumthang hot springs showed a numerous thermotolerant microbes and producers of thermostable enzymes. Several species of *Bacillus*, *Paenibacillus*, *Thermobacillus* remarkably produce hydrolytic thermostable enzyme, which can tolerate temperature up to 65°C (Sahay et al., 2017). The microbial diversity and enzymatic profile from the Western Ghats is represented in Table 11.4.

11.6 Metagenomic diversity of cellulolytic microbes

Over the past three decades, metagenomic studies have revealed an entirely new "micro" universe with the potential of having intense impacts on countless aspects of the society. Metagenomic is a molecular tool to study the microbial communities of various ecological regions, whose pure cultures cannot be obtained, by analyzing extracted DNA directly from the environmental samples. It also plays role in

TABLE 11.4 Microbial diversity of aerobic soils and enzymatic profile in the Western Ghats.

Source	Microorganism	Enzymatic activity		References
Soil sample of Neyyar Wildlife Sanctuary, Kerala, India	*Streptomyces clavifer*	12.6 U/mL		Sugathan (2015)
Soil sample of Bhadra Wildlife Sanctuary, Chikkamagaluru, Karnataka, India		CMCase (U/mL)	FPase (U/mL)	
	Penicillium pallidum	0.059 + 0.001	0.057 + 0.001	Banakar (2012)
	Aspergillus terreus	0.056 + 0.001	0.054 + 0.005	
	Trichoderma viride	0.052 + 0.003	0.054 + 0.002	
Soil sample of Karnataka, India		Amylase-specific activity (μmol/mL/min)		
	Aspergillus niger	4.08 + 0.09		Mukunda et al. (2013)
	Aspergillus sp. 199	4.98 + 0.06		
	Fusarium sp.	2.92 + 0.10		

prospecting hidden genetic traits by discovering novel genes, pathways, and enzymes with better biochemical function. As traditional microbiological techniques have limitations to study very less portion of cultured microbes from environment, metagenomics facilitates the identification of potent application genes from environmental samples through culture-independent methods. With the advancement of the technique, it has been used to investigate novel cellulase-producing microbial communities. Metagenomic libraries of cellulose digesters from mines, coding cellulase was constructed by Healy et al. (1995). Microbes producing cellulose-degrading enzymes reside in a broad range of environmental niches such as rumen/guts of ruminants, termite/insects guts and other extreme conditions like high saline or temperature (Polizeli et al., 2005).

Discovering novel enzymes via metagenomic approach has massive potential for finding a wide range of beneficial biocatalysts (Sleator et al., 2008). Functional metagenomics has become an everyday method for the discovery of industrially significant enzymes from natural and engineered ecosystems. Various cellulase genes have been found by the functional metagenomic approach (Handelsman, 2005; Vieites et al., 2009). Westmann et al. (2018) mined goat rumen metagenome for discovering new microorganisms, and their gene or enzymes coded by the genes. They have done identification, cloning, sequencing of novel sequences of DNA for cellulose-degrading coding genes from the rumen of goat bacterial metagenomes (Westmann et al., 2018). Recently, Ariaeenejad et al. (2020) mined and discovered a novel alkali and thermostable endo-β-1,4-glucanase from rumen of sheep metagenome. Another study by Ferrer et al. (2005) identified nine endoglucanases, 12 esterases and one cyclodextrin in the rumen of cow metagenome (Ferrer et al., 2005). A study in termites (wood-feeding) by Warnecke et al. (2007) reported to identify the presence of multiple cellulase enzyme families such as GH5, GH8, GH9, GH44, GH45, GH51, and GH74 (endoglucanases), GH1 and GH3 (β-glucosidases) (Warnecke et al., 2007). Higher number of cellulolytic microbes are found in the forest soil compared to garden soil or agricultural land (Houfani et al., 2019). Recently, Wilhelm et al.'s (2019) study of coniferous forest's soil by shotgun metagenomics suggested that fungi showed prominent cellulase production, whereas bacteria (Gram negative) are majorly involved in lignin degradation. *Bacillus licheniformis, Bacillus cereus, B. subtilis, Serratia,* and *Pseudomonas* are widely reported bacterial genera as cellulase producers derived from various agricultural or forest wastes (Srivastava et al., 2018). In a recent study, *B. licheniformis* YNP5-TSU via proteo-metagenomic approach showed upregulation of cellulolytic enzymes under magnesium-rich conditions (O'Hair et al., 2020).

11.7 Metaproteomic analysis of cellulolytic microbes

The fundamental objective of metaproteomic investigation has been to depict the effective role of specific microorganisms in the microbial ecosystem of different microbial community. In recent times, it has been proposed that metaproteomics could be utilized for exploration of microbes to inquire for the most potential enzyme to enhance the choice process of industrially significant enzymes. To simplify the intricacy of metaproteomic sample for focused exploration of enzymes, cellulase-producing microbial communities kept on an artificially chemical characterized medium in a dormant metabolic consistent state. From this stage, it tends to be particularly induction of ideal enzyme activity. To discriminate and characterization of the extracellular portion of protein of induced and noninduced fraction from same community can be compared by differential gel electrophoresis. However, the intention of bioprospecting is for enzymes to cloning, production, and characterization for industrial applications, and it was concluded that recognition against public databases was not adequate to recognize the correct gene or protein sequence for cloning of the identified novel enzymes.

However, for the purpose of bioprospecting for enzymes to clone, produce, and characterize for practical applications, it was concluded that identification against public databases was not sufficient to identify the correct gene or protein sequence for cloning of the identified novel enzymes.

11.8 Metatranscriptomic analysis of cellulolytic microbes

Metatranscriptomic analysis involves sequencing of cDNAs procured from the transcripts and consequently allow deep analysis of active function and of targeted genes, have been done on microbiomes of various ecosystem. This approach is extensively utilized to discover microbiomes in aquatic (Sheik et al., 2014; Shi et al., 2009) and terrestrial ecosystem (de Menezes et al., 2012; Maurice et al., 2013). Huang et al. reported that diverse range of thermophilic dynamic CAZymes, enzyme genes facilitating citric acid cycle, Embden-Meyerhof pathway, Kreb cycle, fermentation at higher temperature, 50°C and 62°C (Huang et al., 2017). Differential gene expression study revealed that thermophilic bacteria *Micromonospora* spp. Overexpressed the enzyme from GH family 48, CBM family 2 and 33 domain. This overexpression in thermophilic microbial group derived from small amount of genes, presently absent in protein sequence databases (Simmons et al., 2014). Another metatranscriptomic analysis from cow rumen representing gene encoding

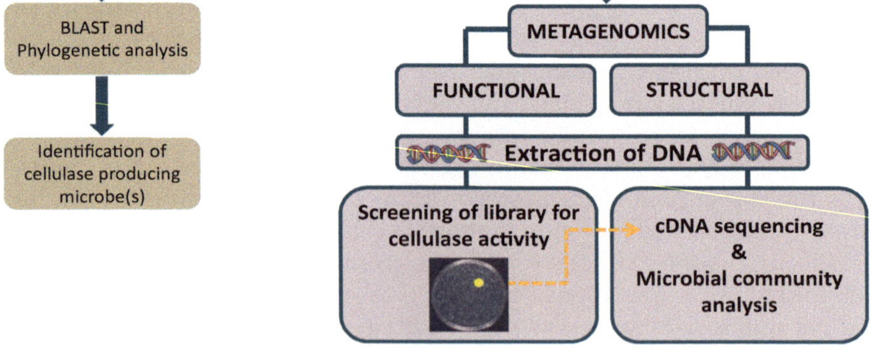

FIGURE 11.5 Approaches in metagenomics for exploring environmental samples for cellulose-producing microbes which cannot be cultured in plate.

cellulases resembled with GH families (i.e., GH5, GH9, GH45, and GH48), producing by *Ruminococcus* sp. and *Fibrobacter* sp. *GH48* cellobiohydrolases transcripts are high in number as compared to other cellulases (Dai et al., 2015). Exploration of metatranscriptomes from microbial community under optimized condition are favorable for bioconversion of lignocellulosic biomass for energy generation is a dominant process for discovery of industrially important novel robust enzymes (Fig. 11.5).

11.9 Conclusion

In this chapter, the cellulolytic microbes, their properties, and characteristics from different various biodiversity hotspots were discussed. The major applications of these ecologically important groups of microbes were also highlighted, specifically the biotechnological potential of cellulolytic enzymes (cellulase, endoglucanase, β-glucosidase, xylanase, etc.). Discussion on culturable and unculturable cellulolytic microbial diversity reveals that more metagenomic and metaproteomic studies are needed to specifically study the diversity and biotechnological importance of the cellulolytic microbes and to establish their role in ecological processes.

References

Abt, B., Foster, B., Lapidus, A., Clum, A., Sun, H., Pukall, R., Lucas, S., del Rio, T. G., Nolan, M., Tice, H., Cheng, J. F., Pitluck, S., Liolios, K., Ivanova, N., Mavromatis, K., Ovchinnikova, G., Pati, A., Goodwin, L., Chen, A., & Klenk, H. P. (2010). Complete genome sequence of *Cellulomonas flavigena* type strain (134 T). *Standards in Genomic Sciences*, *3*(1), 15–25. Available from https://doi.org/10.4056/sigs.1012662.

Akinosho, H., Yee, K., Close, D., & Ragauskas, A. (2014). The emergence of *Clostridium thermocellum* as a high utility candidate for consolidated bioprocessing applications. *Frontiers in Chemistry*, *2*.

Aksenova, H. Y., Rainey, F. A., Janssen, P. H., Zavarzin, G. A., & Morgan, H. W. (1992). *Spirochaeta thermophila* sp. nov., an obligately anaerobic, polysaccharolytic, extremely thermophilic bacterium. *International Journal of Systematic Bacteriology*, *42*(1), 175–177. Available from https://doi.org/10.1099/00207713-42-1-175.

Ariaeenejad, S., Sheykhabdolahzadeh, A., Maleki, M., Kavousi, K., Shahraki, M. F., & Salekdeh, G. H. (2020). A novel high performance metagenome-derived alkali-thermostable endo-β-1, 4-glucanase for lignocellulosic biomass hydrolysis in the harsh conditions. *BMC Biotechnology*, *20*.

Asgher, M., Wahab, A., Bilal, M., & Nasir Iqbal, H. M. (2016). Lignocellulose degradation and production of lignin modifying enzymes by Schizophyllum commune IBL-06 in solid-state fermentation. *Biocatalysis and Agricultural Biotechnology*, *6*, 195–201. Available from https://doi.org/10.1016/j.bcab.2016.04.003.

Azadfar, M., Gao, A. H., & Chen, S. (2015). Structural characterization of lignin: A potential source of antioxidants guaiacol and 4-vinylguaiacol. *International Journal of Biological Macromolecules, 75*, 58–66. Available from https://doi.org/10.1016/j.ijbiomac.2014.12.049.

Banakar, S.P. (2012). Research Article Isolation and Partial Purification of Fungal Cellulases From Forest Soils of Bhadra Wildlife Sanctuary, *Western Ghats of Southern India. 4*, 76–82.

Barabote, R. D., Xie, G., Leu, D. H., Normand, P., Necsulea, A., Daubin, V., Médigue, C., Adney, W. S., Xin, C. X., Lapidus, A., Parales, R. E., Detter, C., Pujic, P., Bruce, D., Lavire, C., Challacombe, J. F., Brettin, T. S., & Berry, A. M. (2009). Complete genome of the cellulolytic thermophile *Acidothermus cellulolyticus* IIB provides insights into its ecophysiological and evolutionary adaptations. *Genome Research, 19*(6), 1033–1042. Available from https://doi.org/10.1101/gr.084848.108.

Behera, S. S., & Ray, R. C. (2016). Solid state fermentation for production of microbial cellulases: Recent advances and improvement strategies. *International Journal of Biological Macromolecules, 86*, 656–669. Available from https://doi.org/10.1016/j.ijbiomac.2015.10.090.

Bhat, A., Riyaz-Ul-Hassan, S., Ahmad, N., Srivastava, N., & Johri, S. (2013). Isolation of cold-active, acidic endocellulase from Ladakh soil by functional metagenomics. *Extremophiles: Life Under Extreme Conditions, 17*(2), 229–239. Available from https://doi.org/10.1007/s00792-012-0510-8.

Blifernez-Klassen, O., Klassen, V., Doebbe, A., Kersting, K., Grimm, P., Wobbe, L., & Kruse, O. (2012). Cellulose degradation and assimilation by the unicellular phototrophic eukaryote *Chlamydomonas reinhardtii*. *Nature Communications, 3*. Available from https://doi.org/10.1038/ncomms2210.

Blumer-Schuette, S. E., Lewis, D. L., & Kelly, R. M. (2010). Phylogenetic, microbiological, and glycoside hydrolase diversities within the extremely thermophilic, plant biomass-degrading genus caldicellulosiruptor. *Applied and Environmental Microbiology, 76*(24), 8084–8092. Available from https://doi.org/10.1128/AEM.01400-10.

Bredholt, S., Mathrani, I. M., & Ahring, B. K. (1995). Extremely thermophilic cellulolytic anaerobes from icelandic hot springs. *Antonie Van Leeuwenhoek, 68*(4), 263–271. Available from https://doi.org/10.1007/BF00874135.

Cazemier, A. E., Den Camp, H. J. M. O., Hackstein, J. H. P., & Vogels, G. D. (1997). Fibre digestion in arthropods. *Comparative Biochemistry and Physiology - A Physiology, 118*(1), 101–109. Available from https://doi.org/10.1016/S0300-9629(96)00443-4.

Costa, A. F. S., Almeida, F. C. G., Vinhas, G. M., & Sarubbo, L. A. (2017). Production of bacterial cellulose by Gluconacetobacter hansenii using corn steep liquor as nutrient sources. *Frontiers in Microbiology, 8*. Available from https://doi.org/10.3389/fmicb.2017.02027.

Cassia Pereira, J., Paganini Marques, N., Rodrigues, A., Brito de Oliveira, T., Boscolo, M., Da Silva, R., Gomes, E., & Bocchini Martins, D. A. (2015). Thermophilic fungi as new sources for production of cellulases and xylanases with potential use in sugarcane bagasse saccharification. *Journal of applied microbiology, 118*(4), 928–939.

Chu, C. Y., Tseng, C. W., Yueh, P. Y., Duan, C. H., & Liu, J. R. (2011). Molecular cloning and characterization of a β-glucanase from Piromyces rhizinflatus. *Journal of bioscience and bioengineering, 111*(5), 541–546.

Dai, X., Tian, Y., Li, J., Su, X., Wang, X., Zhao, S., Liu, L., Luo, Y., Liu, D., Zheng, H., Wang, J., Dong, Z., Hu, S., & Huang, L. (2015). Metatranscriptomic analyses of plant cell wall polysaccharide degradation by microorganisms in the cow rumen. *Applied and Environmental Microbiology, 81*(4), 1375–1386. Available from https://doi.org/10.1128/AEM.03682-14.

De Bhowmick, G., Sarmah, A. K., & Sen, R. (2018). Lignocellulosic biorefinery as a model for sustainable development of biofuels and value added products. *Bioresource Technology, 247*, 1144–1154. Available from https://doi.org/10.1016/j.biortech.2017.09.163.

de Menezes, A., Clipson, N., & Doyle, E. (2012). Comparative metatranscriptomics reveals widespread community responses during phenanthrene degradation in soil. *Environmental Microbiology, 14*(9), 2577–2588. Available from https://doi.org/10.1111/j.1462-2920.2012.02781.x.

Delalibera, I., Handelsman, J., & Raffa, K. F. (2005). Contrasts in cellulolytic activities of gut microorganisms between the wood borer, *Saperda vestita* (Coleoptera: Cerambycidae), and the bark beetles, Ips pini and *Dendroctonus frontalis* (Coleoptera: Curculionidae). *Environmental Entomology, 34*(3), 541–547. Available from https://doi.org/10.1603/0046-225X-34.3.541.

Doi, R. H. (2008). Cellulases of mesophilic microorganisms: Cellulosome and noncellulosome producers. *Annals of the New York Academy of Sciences, 1125*, 267–279. Available from https://doi.org/10.1196/annals.1419.002.

Fatokun, E. N., Nwodo, U. U., & Okoh, A. I. (2016). Classical optimization of cellulase and xylanase production by a marine Streptomyces species. *Applied Sciences, 6*(10), 286.

Ferrer, M., Golyshina, O. V., Chernikova, T. N., Khachane, A. N., Reyes-Duarte, D., Martins Dos Santos, V. A. P., Strompl, C., Elborough, K., Jarvis, G., Neef, A., Yakimov, M. M., Timmis, K. N., & Golyshin, P. N. (2005). Novel hydrolase diversity retrieved from a metagenome library of bovine rumen microflora. *Environmental Microbiology, 7* (12), 1996–2010. Available from https://doi.org/10.1111/j.1462-2920.2005.00920.x.

Gangwar, P., Alam, S. I., Bansod, S., & Singh, L. (2009). Bacterial diversity of soil samples from the western Himalayas, India. *Canadian Journal of Microbiology, 55*(5), 564–577. Available from https://doi.org/10.1139/W09-011.

Gaur, R., & Tiwari, S. (2015). Isolation, production, purification and characterization of an organic-solvent-thermostable alkalophilic cellulase from Bacillus vallismortis RG-07. *BMC Biotechnology, 15*(1), 19.

Gautam, S. P., Bundela, P. S., Pandey, A. K., Awasthi, M. K., & Sarsaiya, S. (2010). Cellulase production by pseudomonas sp. Isolated from municipal solid waste compost. *International Journal of Academic Research, 2*(6).

Gautam, S. P., Bundela, P. S., Pandey, A. K., Jamaluddin., Awasthi, M. K., & Sarsaiya, S. (2012). Diversity of cellulolytic microbes and the biodegradation of municipal solid waste by a potential strain. *International Journal of Microbiology, 2012*. Available from https://doi.org/10.1155/2012/325907.

Goyari, S., Devi, S. S., Kalita, M. C., & Talukdar, N. C. (2014). Population, diversity and characteristics of cellulolytic microorganisms from the Indo-Burma Biodiversity hotspot. *SpringerPlus, 3*(1). Available from https://doi.org/10.1186/2193-1801-3-700.

Guerriero, G., Sergeant, K., Legay, S., Hausman, J. F., Cauchie, H. M., Ahmad, I., & Siddiqui, K. S. (2018). Novel insights from comparative in silico analysis of green microalgal cellulases. *International Journal of Molecular Sciences, 19*(6). Available from https://doi.org/10.3390/ijms19061782.

Gupta, P., Sangwan, N., Lal, R., & Vakhlu, J. (2015). Bacterial diversity of Drass, cold desert in Western Himalaya, and its comparison with Antarctic and Arctic. *Archives of Microbiology, 197*(6), 851–860. Available from https://doi.org/10.1007/s00203-015-1121-4.

Handelsman, J. (2005). Metagenomics: Application of genomics to uncultured microorganisms. *Microbiology and Molecular Biology Reviews, 68*, 195. Available from https://doi.org/10.1128/MMBR.69.1.195.2005.

Healy, F. G., Ray, R. M., Aldrich, H. C., Wilkie, A. C., Ingram, L. O., & Shanmugam, K. T. (1995). Direct isolation of functional genes encoding cellulases from the microbial consortia in a thermophilic, anaerobic digester maintained on lignocellulose. *Applied Microbiology and Biotechnology, 43*(4), 667–674. Available from https://doi.org/10.1007/BF00164771.

Hemansi, G., R. Kuhad, R. C., & Saini, J. K. (2018). Cost effective production of complete cellulase system by newly isolated *Aspergillus niger* RCKH-3 for efficient enzymatic saccharification: Medium engineering by overall evaluation criteria approach (OEC). *Biochemical Engineering Journal, 132*, 182–190. Available from https://doi.org/10.1016/j.bej.2018.01.019.

Houfani, A. A., Větrovský, T., Navarrete, O. U., Štursová, M., Tláskal, V., Beiko, R. G., Boucherba, N., Baldrian, P., Benallaoua, S., & Jorquera, M. A. (2019). Cellulase-hemicellulase activities and bacterial community composition of different soils from algerian ecosystems. *Microbial Ecology, 77*(3), 713–725. Available from https://doi.org/10.1007/s00248-018-1251-8.

Hreggvidsson, G. O., Kaiste, E., Holst, O., Eggertsson, G., Palsdottir, A., & Kristjansson, J. K. (1996). An extremely thermostable cellulase from the thermophilic eubacterium *Rhodothermus marinus*. *Applied and Environmental Microbiology, 62*(8), 3047–3049. Available from https://doi.org/10.1128/0.62.8.3047-3049.1996.

Huang, Y., Yi, Z., Jin, Y., Huang, M., He, K., Liu, D., Luo, H., Zhao, D., He, H., Fang, Y., & Zhao, H. (2017). Metatranscriptomics reveals the functions and enzyme profiles of the microbial community in Chinese Nong-flavor liquor starter. *Frontiers in Microbiology, 8*. Available from https://doi.org/10.3389/fmicb.2017.01747.

Islam, F., & Roy, N. (2018). Screening, purification and characterization of cellulase from cellulase producing bacteria in molasses. *BMC Research Notes, 11*(1), 445.

Karigar, C. S., & Rao, S. S. (2011). Role of microbial enzymes in the bioremediation of pollutants: A review. *Enzyme Research, 2011*(1). Available from https://doi.org/10.4061/2011/805187.

Kim, D. (2018). Physico-chemical conversion of lignocellulose: Inhibitor effects and detoxification strategies: A mini review. *Molecules (Basel, Switzerland), 23*(2). Available from https://doi.org/10.3390/molecules23020309.

Krause, D. O., Denman, S. E., Mackie, R. I., Morrison, M., Rae, A. L., Attwood, G. T., & McSweeney, C. S. (2003). Opportunities to improve fiber degradation in the rumen: Microbiology, ecology, and genomics. *FEMS Microbiology Reviews, 27*(5), 663–693. Available from https://doi.org/10.1016/S0168-6445(03)00072-X.

Kuhad, R. C., Gupta, R., & Singh, A. (2011). Microbial cellulases and their industrial applications. *Enzyme Research, 2011*(1). Available from https://doi.org/10.4061/2011/280696.

Kumbar, S., Laurencin, C., & Deng, M. (2014). *Natural and synthetic biomedical polymers. Natural and synthetic biomedical polymers* (pp. 1–402). Elsevier Inc. Available from https://doi.org/10.1016/C2011-0-07330-1.

Liang, Y. L., Zhang, Z., Wu, M., Wu, Y., & Feng, J. X. (2014). Isolation, screening, and identification of cellulolytic bacteria from natural reserves in the subtropical region of China and optimization of cellulase production by Paenibacillus terrae ME27-1. *BioMed Research International, 2014*.

Liebl, W., Ruile, P., Bronnenmeier, K., Riedel, K., Lottspeich, F., & Greif, I. (1996). Analysis of a *Thermotoga maritima* DNA fragment encoding two similar thermostable cellulases, CelA and CelB, and characterization of the recombinant enzymes. *Microbiology (Reading, England), 142*(9), 2533–2542. Available from https://doi.org/10.1099/00221287-142-9-2533.

Liming, X., & Xueliang, S. (2004). High-yield cellulase production by Trichoderma reesei ZU-02 on corn cob residue. *Bioresource technology, 91*(3), 259–262.

Listyaningrum, N. P., Sutrisno, A., & Wardani, A. K. (2018). Characterization of thermostable cellulase produced by Bacillus strains isolated from solid waste of carrageenan. *E&ES, 131*(1), 012043.

Liu, D., Zhang, R., Yang, X., Wu, H., Xu, D., Tang, Z., & Shen, Q. (2011). Thermostable cellulase production of Aspergillus fumigatus Z5 under solid-state fermentation and its

application in degradation of agricultural wastes. *International Biodeterioration & Biodegradation, 65*(5), 717–725.
Lykidis, A., Mavromatis, K., Ivanova, N., Anderson, I., Land, M., DiBartolo, G., Martinez, M., Lapidus, A., Lucas, S., Copeland, A., Richardson, P., Wilson, D. B., & Kyrpides, N. (2007). Genome sequence and analysis of the soil cellulolytic actinomycete *Thermobifida fusca* YX. *Journal of Bacteriology, 189*(6), 2477–2486. Available from https://doi.org/10.1128/JB.01899-06.
Mandels, M., Andreotti, R., & Roche, C. (1976). Measurement of saccharifying cellulase. *Biotechnology and Bioengineering Symposium, 6.*
Maurice, C. F., Haiser, H. J., & Turnbaugh, P. J. (2013). Xenobiotics shape the physiology and gene expression of the active human gut microbiome. *Cell, 152*(1–2), 39–50. Available from https://doi.org/10.1016/j.cell.2012.10.052.
Moretti, M., Bocchini-Martins, D. A., Silva, R. D., Rodrigues, A., Sette, L. D., & Gomes, E. (2012). Selection of thermophilic and thermotolerant fungi for the productionof cellulases and xylanases under solid-state fermentation. *Brazilian Journal of Microbiology, 43* (3), 1062–1071.
Molinari, H. B. C., Pellny, T. K., Freeman, J., Shewry, P. R., & Mitchell, R. A. C. (2013). Grass cell wall feruloylation: Distribution of bound ferulate and candidate gene expression in *Brachypodium distachyon. Frontiers in Plant Science, 4.* Available from https://doi.org/10.3389/fpls.2013.00050.
Mukunda, S., Onkarappa, R., & Prashith, K. (2013). Isolation and Screening of Industrially Important Fungi from the Soils of Western Ghats of Agumbe and Koppa, Karnataka, India. *Science, Technology and Arts Research Journal, 1*(4), 27. Available from https://doi.org/10.4314/star.v1i4.98816.
Nicholson, M. J., Theodorou, M. K., & Brookman, J. L. (2005). Molecular analysis of the anaerobic rumen fungus *Orpinomyces*—Insights into an AT-rich genome. *Microbiology (Reading, England), 151*(1), 121–133. Available from https://doi.org/10.1099/mic.0.27353-0.
Novotna, Z., Procházka, J., Šimůnek, J., & Fliegerová, K. (2010). Xylanases of anaerobic fungus Anaeromyces mucronatus. *Folia microbiologica, 55*(4), 363–367.
Nunes, C. S., & Kunamneni, A. (2018). *Laccases—properties and applications. Enzymes in human and animal nutrition* (pp. 133–161). Academic Press.
O'Hair JA, Li H, Rangu M, Thapa S, Yong Y, Fish T, et al. (2019). Proteomic Effects of Magnesium Stress on Biofilm Associated Proteins Isolated from Cellulolytic Bacillus licheniformis YNP5-TSU. J Proteomics Bioinform 12:7. https://doi.org/10.35248/0974-276X.19.12.504
Patel, G. B., Khan, A. W., Agnew, B. J., & Colvin, J. R. (1980). Isolation and characterization of an anaerobic, cellulolytic microorganism, Acetivibrio cellulolyticus gen. nov., sp. nov. *International Journal of Systematic and Evolutionary Microbiology, 30*(1), 179–185.
Pawar, K. D., Dar, M. A., Rajput, B. P., & Kulkarni, G. J. (2014). Enrichment and identification of cellulolytic bacteria from the gastrointestinal tract of giant African snail, *Achatina fulica. Applied Biochemistry and Biotechnology, 175*(4), 1971–1980. Available from https://doi.org/10.1007/s12010-014-1379-z.
Peng, X., Qiao, W., Mi, S., Jia, X., Su, H., & Han, Y. (2015). Characterization of hemicellulase and cellulase from the extremely thermophilic bacterium Caldicellulosiruptor owensensis and their potential application for bioconversion of lignocellulosic biomass without pretreatment. *Biotechnology for biofuels, 8*(1), 131.
Pinheiro, G. L., Correa, R. F., Cunha, R. S., Cardoso, A. M., Chaia, C., Clementino, M. M., Garcia, E. S., de Souza, W., & Frasés, S. (2015). Isolation of aerobic cultivable cellulolytic bacteria from different regions of the gastrointestinal tract of giant land snail *Achatina fulica. Frontiers in Microbiology, 6.* Available from https://doi.org/10.3389/fmicb.2015.00860.

Polizeli, M. L. T. M., Rizzatti, A. C. S., Monti, R., Terenzi, H. F., Jorge, J. A., & Amorim, D. S. (2005). Xylanases from fungi: Properties and industrial applications. *Applied Microbiology and Biotechnology, 67*(5), 577–591. Available from https://doi.org/10.1007/s00253-005-1904-7.

Poovaiah, C. R., Nageswara-Rao, M., Soneji, J. R., Baxter, H. L., & Stewart, C. N. (2014). Altered lignin biosynthesis using biotechnology to improve lignocellulosic biofuel feedstocks. *Plant Biotechnology Journal, 12*(9), 1163–1173. Available from https://doi.org/10.1111/pbi.12225.

Pradhan, S., Srinivas, T. N. R., Pindi, P. K., Kishore, H. H., Begum, Z., Singh, P. K., Singh, A. K., Pratibha, M. S., Yasala, A. K., Reddy, G. S. N., & Shivaji, S. (2010). Bacterial biodiversity from Roopkund Glacier, Himalayan mountain ranges, India. *Extremophiles: Life Under Extreme Conditions, 14*(4), 377–395. Available from https://doi.org/10.1007/s00792-010-0318-3.

Riedel, K., Ritter, J., & Bronnenmeier, K. (1997). Synergistic interaction of the *Clostridium stercorarium* cellulases Avicelase I (CelZ) and Avicelase II (CelY) in the degradation of microcrystalline cellulose. *FEMS Microbiology Letters, 147*(2), 239–243. Available from https://doi.org/10.1016/S0378-1097(96)00533-2.

Sadhu, S., Saha, P., Sen, S. K., Mayilraj, S., & Maiti, T. K. (2013). Production, purification and characterization of a novel thermotolerant endoglucanase (CMCase) from Bacillus strain isolated from cow dung. *SpringerPlus, 2*(1), 10.

Sadhu, S., Saha, P., Mayilraj, S., & Maiti, T. K. (2011). Lactose-enhanced cellulase production by *Microbacterium* sp. isolated from fecal matter of Zebra (Equus zebra). *Current Microbiology, 62*(3), 1050–1055. Available from https://doi.org/10.1007/s00284-010-9816-x.

Saini, J. K., Gupta, R., Verma, A., Gaur, P., Saini, R., Shukla, R., & Kuhad, R. C. (2019). *Integrated lignocellulosic biorefinery for sustainable bio-based economy. In* Sustainable approaches for biofuels production technologies (pp. 25–46). Cham: Springer.

Sahay, H., Babu, B. K., Singh, S., Kaushik, R., Saxena, A. K., & Arora, D. K. (2013). Cold-active hydrolases producing bacteria from two different sub-glacial Himalayan lakes. *Journal of Basic Microbiology, 53*(8), 703–714. Available from https://doi.org/10.1002/jobm.201200126.

Sahay, H., Yadav, A. N., Singh, A. K., Singh, S., Kaushik, R., & Saxena, A. K. (2017). Hot springs of Indian Himalayas: Potential sources of microbial diversity and thermostable hydrolytic enzymes. *3 Biotech, 7*(2). Available from https://doi.org/10.1007/s13205-017-0762-1.

Salwan, R., Gulati, A., & Kasana, R. C. (2010). Phylogenetic diversity of alkaline protease-producing psychrotrophic bacteria from glacier and cold environments of Lahaul and Spiti, India. *Journal of Basic Microbiology, 50*(2), 150–159. Available from https://doi.org/10.1002/jobm.200800400.

Sethi, S., Datta, A., Gupta, B. L., & Gupta, S. (2013). Optimization of cellulase production from bacteria isolated from soil. *International Scholarly Research Notices*, 2013.

Sheik, C. S., Jain, S., & Dick, G. J. (2014). Metabolic flexibility of enigmatic SAR324 revealed through metagenomics and metatranscriptomics. *Environmental Microbiology, 16*(1), 304–317. Available from https://doi.org/10.1111/1462-2920.12165.

Shi, Y., Tyson, G. W., & Delong, E. F. (2009). Metatranscriptomics reveals unique microbial small RNAs in the oceans water column. *Nature, 459*(7244), 266–269. Available from https://doi.org/10.1038/nature08055.

Shivaji, S., Pratibha, M. S., Sailaja, B., Hara Kishore, K., Singh, A. K., Begum, Z., Anarasi, U., Prabagaran, S. R., Reddy, G. S. N., & Srinivas, T. N. R. (2011). Bacterial diversity of soil in the vicinity of Pindari glacier, Himalayan mountain ranges, India, using culturable bacteria and soil 16S rRNA gene clones. *Extremophiles: Life Under Extreme Conditions, 15*(1), 1–22. Available from https://doi.org/10.1007/s00792-010-0333-4.

Siqueira, J. G. W., Rodrigues, C., de Souza Vandenberghe, L. P., Woiciechowski, A. L., & Soccol, C. R. (2020). Current advances in on-site cellulase production and application on lignocellulosic biomass conversion to biofuels: A review. *Biomass and Bioenergy, 132,* 105419.

Simmons, C. W., Reddy, A. P., D'haeseleer, P., Khudyakov, J., Billis, K., Pati, A., Simmons, B. A., Singer, S. W., Thelen, M. P., & Vandergheynst, J. S. (2014). Metatranscriptomic analysis of lignocellulolytic microbial communities involved in high-solids decomposition of rice straw. *Biotechnology for Biofuels, 7*(1). Available from https://doi.org/10.1186/s13068-014-0180-0.

Sleator, R. D., Shortall, C., & Hill, C. (2008). Metagenomics. *Letters in Applied Microbiology, 47*(5), 361–366. Available from https://doi.org/10.1111/j.1472-765X.2008.02444.x.

Soares, F. L., Melo, I. S., Dias, A. C. F., & Andreote, F. D. (2012). Cellulolytic bacteria from soils in harsh environments. *World Journal of Microbiology and Biotechnology, 28*(5), 2195–2203. Available from https://doi.org/10.1007/s11274-012-1025-2.

Srinivas, T. N. R., Singh, S. M., Pradhan, S., Pratibha, M. S., Kishore, K. H., Singh, A. K., Begum, Z., Prabagaran, S. R., Reddy, G. S. N., & Shivaji, S. (2011). Comparison of bacterial diversity in proglacial soil from Kafni Glacier, Himalayan Mountain ranges, India, with the bacterial diversity of other glaciers in the world. *Extremophiles: Life Under Extreme Conditions, 15*(6), 673–690. Available from https://doi.org/10.1007/s00792-011-0398-8.

Srivastava, N., Srivastava, M., Mishra, P. K., Gupta, V. K., Molina, G., Rodriguez-Couto, S., Manikanta, A., & Ramteke, P. W. (2018). Applications of fungal cellulases in biofuel production: Advances and limitations. *Renewable and Sustainable Energy Reviews, 82,* 2379–2386. Available from https://doi.org/10.1016/j.rser.2017.08.074.

Sugathan, S. (2015). *Cellulolytic Activity of Streptomyces Clavifer TBG-MNR13 (MTCC 4150), Isolated From Neyyar Wild Life Sanctuary of Kerala, India.* **13**(September).

Sunna, A., Moracci, M., Rossi, M., & Antranikian, G. (1997). Glycosyl hydrolases from hyperthermophiles. *Extremophiles: Life Under Extreme Conditions, 1*(1), 2–13. Available from https://doi.org/10.1007/s007920050009.

Svetlichnii, V. A. (1988). *Dictyoglomus turgidus* sp. nov., a new extremely thermophilic eubacterium isolated from hot springs of the Uzon volcano caldera. *Mikrobiologiya, 57,* 435–441.

Thakur, V., Kumar, V., Kumar, S., & Singh, D. (2018). Diverse culturable bacterial communities with cellulolytic potential revealed from pristine habitat in indian trans-Himalaya. *Canadian Journal of Microbiology, 64*(11), 798–808. Available from https://doi.org/10.1139/cjm-2017-0754.

Truong, L. V., Tuyen, H., Helmke, E., Binh, L. T., & Schweder, T. (2001). Cloning of two pectate lyase genes from the marine Antarctic bacterium *Pseudoalteromonas haloplanktis* strain ANT/505 and characterization of the enzymes. *Extremophiles: Life Under Extreme Conditions, 5*(1), 35–44. Available from https://doi.org/10.1007/s007920000170.

Venkatachalam, S., Gowdaman, V., & Prabagaran, S. R. (2015). Culturable and culture-independent bacterial diversity and the prevalence of cold-adapted enzymes from the Himalayan mountain ranges of India and Nepal. *Microbial Ecology, 69*(3), 472–491. Available from https://doi.org/10.1007/s00248-014-0476-4.

Ventorino, V., Ionata, E., Birolo, L., Montella, S., Marcolongo, L., de Chiaro, A., Espresso, F., Faraco, V., & Pepe, O. (2016). Lignocellulose-adapted endo-cellulase producing Streptomyces strains for bioconversion of cellulose-based materials. *Frontiers in Microbiology, 7,* 2061.

Vieites, J. M., Guazzaroni, M. E., Beloqui, A., Golyshin, P. N., & Ferrer, M. (2009). Metagenomics approaches in systems microbiology. *FEMS Microbiology Reviews, 33*(1), 236–255. Available from https://doi.org/10.1111/j.1574-6976.2008.00152.x.

Warnecke, F., Luginbühl, P., Ivanova, N., Ghassemian, M., Richardson, T. H., Stege, J. T., Cayouette, M., McHardy, A. C., Djordjevic, G., Aboushadi, N., Sorek, R., Tringe, S. G., Podar, M., Martin, H. G., Kunin, V., Dalevi, D., Madejska, J., Kirton, E., Platt, D., &

Leadbetter, J. R. (2007). Metagenomic and functional analysis of hindgut microbiota of a wood-feeding higher termite. *Nature, 450*(7169), 560–565. Available from https://doi.org/10.1038/nature06269.

Westmann, C. A., Alves, L. d F., Silva-Rocha, R., & Guazzaroni, M. E. (2018). Mining novel constitutive promoter elements in soil metagenomic libraries in *Escherichia coli. Frontiers in Microbiology, 9*. Available from https://doi.org/10.3389/fmicb.2018.01344.

Wilhelm, R. C., Singh, R., Eltis, L. D., & Mohn, W. W. (2019). Bacterial contributions to delignification and lignocellulose degradation in forest soils with metagenomic and quantitative stable isotope probing. *ISME Journal, 13*(2), 413–429. Available from https://doi.org/10.1038/s41396-018-0279-6.

Willis, J. D., Oppert, C., & Jurat-Fuentes, J. L. (2010). Methods for discovery and characterization of cellulolytic enzymes from insects. *Insect Science, 17*(3), 184–198. Available from https://doi.org/10.1111/j.1744-7917.2010.01322.x.

Yin, L. J., Huang, P. S., & Lin, H. H. (2010). Isolation of cellulase-producing bacteria and characterization of the cellulase from the isolated bacterium Cellulomonas sp. YJ5. *Journal of Agricultural and Food Chemistry*, 58(17), 9833–9837.

Yang, M., Zhao, J., Yuan, Y., Chen, X., Yang, F., & Li, X. (2020). Comparative metagenomic discovery of the dynamic cellulose-degrading process from a synergistic cellulolytic microbiota. *Cellulose, 28*(4), 2105–2123.

Zhang, H. B., Yang, M. X., & Tu, R. (2008). Unexpectedly high bacterial diversity in decaying wood of a conifer as revealed by a molecular method. *International Biodeterioration and Biodegradation, 62*(4), 471–474. Available from https://doi.org/10.1016/j.ibiod.2008.06.001.

Zhou, X., Smith, J. A., Oi, F. M., Koehler, P. G., Bennett, G. W., & Scharf, M. E. (2007). Correlation of cellulase gene expression and cellulolytic activity throughout the gut of the termite *Reticulitermes flavipes. Gene, 395*(1–2), 29–39. Available from https://doi.org/10.1016/j.gene.2007.01.004.

CHAPTER

12

Biodiversity of cold-adapted extremophiles from Antarctica and their biotechnological potential

Laxmi Jadhav[1], *Vrushali Phalke*[1], *Stutee Panse*[2], *Smita Patil*[3] *and Ashok Bankar*[1]

[1]Department of Microbiology, MES Abasaheb Garware College, Savitribai Phule Pune University, Pune, India, [2]Department of Microbiology, PES Modern College of Arts, Science and Commerce, Savitribai Phule Pune University, Ganeshkhind, Pune, India, [3]Department of Microbiology, Camp Education Society's Arvind B. Telang Senior College of Arts, Commerce and Science, Savitribai Phule Pune University, Pune, India

12.1 Introduction

Most ecosystems on the planet are adapted to temperatures less than 5°C. This is because about 70% surface of earth is occupied by oceans having continuous lower level of temperature (5°C). Further, it is observed that about 18% earth is occupied by regions such as polar, glacier, and alpine. The Antarctic environment is the coldest region with the lowest temperature −89.2°C. Organisms are continuously exposed to the stress conditions such as limited nutrients, desiccation, adverse solar radiations, high salinity, and low biochemical activity (Pearce et al., 2009). It has been seen that about 98% of Antarctica has covered with ice and very less free water is available. Therefore microbes that live in these harsh environmental conditions have developed adaptation strategies such as the production of some specific

Microbial Diversity and Ecology in Hotspots
DOI: https://doi.org/10.1016/B978-0-323-90148-2.00013-4

compounds like antifreeze proteins such as gu3A, gu3B, and afp5A and antifreezing lipopolysaccharides (Muñoz et al., 2017; Pearce et al., 2009). Seas around Antarctica are highly diverse and productive for the growth of phytoplankton that supports the marine food web. Sufficient amounts of nutrients are released into water due to several physical processes. Due to availability of sufficient nutrients in water that supports the growth of diverse groups of organisms. Therefore the Antarctic ecosystem is considered to be one of the most stable ecosystems with high biodiversity (Laybourn-Parry, 2009). Different types of microorganisms that have the ability to thrive in such extreme environments of Antarctica are used to understand the eco-physiological and biochemical characteristics of living forms. Different types of microorganism from all other sources around the globe continuously arrive and adapt in the Antarctic region (Pearce, 2009). The potential of microbes to become Antarctica microflora mostly depend on their competition with the existing microflora and variable evolution rate. Thus microorganisms of Antarctic continent are well known for their ability to thrive and survive in the extreme environmental conditions (Hughes et al., 2007; Pearce, 2009; Pearce et al., 2010).

Due to the small size of bacteria which are not able to insulate themselves and thus follow different adaptation strategies (Huston, 2008). These microbes have some special type of adaptive protein sequence and folding, and membrane lipid composition. The newly formed nascent polypeptide must be properly folded in order to remain functional under cold conditions. These microbes have some special type of adaptive protein sequence and folding. It has seen that psychrophiles having a higher level of serine and a less proline as well as acidic residues in cold adaptive proteins (Isabelle et al., 2018). It has been also reviewed that the bacterial molecular chaperons such as trigger factor, GroEL, Hsc66, CaseinoLytic proteases, and GroES have a crucial role in cold adaptation. These chaperones and nascent polypeptides interact with each other in order to regulate the cotranslational protein folding (Singh et al., 2015). These microbes also have some modified membrane lipid compositions in order to adapt to cold conditions. In bacteria, the alteration in fatty acids composition may change the bacterial membrane fluidity. At low temperature, bacteria can change the fluidity of their membrane by enhancing concentration of *cis*-vaccenic acid with palmitoleic acid elongation process (Garwin & Cronan, 1980). Incorporation of branched-chain fatty acids into lipid bilayer also alters the fluidity of the membrane. At low temperature, level of anteiso-branched-chain fatty acids have been enhanced while iso-branched-chain fatty acids have been reduced (Suutari & Laakso, 1992).

Because of different adaptation strategies and broad metabolic capabilities of these microbes, this has made them useful in bioremediation and production of several enzymes (Pereira et al., 2017). Thus cold adaptive proteins or enzymes like cellulases, lipases, and proteases obtained from different psychrophiles are of high importance in the chemical industry, food industry, and textile industry (Pearce, 2009; Shahidi & Janak Kamil, 2001). In this chapter, we have discussed the potential of psychrophiles for various biotechnological applications.

12.2 Biodiversity of extremophiles from Antarctica

The continent of Antarctica having size of 5.5 million square miles is measured as the largest part of the earth. It consists different regions having varied physiological conditions. Therefore microbial diversity is commonly observed among the Antarctic ecosystems (Laybourn-Parry, 2009). There are several factors that contribute to the isolation of Antarctic microorganisms from the rest of the biosphere. The dominant microbial species are highly diverse. The microbiological study of Antarctic soil and air has been studied (Ekelöf, 1908) by several researchers (Gazert, 1901; Pirie, 1904; Tsiklinsky, 1908). Initially, identification of these microorganisms was difficult only based on the culture-dependent methods. However, initially identification of microbes by rRNA gene sequencing that helps to discover new species by culture-independent manner. Further, several groups of microorganisms have identified in Antarctica including eubacteria, cyanobacteria, fungus, lichens, and viruses (Pearce, 2017).

12.2.1 Bacteria

Bacterial population obtained from Antarctica is classified into different categories like obligate psychrophilic bacteria, psychrotolerant bacteria, and psychrotrophs based on the growth conditions observed. The bacterial diversity of several habitats like water, ice core, soil, and sediments have been studied by using noncultivable methodologies like rRNA gene sequencing reported previously (Bottos et al., 2014; Shivaji & Reddy, 2010; Shivaji et al., 2017). Several studies revealed that the Antarctic soils consist of diverse groups of microbes such as Actinobacteria, Proteobacteria, and Bacteroidetes reported earlier (Bottos et al., 2014). The bacterial cell count is influenced by total water, nitrogen, and organic content of the soil (Bailey & Wynn-Williams, 1982). Other soil properties such as pH, moisture, nutrients, carbon, copper, and lead content also have an influence on bacterial populations as reported previously (Aislabie et al., 2009; Barrett et al., 2006; Chong et al., 2009; Wynn-Williams, 1996). Researchers have studied the biodiversity of soil prokaryotes from 15 different regions of Signy Isaland, Maritime Antarctic (Chong et al., 2009). The study showed that most of dominant species belong to the phylum Proteobacteria, Bacteroidetes, and Actinobacteria. Some rare phyla belong to Planctomycetes, Firmicutes, and Spirochetes reported previously (Papale et al., 2019). After a comparative study of different habitats (table can be more appropriate), it was observed that soil, water, and algal mats are found to be more diverse when compared with sediments and ice (Shivaji et al., 2017).

12.2.2 Viruses

Viruses are found in almost all habitats on the earth and thus they are also found in polar aquatic systems (De Cárcer et al., 2015; Säwström et al., 2007). However, viruses are less abundant when compared to the bacterial population (Dickinson et al., 2016). In Antarctic waters, small circular ssDNA viruses and dsDNA viruses have been identified. There are some viruses having potential to change their genetic material from a ssDNA to dsDNA when ice-covered lakes are converted to an open-water lake in summer season (López-Bueno et al., 2009; Zawar-Reza et al., 2014). Along with the aquatic ecosystems, some of the Aquatic penguin species have been screened for identification of viruses studied previously (Wille et al., 2019, 2020).

12.2.3 Fungi

Some fungal species are very well adjusted with the extreme conditions like low water, low nutrients, cold temperature, and low rates of metabolite transfer reported earlier (Jansson, 2014; Steven et al., 2006). It is studied that fungi inhabiting the permafrost are mostly present in dormant state and thus they may extend the survival periods in permafrost (Ozerskaya et al., 2009). Researchers have identified about 213 and 351 fungi were isolated from permafrost and Maritime Antarctica, respectively. Dominant groups of fungi such as *Penicillium, Oidiodendron, Pseudogymnoascus, Bionectriaceae, Helotiales, Mortierellaceae,* and *Pseudeurotium* have been identified (da Silva et al., 2020). Some other genera also found in Antarctica are *Codophora, Neurospora, Aspergillus,* and *Pseudodeflectus* (Albores et al., 2018). The yeast strains such as *Mrakia blollopis, M. robertii,* and *Mrakiella niccombsii* are found in the Antarctic glaciers (Thomas-Hall et al., 2010). Culturable microfungal diversity has been well understood from the water column of ice-covered lake in Taylor Valley. It is observed that the abundance of culturable fungi decreases with increasing depth of water with concurrent shift in diversity. This revealed that the spatial discontinuities of fungi have been observed within oligotrophic aquatic environments (Connell et al., 2018).

12.2.4 Cyanobacteria

It has been reported that Cyanobacteria plays a very valuable role in carbon and nitrogen fixation in the soils of Antarctica. The cold-adapted microorganisms are recognized for their ability of coloring snow.

Ankistrodesmus, Chlamydomonas nivalis, Chloromonas, Raphidonema, Mycanthococcus, Oscillatoria priestleyi, Phormidium murrayi, and some dinoflagellates are well known for their ability to color the snow in Antarctica as reported previously (Prescott, 1978; Quesada & Vincent, 1997). It is also reported that biological pigments are important for cyanobacteria to survive in the harsh conditions of Antarctica. Some species of cyanobacteria are reported for their potential to produce capsular polysaccharides and released polysaccharides reported in previous study (Nicolaus et al., 2000). Cyanobacteria are well adapted to the cold environment that have the ability of nitrogen fixation, nitrate uptake, ammonium-uptake, nitrate reduction, and photosynthesis at cold conditions (Pandey et al., 2004). Thus the global glaciations have been affected due to the cyanobacterial populations.

12.2.5 Archaea

Archaeal species belonging to the phylum Crenarchaeota, Korarchaeota, and Euryarchaeota are found in Antarctic habitats (Karr et al., 2006). The archaeal population is classified into two groups, namely, methanogens and halophiles. The methanogenic and methanotrophic archaea are found abundantly in the Antarctic habitats. Commonly occurring methanogens such as *Methanococcoides burtonii, Methanosarcina mazei, Methanosarcina barkeri, Methanosarcina acetivorans, Methanobrevibacter smithii*, and *Methanogenium frigidum* have been reported previously (Franzmann et al., 1997, 1988). Some halophiles such as *Halogeometricum borinquense, Haloarcula marismortui, Natronomonas pharaonis, Halobaculum* spp., and *Halorhabdus* spp. are also been found to be inhabitants of Antarctica (Papale et al., 2019). Psychrophilic archaea such as *Sphingopyxis alaskensis, Cenarchaeum symbiosum*, and *Halorubrum lacusprofundi* are also well studied earlier (Cavicchioli et al., 2003; Franzmann et al., 1988; Schleper et al., 2005).

12.3 Potential hotspots in Antarctica

The psychrophilic microbes have been widely spread around different regions of the world like ponds, mountains, and ice shelves. In some regions, it was observed that psychrophilic microbes are found to be more prevalent than others. Such international locations are well known such as Hindkush Karakoram Himalaya, European Alps, Arctic Ocean, and Svalbard (Hassan et al., 2016). Antarctic microbial hotspots are spread over the continent. In Table 12.1, some prevalent areas are listed as well as their locations has been shown in the map (Fig. 12.1).

It has been seen that East Antarctica, Antarctic peninsula, and Victoria Land are the most diverse and prevalent hotspots in the continent. It was

TABLE 12.1 Distribution of psychrophilic microorganisms in different regions of Antarctica.

Sr no.	Region/location	Psychrophilic microbe	Genus/species	Reference
1	Schirmacher Oasis	Fungi	Microbes belonging to genera—*Acremonium, Aspergillus, Cladosporium, Fusarium*, and *Trichoderma*	Singh et al. (2006)
2	Skarvsnes ice-free area	Fungi	Microbes from genera Ascomycetous, basidiomycetous. Including *Embellisia, Phoma, Geomyces, Mrakia, Cryptococcus, Dioszegia, Rhodotorula*	Tsuji et al. (2013)
3	Lakes in Stornes Peninsula, Larsemann Hills	Bacteria	Members of *Pseudomonas, Staphylococcus, Salmonella* species	Chauhan et al. (2015)
4	Davis Mossel Lake area	Bryophytes associated fungi	*Pseudogymnoascus pannorum, Phoma herbarum*	Kerry (1990)
5	Ace Lake, Vestfold Hills	Archaea	*Methanogenium frigidum*	Franzmann et al. (1997)
6	Vestfold Hills, Mawson Rock, and nearby localities	Algae	*Chroococcidiopsis* sp., *Myxosarcina* sp., *Gloeotheca* sp., *Trochiscia* sp., *Chlorella* sp.	Broady (1981)
7	Mitchell peninsula, East Antarctica	Bacteria	40 Phyla including 111 classes including *Chloroflexi, Candidate* division *WPS-2, Bacteroidetes, Firmicutes*	
8	Mitchell peninsula, East Antarctica	Fungi	5 Phyla with 22 classes including Ascomycota, Basidiomycota	Ji et al. (2016)
9	Antarctic continental shelf area within the Mertz Glacier Polynya (MGP) region	Bacteria	Gamma-Proteobacteria, delta- and epsilon-Proteobacteria, Flavobacteria, green sulfur bacteria	Bowman and McCuaig (2003)
10	Antarctic continental shelf area within the Mertz Glacier Polynya (MGP) region	Archaebacteria	Euryarchaeota, Crenarchaeota	Bowman and McCuaig (2003)
11	Northern Victoria Land	Algae	*Chlorophyta, Cyanobacteria, Xanthophyta, Bacillariophyta*	Cavacini (2001)
12	Balleny Islands	Algae	*Phormidium priestley, Chlamydomonas hallenyana* KDl sp. nov., *Bracteacoccus minor*	Kol and Flint (1968)

(*Continued*)

TABLE 12.1 (Continued)

Sr no.	Region/location	Psychrophilic microbe	Genus/species	Reference
13	Brine pockets in Tarn Flat area, Northern Victoria Land	Algae	Species belonging to Proteobacteria (genus *Thiomicrospora, Psychrobacter*), Bacteroidetes (genus *Ulvibacter*), Actinobacteria, Spirochetes, Firmicutes, Planctomycetes	Papale et al. (2019)
14	Brine pockets in Tarn Flat area, Northern Victoria Land	Archaeabacteria	Members of Euryarchaeota, Crenarchaeota, Korarchaeota	Papale et al. (2019)
15	Brine pockets in Tarn Flat area, Northern Victoria Land	Virus	Virus-like particles	Papale et al. (2019)
16	Ponds in McMurdo Iceshelf	Algae	Benthic cyanobacteria	Nadeau and Castenholz (2000)
17	Lake Wellman area, Darwin Mountains	Bacteria	Members of genera Arthrobacter, Adhaeribacter, Pontibacter	Aislabie et al. (2013)
18	Northern shore of Miers Lake (ML)	Bacteria	Members of Acidobacteria, Actinobacteria	Niederberger et al. (2015)
19	Victoria Land, Continental Antarctica	Fungi	33 Fungal strains including *Verticillium, Phoma, Aspergillus, Geomyces*	Fenice et al. (1997)
20	Shackleton's Historic Hut on Cape Royds	Fungi	*Cladosporium, Chaetomium, Pezizomycotina, Onygenales, Nectriaceae*	Blanchette et al. (2010)
21	La Gorce Mountains	Algae	*Phormidium autumnale, Leptolyngbya fragilis, Desrnococcus* cf. *alivaceus*	Broady and Weinstein (1998)
22	Deception Island Volcano	Bacteria	*Arthrobacter, Flavobacterium, Sphingomonas, Psychrobacter*	Bendia et al. (2018)
23	Island Livingston	Fungi	Filamenous fungi including *Mucor, Cladosporium, Alternaria, Aspergillus, Penicillium*	Kostadinova et al. (2009)
24	Greenwich Island, Troub	Fungi	*Penicillium rubens, Penicillium tardochrysogenum, Penicillium chrysogenum*	de Menezes et al. (2020)
25	Barrientos Island	Actinomycetes	*Actinomycete, Streptomyces, Leifsonia, Frankineae, Rhodococcus*, and *Mycobacterium*	Learn-Han et al. (2012)
26	King George Island	Bryophytes associated fungi	*Mortierella gamsii, Phoma* sp., *Chalara* sp., *Phialophora malorum, Monascella* sp., *Pycnostysanus* sp., *Chaunopycnis alba*	Möller and Dreyfuss (1996)

(*Continued*)

TABLE 12.1 (Continued)

Sr no.	Region/location	Psychrophilic microbe	Genus/species	Reference
27	Signy Island terrestrial soils	Actinomycetes	*Demetria*, *Glaciibacter*, *Kocuria*, *Marmoricola*, *Nakamurella*, and *Tsukamurella*	Pan et al. (2013)
28	Collins Glacier	Bacteria	Beta-Proteobacteria, Gamma-Proteobacteria, Alpha-Proteobacteria, *Cytophaga-9*, *Flavobacterium*	García-Echauri et al. (2011)
29	Ellsworth Mountains	Fungi	Members of *Aspergillus*, *Debaryomyces*, *Cladosporium*, *Pseudogymnoascus*, *Penicillium*, and *Hypocreales*	Godinho et al. (2015)

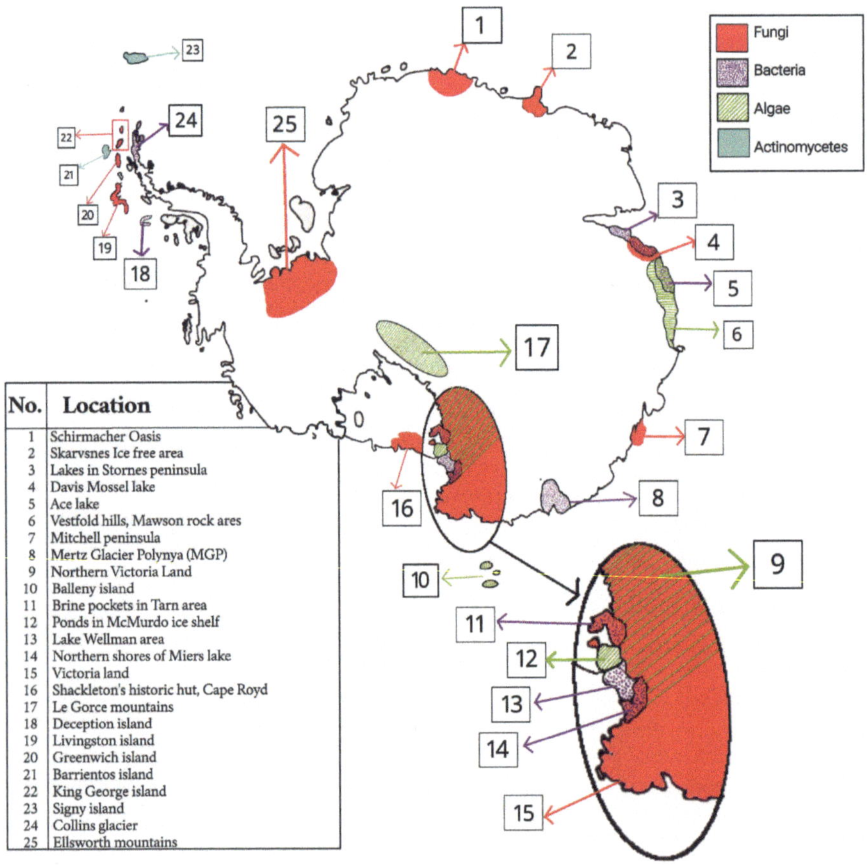

FIGURE 12.1 Distribution of psychrophilic microbes in different regions of Antarctica.

also seen that fungi, bacteria, actinomycetes, and algae are abundant in some of these spots as shown in Table 12.1 (Singh et al., 2015).

12.4 Cold adaption mechanisms in psychrophiles

Studies on different mechanisms of cold-adapted microbes are of high importance in order to exploit them for several biotechnological applications (Wang et al., 2017). The cell membrane composition plays a vital role to survive under different temperature ranges. The psychrophiles have different cell membrane composition as compared to mesophiles and thermophiles. These psychrophiles have the ability to produce a variety of fatty acids like C15:1 iso G, C15:0 iso, C15:0 anteiso, C15:1 ω6c, C15:0 iso 3 OH, C16:0 iso 3OH, and C17:0 iso 3 OH. Thus branched and unsaturated fatty acids are commonly present in the cell membranes of pyschrophiles (Králová, 2017). The psychrophiles adapted in ice-rich area are continuously subjected to the additional stresses. These psychrophiles are polysaccharide-rich extracellular polymeric substances produced in order to protect mechanical disruption due to ice particles (Junge et al., 2004; Thomas & Dieckmann, 2002).

The psychrophiles also showed the potential to survive in the extreme cold environments by regulating their energy conservation and biosynthetic pathways (Amato & Christner, 2009). The cold adaptive classes of enzymes, such as glucanases, hydrogenases, isomerases, hydrolases, and nucleic acid-modifying enzymes, are found functional at low temperature (Adams & Kelly, 1994; Demirjian et al., 2001; Siddiqui & Cavicchioli, 2006). Thus proteins of psychrophiles also play a crucial role in cold adaptation mechanisms. It is also reported that cold shock proteins (Csps) are constitutively secreted by psychrophiles and thus known as cold adaptive proteins (D'Amico et al., 2006). Study also revealed that the overexpression of Csps gene shows the increased cold resistance (Jung et al., 2010). Thus it is studied that Csps along with some additional small proteins acts as RNA chaperons that have a crucial role in stabilizing the secondary structure of transcripts (Jones & Inouye, 1994; Williams et al., 2010). Thus there are different strategies seen in the psychrophiles that have the ability to thrive in cold conditions as shown in Fig. 12.2 (De Maayer et al., 2014).

12.5 Biotechnological applications of psychrophiles from Antarctica

Due to cold tolerance ability of Antarctic psychrophiles, they have several biotechnological applications (Buzzini et al., 2012). Psychrophilic

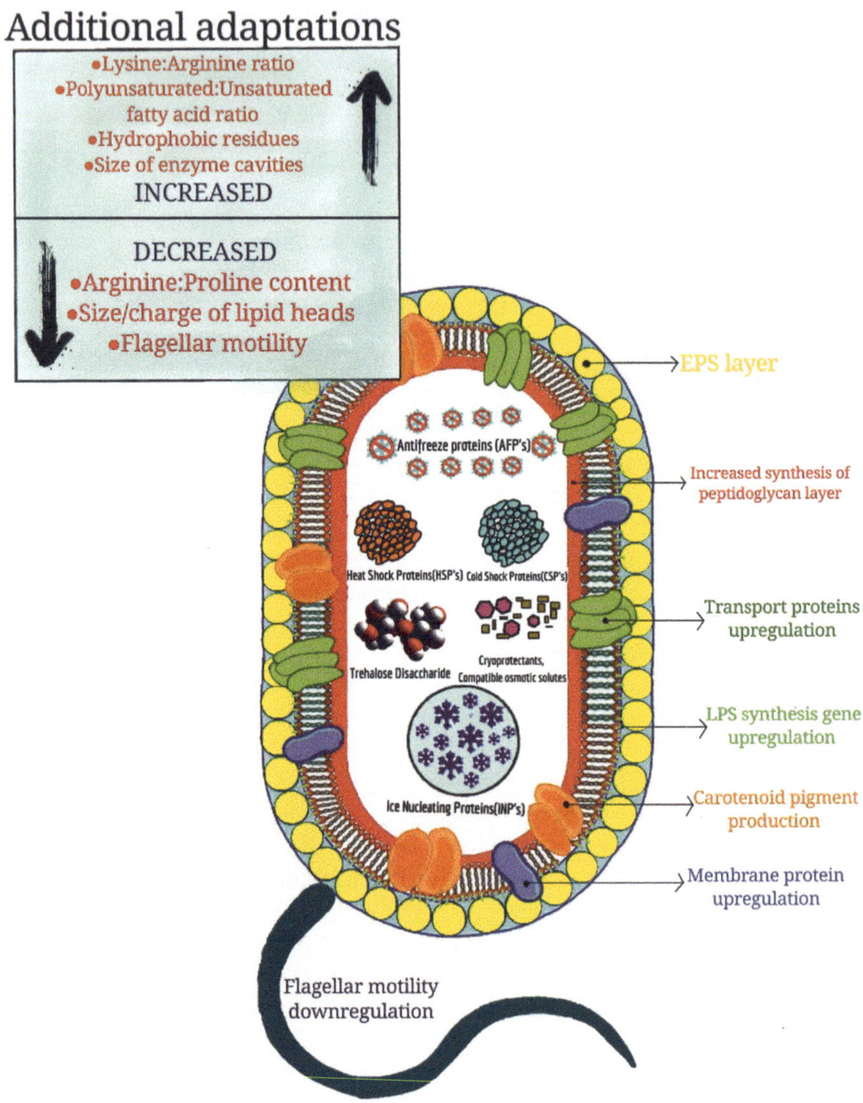

FIGURE 12.2 Different cold adaptation mechanisms in psychrophiles.

bacteria, yeast, and fungi have been used for their biotechnological applications as shown in Fig. 12.3.

Psychrophiles are considered to be more stable than (why do you think they are more stable than other extremophiles) thermophiles and mesophiles and thus preferred for production of several industrial value-added products. Some major fields of interest are as follows: (1)

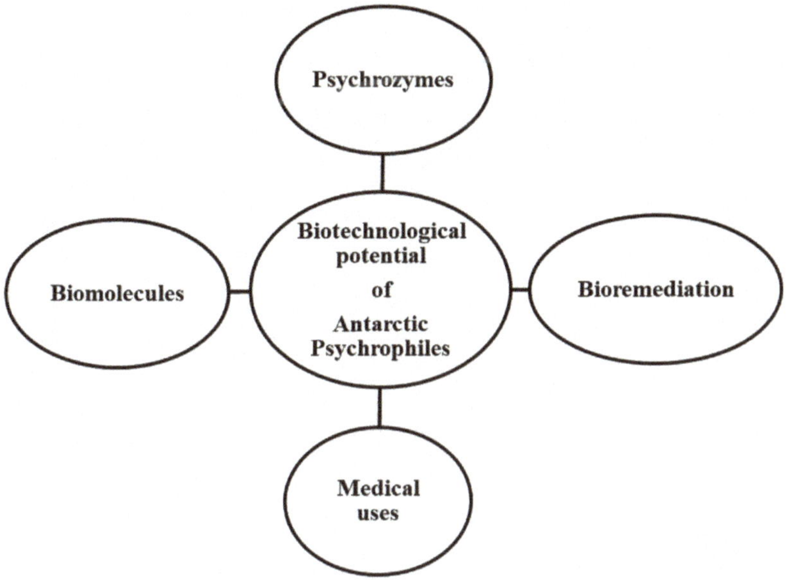

FIGURE 12.3 Biotechnological applications of psychrophiles isolated from Antarctic regions.

Enzymes: Psychrophilic enzymes are commonly used in many industrial processes including food processing and laboratory reagents preparations. (2) *Biomolecules*: polyunsaturated fatty acids (PUFAs) antifreeze proteins obtained from psychrophiles are mostly employed as food additives and dietary supplements. (3) *Pharmaceuticals and medical uses*: Many antibiotics and anticancer drugs have been obtained from Antarctic psychrophiles. Thus they have high demands in cosmetics and nutraceutical companies. (4) *Bioremediation*: Cold-adapted microbes such as *Arthrobacter psychrolactophilus*, *Rhodococcus* spp., and *Rhodotorula psychrophenolica* have been used for bioremediation of oil spills in frozen oceans that have been reviewed earlier (Margesin & Feller, 2010). Products obtained from Antarctic psychrophiles have been enlisted in Table 12.2.

12.5.1 Enzymes (psychrozymes)

Enzymes obtained from the psychrophiles are called psychrozymes. The psychrophiles have been reported for their potential to produce different types of enzymes such as amylases, lipases, proteases, and cellulases in order to survive in cold environments. These psychrozymes have several applications in chemical, biochemical, agricultural, and

TABLE 12.2 Cold-active enzymes/products of psychrophiles isolated from Antarctic cold environments.

Sr no.	Cold adaptive enzymes/products	Name of organisms	Origin of source	Applications	Reference
1	Pectinase	*Geomyces* sp. strain F09-T3–2	Marine sponges in Antarctica	Food and beverage industries	Poveda et al. (2018)
2	β-Galactosidase	*Pseudoalteromonas haloplanktis* TAE 79	Antarctic marine	Lactose hydrolysis at cold conditions during milk storage	Hoyoux et al. (2001)
3	β-Galactosidase	*Tausonia pullulans* 17-1	Antarctic cold environment	Production of lactose-free or low-lactose products	Pulicherla et al. (2011), Zhang et al. (2012)
4	Protease	*Pseudoalteromonas* sp. NJ276	Antarctic marine	Food processing at low temperature	Wang et al. (2008)
5	Protease	*Colwellia* sp. NJ341	Antarctic sea-ice	Food industry	Wang et al. (2005)
6	Cellulase	*Pseudoalteromonas haloplanktis*	Antarctic seawater	Additive in detergent industry	Violot et al. (2005)
7	Xylanase	*Cladosporium* sp.	Marine sponges of King George Island, Antarctica	Several applications in probiotic inclusion as well as a potential therapeutic agent in food	Gomes et al. (2000), Petrescu et al. (2000)
8	α-Amylase M175	*Pseudoalteromonas* sp.	Antarctic sea-ice	Commercial detergents	Wang et al. (2018)
9	Lipase	*Bacillus pumilus* ArcL5	Arctic seawater	Detergent additives used at low temperatures and biocatalysts for the biotransformation of heat-labile compounds	Wi et al. (2014)
10	Antifreeze protein	*Pyramimonas gelidicola*	Antarctic marine	Food industries	Jung et al. (2014)
11	Polyunsaturated fatty acids	*Colwellia* sp.	Coastal Antarctica	Production of docosahexaenoic acid (omega 3)	Bowman et al. (1998)
12	Cell biomass	*Aspergillus fumigatus*	Antarctic cold environment	Phenol and its derivatives degradation	Gerginova et al. (2013)

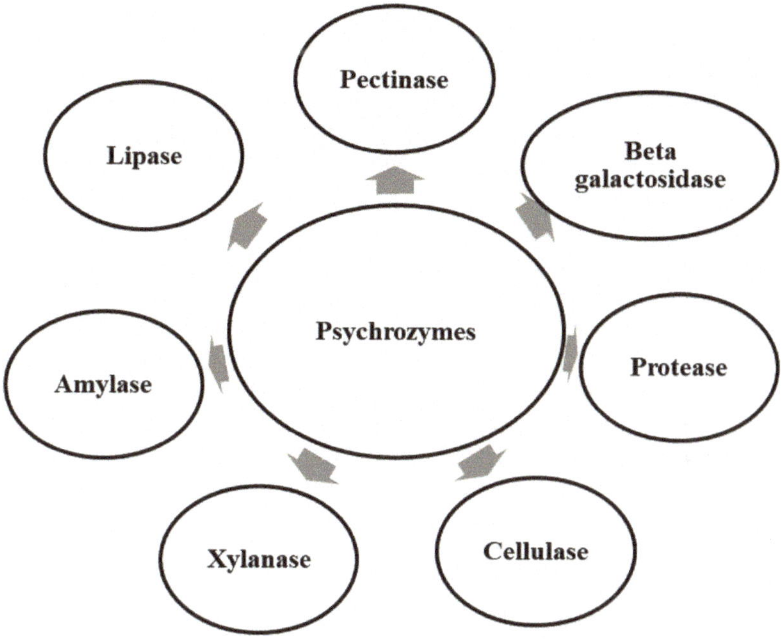

FIGURE 12.4 Psychrozyme production from the psychrophiles isolated from Antarctic environments.

biotechnological processes (Dumorné et al., 2017). In psychrozymes, the weak interactions such as hydrogen bonds, aromatic interactions, and helix dipoles are absent and therefore these enzymes are more stable and found useful for industrial processes. The psychrozymes have high specific activity at cold conditions, easy to inactivate, easy to recover, and less chances of contamination. Due to all these special properties, **psychrozymes become** promising **candidates** for several biotechnological applications (Pulicherla et al., 2011). There are some industrially important psychrozymes as shown in Fig. 12.4.

12.5.1.1 *Pectinases*

The pectinases are enzymes that degrade pectic substances. Therefore pectinases are mostly useful in many processes like fruit juice preparation, coffee, and tea processing as well as vegetable oil extraction. Raw fruit juices are treated with pectinases for removal of insoluble pectic particles. Thus degradation of the pectic particles to reduce the viscosity of the juice is the best strategy used in juice industries. The fruit juice processing is mostly performed at low temperature and therefore cold adaptive pectinases obtained from psychrophiles are preferred than the mesophilic pectinases (Poveda et al., 2018; Pulicherla et al., 2011). The

Geomyces sp. strain F09-T3−2 with pectinase activity was obtained from marine sponges of Antarctica. This strain has been used in beverage as well as food industries (Poveda et al., 2018). Antarctic fungal species such as *Arthrobotrys*, *Cladosporium*, *Leucosporidium*, and *Aureobasidium* showed pectinase activities (Aislabie et al., 2013; Bendia et al., 2018; Blanchette et al., 2010; Bowman & McCuaig, 2003; Broady & Weinstein, 1998; Broady, 1981; Cavacini, 2001; Chauhan et al., 2015; de Menezes et al., 2020; Fenice et al., 1997; Franzmann et al., 1997, 1988; García-Echauri et al., 2011; Godinho et al., 2015; Ji et al., 2016; Kerry, 1990; Kol & Flint, 1968; Kostadinova et al., 2009; Learn-Han et al., 2012; Möller & Dreyfuss, 1996; Nadeau & Castenholz, 2000; Niederberger et al., 2015; Pan et al., 2013; Papale et al., 2019; Singh et al., 2006; Tsuji et al., 2013). Thus they have been employed in process of fruit juice clarification. They are also found useful in wastewater treatment process and also used in improvement of wine quality (Divya & Padma, 2015). Thus not only bacteria, but different fungi and sponges also have the ability to produce cold-adapted pectinases.

12.5.1.2 β-Galactosidase or lactase

β-Galactosidase is a glycoside hydrolase enzyme that converts lactose into galactose and glucose. About 75% of human populations is considered to be lactose intolerant and therefore lactose-free dairy products have high demand (Margesin & Feller, 2010). Lactose is degraded by treating with β-galactosidases enzymes that are obtained from psychrophiles. This helps to reduce crystallization of milk as well as increases the sweetness too (Rao et al., 1998). The cold-adapted lactase can hydrolyze the lactose at cold condition (4°C) (Hamid et al., 2014). The mesophilic *Kluyveromyces lactis* was commonly employed for lactose hydrolysis, but it worked poorly at low temperature and therefore was replaced by psychrophilic β-galactosidases (Karasova et al., 2002). A cold-adapted β-galactosidase enzyme produced by a marine bacterium *Pseudoalteromonas haloplanktis* TAE 79 can hydrolyze the lactose at low temperature during milk storage (Hoyoux et al., 2001). This enzyme is commonly used in tagatose production. Tagatose is a well-known monosaccharide having low caloric values and glycemic index (Van De Voorde et al., 2014). Many Antarctic yeasts are studied for their ability for production of β-galactosidase enzymes. For example, *Guehomyces pullulans* 17-1 isolated from the Antarctic sea sediment was used for production of both extracellular as well as cell-bound β-galactosidases enzymes (Song et al., 2010). A recombinant cold-active β-galactosidase enzyme was produced by an Antarctica bacterium *Pseudoalteromonas* sp., an Antarctic bacterium *Pseudoalteromonas* sp. This enzyme showed about 90% milk lactose hydrolysis after 6 hours at 30°C and 28 hours at 15°C, respectively

(Turkiewicz et al., 2003). An Antarctic bacterium, *P. haloplanktis* TAE 79 has been reported for cold-active lactase production (Gerday et al., 2000). This enzyme exhibited milk lactose hydrolysis at cold conditions reported earlier (Hoyoux et al., 2001). Further, an Antarctic yeast, *Tausonia pullulans* 17-1 has been reported for production of β-galactosidase (17.2 U/mL) with highest lactose hydrolysis activity (Zhang et al., 2012). Thus psychrophilic lactases are in high demand to obtain lactose-free or low-lactose-containing products like low-lactose yoghurt, low-lactose milk, and low-lactose ice-cream (Pulicherla et al., 2011).

12.5.1.3 Proteases

Proteases have the ability to hydrolyze the peptide linkages of proteins as well as polypeptides. These enzymes are of high demand and account for about 60% of total global sale of enzymes (Rao et al., 1998). Many diseases such as cancer and cardiovascular disorders have been treated by protease-based drugs (Turk, 2006). Cold-active proteases have several applications in food, cosmetics, bakery products, detergents, beverages, bioremediation process, and in textile industry (Anwar & Saleemuddin, 1998; Baghel et al., 2005; Gupta et al., 2002; Hamamoto et al., 1994). These proteases are found to be beneficial to save energy and also replace the toxic chemicals that are commonly used in hides and skin dehairing process (Pulicherla et al., 2011). These proteases are commonly used in detergent in order to remove protein-based stains such as human sweat, egg, grass, and blood on clothes (Banerjee et al., 1999). A variety of psychrophilic bacteria, yeasts, and fungi are screened from Antarctic cold environments. Antarctic fungi belonging to the genera *Acremonium, Trichoderma, Vanrija, Candida, Rhodotorula,* and *Leucosporidium* have been identified earlier (Lario et al., 2015; Nascimento et al., 2015; Pazgier et al., 2003; Ray et al., 1992). An Antarctic marine yeast, *Rhodotorula mucilaginosa* L7 has been reported for protease enzyme production. This purified enzyme showed maximum enzyme activity at 50°C and pH 5.0 (Lario et al., 2015). Similarly, a cold-active protease was purified from *Janthinobacterium lividum* (Kim et al., 2018) and *Candida humicola* isolated from Antarctic environments. This protease enzyme revealed its activity in the temperature range 0°C–45°C (Ray et al., 1992).

Antarctic bacterial genus like *Pseudoalteromonas* sp., *Psychrobacter* sp., *Marinobacter* sp., and *Polaribacter* sp. have been reported for their ability to produce cold-active proteases having several biotechnological applications. All these bacteria revealed the maximum enzyme activity at low temperature as compared to mesophilic bacteria (Acevedo et al., 2013). The serine protease produced by *Glaciozyma antarctica* 107 exhibited maximum activity at 25°C (Pazgier et al., 2003). The maximum enzyme activity by *Leucosporidium antarcticum* was displayed in the temperature range 0°C–25°C (Pazgier et al., 2003). Several bacteria isolated from different Antarctic marine habitats have been reported for their ability to produce protease enzyme. These isolates were identified as *Colwellia demingae* ACAM 607, *Pseudoalteromonas* sp., and *Cytophaga* like strain IC166 (Buia, 1997). Other bacterial isolates from Potter cove in Antarctica were identified as *Psychrobacter, Salinibacterium, Colwellia, Shewanella,* and *Pseudoalteromonas* sp. (Tropeano et al., 2012). The extracellular cold adaptive protease enzyme was obtained from the marine psychrophilic bacterium *Pseudoalteromonas* sp. NJ276. This enzyme showed high catalytic activity and broad substrate specificity and thus used in food processing at cold condition (Wang et al., 2008). Therefore cold-active proteases produced by different yeasts, fungi, and bacteria displayed several applications in many industrial processes. Antarctic *Colwellia* sp. NJ341 isolated from sea-ice showed maximum enzyme activity at 0°C. This protease enzyme also displayed higher thermostability when compared to other cold-active proteases. This characteristic makes them a special enzyme for use in several industrial processes with high risk of microbial contamination seen in many products (Wang et al., 2017).

12.5.1.4 Cellulases

Cellulases have the ability to hydrolyze α-1,4-glycosidic bonds of cellulose. The enzyme that hydrolyzes a hemicellulose is known as hemicellulases. Cellulose is degraded and converted into different value-added products like glucose, cellobiose, and cello-oligosaccharides (Sukumaran et al., 2005). Cellulases are mostly utilized in laundry detergents that display color brightness and also give softness to the product. Pumice stones are extensively employed for production of "Faded jeans." However, recently pumice stones are being replaced by cellulases due to their high specificity (Pulicherla et al., 2011). These enzymes are also employed for beans drying in food industry. Cold-active cellulases displayed maximum specific activity at cold condition as well as moderate temperatures. Thus these enzymes are found to be more useful than enzymes obtained from mesophiles used earlier.

Several fungi isolated from Antarctica exhibited cellulase activities. These fungi were isolated from different environments such as wood, soil, ornithogenic soil, marine sponges, and other organic materials (Duarte et al., 2018). Cellulase-producing filamentous fungi are belonging to genera *Penicillium, Cadophora, Cladosporium,* and *Geomyces* (Pulicherla et al., 2011). Antarctic yeasts belonging to genera *Cryptococcus* and *Mrakia* are reported for their ability to produce cellulase. Similarly, cellulase-producing Antarctic bacterial genera such as *Colwellia, Psychromonas, Pseudoalteromonas,* and *Polaribacter, Firmicutes,*

Actinobacter, Gamma-proteobacter, and *Bacteroidetes* have been reported earlier (Herrera et al., 2017; Loperena et al., 2012; Melo et al., 2014; Soares et al., 2012; Tropeano et al., 2012). Cold-active cellulases are found to be tolerant to many organic solvents and thus they can be beneficial to produce different volatile heat-sensitive products such as perfumes, flavors, and fragrances (Maharana & Ray, 2015). Cold-active cellulases are found useful in paper recycling, textile, cotton processing, paper, detergent production, and also for biofuel production reported previously (Duarte et al., 2018).

12.5.1.5 *Xylanases*

Xylan is a major part of plant hemicellulose based on a β-1,4-linked D-xylosyl backbone. Xylanases are enzymes that hydrolyze xylan into xylose (Hamid et al., 2014). These enzymes are well known for their biotechnological applications such as fruit juice clarification, pentose production, digestion improvement, biotransformation of lignocellulosic materials into fuels and many chemicals (Garg et al., 1998; Srinivasan & Rele, 1995). There are several Antarctic fungi identified from genera *Cryptococcus, Candida, Alternaria, Naganishia, Penicillium, Tausonia, Phoma,* and *Trichoderma* which are reported for xylanase production (Carrasco et al., 2012; Duarte et al., 2018; Gomes et al., 2000; Petrescu et al., 2000). Cold-active xylanases have been produced by different Antarctic fungi as novel and promising source. It was also observed that xylanase obtained from *Cladosporium* sp. displayed highest enzyme activity at low temperatures (Del-Cid et al., 2014). Similar enzyme activity was also observed in marine Antarctic bacteria, *Flavobacterium frigidarium* that was isolated from marine sediment of shallow waters (Humphry et al., 2001). Almost 38 Antarctic fungi were reported for xylanase activity that were obtained from marine sponges of King George Island. Among all these fungi, *Cladosporium* sp. displayed the highest enzyme activity in cold condition (Del-Cid et al., 2014).

Similarly, the yeast *Cryptococcus adeliensis* isolated from Antarctica produced cold-active xylanase with high catalytic activity and low activation energy (0°C–20°C). Such cold adaptive yeast has several applications in probiotic inclusion as well as a potential therapeutic agent in food (Gomes et al., 2000; Petrescu et al., 2000; Shivaji & Prasad, 2009). Further, it has been studied among the 189 cold adaptive bacterial isolates, about 20.1% bacterial isolates displayed xylanase activity. These bacterial isolates were identified from the genera *Psychromonas* sp., *Pseudoalteromonas* sp., *Polaribacter* sp., and *Colwellia* sp. (Tropeano et al., 2012). The cold-adapted xylanases have been used in bakery, biofuel production, chemical generation of lignocellulose, and also in bioremediation processes (Gomes et al., 2000).

12.5.1.6 Amylases

Amylases are enzymes which catalyzes the starch hydrolysis into sugars. Based on their action, they are classified into amylases, β-amylase and γ-amylase. Amylases hydrolyze-1,4-glycosidic bonds of starch and transform into glucose and deutrins (Rani et al., 2015). These enzymes are obtained from various microbial sources like bacteria, yeasts, and fungi. These enzymes have several applications in many industries because of their thermostability and high productivity. Amylases are well known as one of the most useful industrial enzymes. These enzymes have several applications in different industries or processes such as detergent industry, pulp industry, paper industry, textile industry, bread making process, and starch to sugar syrup conversion process (Pulicherla et al., 2011). Thus amylases enzymes have variety of applications in different industries and processes. There is a report on purification and characterization of amylase obtained from cold adaptive Antarctic yeast *Candida antarctica* CBS 6678 (De Mot & Verachtert, 1987).

A cold-active α-amylase has been characterized that was produced by psychrophilic yeast, *G. antarctica* PI12 (Ramli et al., 2013). Further, the structure of this enzyme has been studied in detail (Ramli et al., 2013). It has been studied that about 33 fungal strains were isolated from Victoria Land (Continental Antarctica). Out of these 33 fungal strains, 18 isolates have displayed amylase activity. Higher amylase producer strain was identified as *Geomyces pannorum* var. *pannorum* which is well known for its several applications in industry (Fenice et al., 1997). Antarctic bacteria, *Pseudoalteromonas* sp. M175 reported for their ability to produce a cold-active α-amylase. This bacteria showed resistance toward several commercial detergents. Thus enzyme from this bacteria was used in detergent to increase the stain removal efficiency (Wang et al., 2018). A cold-adapted yeast, *Sporobolomyces ruberrimus* CRUB1141 has been isolated from Antarctica which displayed amylase activity (Vaz et al., 2011). Many bacterial isolates such as *Planococcus, Psychrobacter, Pseudoalteromonas, Arthtobacter, Shewanella,* and *Salinibacterium* are identified as amylase producers (Tropeano et al., 2012). Heat-sensitive α-amylases showed highest amylase activity at low temperature (Chessa et al., 1999; Feller et al., 1999).

12.5.1.7 Lipases

Lipases have the ability to hydrolyze long-chain triglycerides. Lipases have many biotechnological applications and are thus used in different industries (Joseph et al., 2007). Demand for lipase production is continuously increasing and acquired fifth position in the total enzyme market (Szczęsna-Antczak et al., 2014). Cold-active lipases have been reported for their wide variety of applications including additives in the food industry, detergent formulations, bioremediation, molecular biology, and in biotransformation (Joseph et al., 2008). The temperature stability of lipases is the most crucial parameter that is considered during selection of lipases at an industrial level. Cold adaptive lipases are of high significance due to its high enzyme activity at cold conditions and thus used in production of organic compounds. Recently, cold-active lipases are favored by biophysicists, pharmacists, organic chemists, biotechnologists, and microbiologists (Szczęsna-Antczak et al., 2014).

Antarctic fungi belonging to *Mrakia*, *Leucosporidium*, *Geomyces*, *Beauveria*, *Penicillium*, and *Candida* have been identified as lipase producers (Duarte et al., 2015; Florczak et al., 2013; Mohammed et al., 2013; Tsuji et al., 2013; Yang et al., 2013). Antarctic marine yeast *Leucosporidium scottii* L117 displayed the ability to produce lipase enzyme at 20°C in media containing corn steep and olive oil (Duarte et al., 2015). Similarly, Antarctic filamentous fungi *Geomyces* sp. P7 has been reported for their lipase activity at low temperature and later on this enzyme was expressed in *Saccharomyces cerevisiae* (Florczak et al., 2013). In 33 fungal strains isolated from Victoria Land (Antarctica), all except *Arthrobacter ferox* displayed production of lipases (Fenice et al., 1997). Thus lipases have several applications to obtain many value-added products such as medicines, detergents, polymers, fine chemicals, and biodiesel (Szczęsna-Antczak et al., 2014).

12.5.2 Biomolecules

Antarctic psychrophiles have the ability to synthesize a variety of biomolecules which are valuable in biotechnological processes. PUFA and antifreezing proteins (AFPs) are of high importance and extensively used as cold-adapted biomolecules having several biotechnological applications.

12.5.2.1 *Antifreeze proteins*

Production of AFP is considered as one of the important adaptive strategies of many psychrophiles including *Pseudomonas fluorescens*, *Pseudomonas putida*, and *Moraxella* sp. reported previously (Pulicherla et al., 2011). It has been studied that AFPs can adhere to ice crystals and stop their growth inside the cell. This leads to inhibition of ice crystallization in ice creams as well as in frozen yoghurts (Margesin & Feller, 2010). In further study, the AFP obtained from *Typhula ishikariensis* was compared to the novel AFP obtained from an Antarctic fungi *Antarctomyces psychrotrophicus* (Xiao et al., 2010). A calcium-dependent AFP obtained from Antarctic bacterial strains such as *Marinomonas primoryensis* and *Moraxella* sp. were reported earlier (Gilbert et al., 2005;

Yamashita et al., 2002). Similarly, the AFPs are also obtained from an Antarctic marine microalga, *Pyramimonas gelidicola* studied previously (Jung et al., 2010). These psychrophilic AFPs have been used in several food industries in order to enhance the quality of frozen food, flavor the food, as well as to preserve the texture and reduce microbial load in frozen food. It was also seen that direct injection of AFPs in stored fish also increases shelf life. Thus along with the food industry, AFPs are also found beneficial in cryopreservation and cryosurgery of cells (Pulicherla et al., 2011).

12.5.2.2 Production of polyunsaturated fatty acids

PUFAs are made of long-chain fatty acids which have more than one double bond in the backbone. These fatty acids include ω-3, ω-6, and also some other essential fatty acids. Such fatty acids are beneficial for central nervous system for its functioning and good development. Deficiency of omega -3-fatty acids may develop several neurocognitive disorders like autism, dyslexia, and attention deficit hyperactivity disorder (Schuchardt et al., 2010). The most traditional source to get long-chain PUFA is fish oils. However, the production of traditional fish oils come with many risks and problems at purification level. It is also observed that commercial stock of fish will decrease with time. Thus bacterial PUFA sources are getting much more attention (Pulicherla et al., 2011).

Several algal species have also been used for PUFA production. However, it needs higher investment as compared to the bacterial fermentation. In addition to this, bacteria also have the ability to produce only a single type of PUFA, whereas fish and algae produce a mixed type of PUFA (Pulicherla et al., 2011). An Antarctic alga, *Chlamydomonas* sp. ICE-L has been reported for maximum PUFA production at 0°C–5°C (An et al., 2013; Mou et al., 2012). Similarly, Antarctic bacterial genus such as *Shewanella* and *Colwellia* have produced a large amount of PUFAs including eicosapentaenoic acid and docosahexaenoic acid. Different species of *Colwellia* isolated from coastal Antarctica are reported for their ability to produce docosahexaenoic acid (Bowman et al., 1998). Similarly, *Shewanella* sp. GA-22 has been reported for PUFA production. Three types of PUFA such as arachidonic, linoleic, and eicosapentaenoic acids were produced by AG-22 strain reported previously (Gentile et al., 2003).

12.6 Bioremediation

Waste materials are continuously generated due to increasing industrialization and civilization. These wastes are more complex and toxic and regularly discharged in soil, water, and air that leads to environmental pollution. This may be harmful to environment and human health (Azubuike et al., 2016; Gavrilescu et al., 2015). There are several conventional physiochemical methods for remediation of these waste materials. However, these methods have some drawbacks or disadvantages such as they are complex in nature, costly, less effective, and may produce secondary toxic pollutants (Chen et al., 2015). Thus bioremediation processes have received much more attention due to its several advantages over conventional techniques. Several microbes such as bacteria, algae, yeasts, and fungi have been extensively used to remediate the polluted environment in bioremediation process. Cold environments are being extensively exposed to petroleum transport, extraction, and exploration that may result into oil spills. Thus biostimulation and bioaugmentation strategies are extensively used for bioremediation of hydrocarbons (Margesin & Feller, 2010). Psychrophilic microbes are found to be useful for soil remediation and wastewater treatment in winter season when temperature drop down.

Thus use of psychrophiles are found to be highly beneficial in cold environments. An Antarctic fungi, *Mortierella* sp., is reported as a promising green source for treatment of hydrocarbons contaminated environments (Hughes et al., 2007). Phenol and its derivatives are well known as toxic chemicals that are contaminating the different cold regions. Thus cold-adapted microbes are promising green candidates used in bioremediation process. An Antarctic fungal strain of *Aspergillus fumigatus* has been studied for phenol and its derivatives degradation (Gerginova et al., 2013). Toluene-*o*-xylene monooxygenase gene is expressed in the recombinant Antarctic bacterial strain *P. haloplanktis* TAC125 which converted several toxic aromatic compounds into less toxic catechols under wide range of temperature (Papa et al., 2009). Not only microbes are used in bioremediation processes, but currently some psychrozymes such as amylases, lipolytic enzymes, pectinases, proteases, proteinases, and xylanases are being used in bioremediation process (Gerday et al., 2000).

12.7 Pharmaceutical and medical applications

Several Antarctic psychrophiles are reported for their antimicrobial, antifungal, and antiprotozoal activities.

12.7.1 Antimicrobials

Microbes have the ability to colonize and remain dominant in the particular environment by killing or inhibiting their competitors or predators. Several most volatile and soluble secondary metabolites are produced by microbes in order to grow in adverse conditions (Mojib et al., 2010). Several Antarctic microbes have been reported for their antagonistic effects in aforementioned environments. A cold-adapted bacterial strain belonging to phylum Firmicutes has been isolated from King George Island that displayed antagonistic effect against methicillin resistant *Staphylococcus aureus* (Vollú et al., 2014). Most bioactive microbial compounds such as antibiotics, antiviral, antiprotozoal, antifungal, and anthelmintic have been obtained from the Actinomycetes group (Baltz, 2007). Further, an Antarctic actinomycetes, *Nocardioides* sp. A-1 has been reported for their ability to produce many antifungal and antimicrobial agents like lipopeptides and glycolipids. These antimicrobial agents displayed its activity against gram-positive bacteria such as *Sarcina lutea, Bacillus subtilis,* and *S. aureus*. Antimicrobial activity was also observed against the yeast, *Candida tropicalis* and fungi, *Xanthomonas oryzae* (Gesheva & Vasileva-Tonkova, 2012).

Antarctic lichen such as *Stereocaulon alpinum* is known to produce some antimicrobial compounds like lobastin and lobaric acid. They showed inhibitory activity against gram-positive bacteria such as *S. aureus* and *B. subtilis* (Bhattarai et al., 2013). Further, it was seen that *Mrakia frigida* produced a killer toxin against *Candida albicans, C. tropicalis,* and *Metschnikowia bicuspidate* reported previously (Hua et al., 2010). Most antimicrobial activity exhibited by microbes are isolated from Antarctic soils. However, recently Antarctic marine microbes are well known for their potential antimicrobial activity. Similar antimicrobial activity has been reported in Antarctic bacteria *Pseudoalteromonas* against *Burkholderia cepacia* (Maida et al., 2015; Papaleo et al., 2012).

12.7.2 Antioxidants, antitumorals, and other potential compounds

Along with antimicrobial activities, psychrophilic microbes can be used as promising agents for the treatment of noninfectious yet fatal diseases like cancer, diabetes, heart related diseases, and many more. An antioxidant compound has been identified in Antarctic lichen *Ramalina terebrata* showed good antioxidant activity in vitro and in vivo (Paudel et al., 2011). Antioxidants like torulene, ergosterol, beta-carotene, and torularhodin are purified from Antarctic yeast, *Sporobolomyces salmonicolor* AL1. All these antioxidants with exoglucomannan displayed UV-A protection and are being used in sunscreens formulation (Dimitrova et al., 2010, 2013). Photoprotectors have the property to prevent the skin cancer caused due to UV-B radiation. It is studied that a freshwater bacteria, *Lysobacter oligotrophicus,* displayed a photoprotective compound-melanin (Fukuda et al., 2013; Kimura et al., 2015). Violacein like purple violet pigment also displayed anticancer activity. This compound was isolated and purified from an Antarctic.

Janthinobacterium sp. has been reported to display inhibitory effect on murine fibrosarcoma cells. This compound arrested the cell cycle of fibrosarcoma cells and induced their apoptosis. An antiproliferative effect on melanoma cells was also reported without any adverse effect on normal human cells. Thus such pigment can be used as a potential chemotherapeutic agent (Mojib et al., 2013).

12.8 Conclusions

Antarctica is well known a region where several microbes are exposed continuously to different extreme conditions such as limited nutrients, desiccation, low biochemical activity, high salinity, and adverse solar radiation. Different types of microbes like bacteria, yeasts, fungi, and viruses have developed adaption mechanisms in order to survive in extreme cold conditions of Antarctic continent, and thus microbes are known as extremophiles. It is also well known that highest microbial diversity occurs in extremophiles that survive in cold and frozen ecosystems. The psychrophiles have been studied for production of several industrially important enzymes (psychrophiles) like proteases, lipases, pectinases, β-cellulases, amylases, -galactosidase or lactase, and xylonases. These psychrophiles are extensively used to obtain several valuable biomolecules like antifreezing proteins and polyunsaturated acids. They are also used in bioremediation process feasible at cold conditions. Thus it is concluded that psychrophiles isolated from Antarctica regions have several biotechnological applications.

Acknowledgments

All authors are thankful to DST-SERB for providing financial support in form of a major research project to Dr. Ashok Bankar. Authors are also thankful to the Principal, MES Abasaheb Garware College, Pune, Maharashtra, India, for their valuable support and encouragement.

Conflict of interest

All authors declared that there is no conflict of interest.

References

Acevedo, J. P., Rodriguez, V., Saavedra, M., Muñoz, M., Salazar, O., Asenjo, J. A., & Andrews, B. A. (2013). Cloning, expression and decoding of the cold adaptation of a new widely represented thermolabile subtilisin-like protease. *Journal of Applied Microbiology, 114*(2), 352–363. Available from https://doi.org/10.1111/jam.12033.

Adams, M. W. W., & Kelly, R. M. (1994). Thermostability and thermoactivity of enzymes from hyperthermophilic archaea. *Bioorganic and Medicinal Chemistry, 2*(7), 659–667. Available from https://doi.org/10.1016/0968-0896(94)85015-1.

Aislabie, J., Jordan, S., Ayton, J., Klassen, J. L., Barker, G. M., & Turner, S. (2009). Bacterial diversity associated with ornithogenic soil of the Ross Sea region, Antarctica. *Canadian Journal of Microbiology, 55*(1), 21–36. Available from https://doi.org/10.1139/W08-126.

Aislabie, J. M., Lau, A., Dsouza, M., Shepherd, C., Rhodes, P., & Turner, S. J. (2013). Bacterial composition of soils of the Lake Wellman area, Darwin Mountains, Antarctica. *Extremophiles: Life Under Extreme Conditions, 17*(5), 775–786. Available from https://doi.org/10.1007/s00792-013-0560-6.

Albores, S., Sanguinedo, P., Held, B. H., Cerdeiras, M. P., & Blanchette, R. A. (2018). Biodiversity and antimicrobial activity of antarctic fungi from the fildes peninsula, king george island. *Sydowia, 70*, 185–192. Available from https://doi.org/10.12905/0380.sydowia70-2018-0185.

Amato, P., & Christner, B. C. (2009). Energy metabolism response to low-temperature and frozen conditions in Psychrobacter cryohalolentis. *Applied and Environmental Microbiology, 75*(3), 711–718. Available from https://doi.org/10.1128/AEM.02193-08.

An, M., Mou, S., Zhang, X., Ye, N., Zheng, Z., Cao, S., Xu, D., Fan, X., Wang, Y., & Miao, J. (2013). Temperature regulates fatty acid desaturases at a transcriptional level and modulates the fatty acid profile in the Antarctic microalga *Chlamydomonas* sp. ICE-L. *Bioresource Technology, 134*, 151–157. Available from https://doi.org/10.1016/j.biortech.2013.01.142.

Anwar, A., & Saleemuddin, M. (1998). Alkaline proteases: A review. *Bioresource Technology, 64*(3), 175–183. Available from https://doi.org/10.1016/S0960-8524(97)00182-X.

Azubuike, C. C., Chikere, C. B., & Okpokwasili, G. C. (2016). Bioremediation techniques—classification based on site of application: Principles, advantages, limitations and prospects. *World Journal of Microbiology and Biotechnology, 32*(11).

Baghel, V. S., Tripathi, R. D., Ramteke, P. W., Gopal, K., Dwivedi, S., Jain, R. K., Rai, U. N., & Singh, S. N. (2005). Psychrotrophic proteolytic bacteria from cold environment of Gangotri glacier, Western Himalaya, India. *Enzyme and Microbial Technology, 36*(5–6), 654–659. Available from https://doi.org/10.1016/j.enzmictec.2004.09.005.

Bailey, A. D., & Wynn-Williams, D. D. (1982). Soil microbiological studies at Signy Island. *South Orkney Islands. British Antarctic Survey Bulletin, 51*, 167–191.

Baltz, R. H. (2007). Antimicrobials from actinomycetes: Back to the future. *Microbe, 2*(3), 125–131. Available from http://www.asm.org/ASM/files/ccLibraryFiles/Filename/000000002920/znw00307000125.pdf.

Banerjee, U. C., Sani, R. K., Azmi, W., & Soni, R. (1999). Thermostable alkaline protease from *Bacillus brevis* and its characterization as a laundry detergent additive. *Process Biochemistry, 35*(1–2), 213–219. Available from https://doi.org/10.1016/S0032-9592(99)00053-9.

Barrett, J. E., Virginia, R. A., Wall, D. H., Cary, S. C., Adams, B. J., Hacker, A. L., & Aislabie, J. M. (2006). Co-variation in soil biodiversity and biogeochemistry in northern and southern Victoria Land, Antarctica. *Antarctic Science, 18*(4), 535–548. Available from https://doi.org/10.1017/S0954102006000587.

Bendia, A. G., Araujo, G. G., Pulschen, A. A., Contro, B., Duarte, R. T. D., Rodrigues, F., Galante, D., & Pellizari, V. H. (2018). Surviving in hot and cold: Psychrophiles and thermophiles from Deception Island volcano, Antarctica. *Extremophiles: Life Under Extreme Conditions, 22*(6), 917–929. Available from https://doi.org/10.1007/s00792-018-1048-1.

Bhattarai, H. D., Kim, T., Oh, H., & Yim, J. H. (2013). A new pseudodepsidone from the Antarctic lichen *Stereocaulon alpinum* and its antioxidant, antibacterial activity. *Journal of Antibiotics, 66*(9), 559–561. Available from https://doi.org/10.1038/ja.2013.41.

Blanchette, R. A., Held, B. W., Arenz, B. E., Jurgens, J. A., Baltes, N. J., Duncan, S. M., & Farrell, R. L. (2010). An Antarctic hot spot for fungi at Shackleton's historic hut on

Cape Royds. *Microbial Ecology, 60*(1), 29–38. Available from https://doi.org/10.1007/s00248-010-9664-z.

Bottos, E. M., Woo, A. C., Zawar-Reza, P., Pointing, S. B., & Cary, S. C. (2014). Airborne bacterial populations above desert soils of the McMurdo Dry Valleys, Antarctica. *Microbial Ecology, 67*(1), 120–128. Available from https://doi.org/10.1007/s00248-013-0296-y.

Bowman, J. P., & McCuaig, R. D. (2003). Biodiversity, community structural shifts, and biogeography of prokaryotes within Antarctic continental shelf sediment. *Applied and Environmental Microbiology, 69*(5), 2463–2483. Available from https://doi.org/10.1128/AEM.69.5.2463-2483.2003.

Bowman, J. P., Gosink, J. J., McCammon, S. A., Lewis, T. E., Nichols, D. S., Nichols, P. D., Skerratt, J. H., Staley, J. T., & McMeekin, T. A. (1998). *Colwellia demingiae* sp. nov., *Colwellia hornerae* sp. nov., *Colwellia rossensis* sp. nov. and *Colwellia psychrotropica* sp. nov.: Psychrophilic Antarctic species with the ability to synthesize docosahexaenoic acid (22:6ω3). *International Journal of Systematic Bacteriology, 48*(4), 1171–1180. Available from https://doi.org/10.1099/00207713-48-4-1171.

Broady, P. A. (1981). The ecology of chasmolithic algae at coastal locations of Antarctica. *Phycologia, 20*(3), 259–272.

Broady, P. A., & Weinstein, R. N. (1998). Algae, lichens and fungi in La Gorce Mountains, Antarctica. *Antarctic Science, 10*(4), 376–385. Available from https://doi.org/10.1017/s0954102098000467.

Buia, A. (1997). *Psychrophilic enzymes of Antarctic sea-ice bacteria*. BSc (Hons) Thesis.

Buzzini, P., Branda, E., Goretti, M., & Turchetti, B. (2012). Psychrophilic yeasts from worldwide glacial habitats: Diversity, adaptation strategies and biotechnological potential. *FEMS Microbiology Ecology, 82*(2), 217–241. Available from https://doi.org/10.1111/j.1574-6941.2012.01348.x.

Carrasco, M., Rozas, J. M., Barahona, S., Alcaíno, J., Cifuentes, V., & Baeza, M. (2012). Diversity and extracellular enzymatic activities of yeasts isolated from King George Island, the sub-Antarctic region. *BMC Microbiology, 12*. Available from https://doi.org/10.1186/1471-2180-12-251.

Cavacini, P. (2001). Soil algae from northern Victoria Land (Antarctica). *Polar Biosciences, 14*, 45–60.

Cavicchioli, R., Ostrowski, M., Fegatella, F., Goodchild, A., & Guixa-Boixereu, N. (2003). Life under nutrient limitation in oligotrophic marine environments: An eco/physiological perspective of *Sphingopyxis alaskensis* (formerly *Sphingomonas alaskensis*). *Microbial Ecology, 45*(3), 203–217. Available from https://doi.org/10.1007/s00248-002-3008-6.

Chauhan, A., Bharti, P. K., Goyal, P., Varma, A., & Jindal, T. (2015). Psychrophilic pseudomonas in antarctic freshwater lake at stornes peninsula, larsemann hills over east Antarctica. *SpringerPlus, 4*(1). Available from https://doi.org/10.1186/s40064-015-1354-3.

Chen, M., Xu, P., Zeng, G., Yang, C., Huang, D., & Zhang, J. (2015). Bioremediation of soils contaminated with polycyclic aromatic hydrocarbons, petroleum, pesticides, chlorophenols and heavy metals by composting: Applications, microbes and future research needs. *Biotechnology Advances, 33*(6), 745–755. Available from https://doi.org/10.1016/j.biotechadv.2015.05.003.

Chessa, J. P., Feller, G., & Gerday, C. (1999). Purification and characterization of the heat-labile α-amylase secreted by the psychrophilic bacterium TAC 240B. *Canadian Journal of Microbiology, 45*(6), 452–457. Available from https://doi.org/10.1139/w99-021.

Chong, C. W., Dunn, M. J., Convey, P., Tan, G. Y. A., Wong, R. C. S., & Tan, I. K. P. (2009). Environmental influences on bacterial diversity of soils on Signy Island, maritime antarctic. *Polar Biology, 32*(11), 1571–1582. Available from https://doi.org/10.1007/s00300-009-0656-8.

Connell, L., Segee, B., Redman, R., Rodriguez, R. J., & Staudigel, H. (2018). Biodiversity and abundance of cultured microfungi from the permanently ice-covered Lake Fryxell, Antarctica. *Life (Chicago, Ill.: 1978)*, *8*(3). Available from https://doi.org/10.3390/life8030037.

D'Amico, S., Collins, T., Marx, J. C., Feller, G., & Gerday, C. (2006). Psychrophilic microorganisms: Challenges for life. *EMBO Reports*, *7*(4), 385–389. Available from https://doi.org/10.1038/sj.embor.7400662.

da Silva, T. H., Silva, D. A. S., de Oliveira, F. S., Schaefer, C. E. G. R., Rosa, C. A., & Rosa, L. H. (2020). Diversity, distribution, and ecology of viable fungi in permafrost and active layer of Maritime Antarctica. *Extremophiles: Life Under Extreme Conditions*, *24*(4), 565–576. Available from https://doi.org/10.1007/s00792-020-01176-y.

De Cárcer, D. A., López-Bueno, A., Pearce, D. A., & Alcamí, A. (2015). Biodiversity and distribution of polar freshwater DNA viruses. *Science Advances*, *1*(5). Available from https://doi.org/10.1126/sciadv.1400127.

De Maayer, P., Anderson, D., Cary, C., & Cowan, D. A. (2014). Some like it cold: Understanding the survival strategies of psychrophiles. *EMBO Reports*, *15*(5), 508–517. Available from https://doi.org/10.1002/embr.201338170.

De Mot, R., & Verachtert, H. (1987). Purification and characterization of extracellular α-amylase and glucoamylase from the yeast Candida antarctica CBS 6678. *European journal of biochemistry*, *164*(3), 643–654.

de Menezes, G. C. A., Porto, B. A., Amorim, S. S., Zani, C. L., de Almeida Alves, T. M., Junior, P. A. S., Murta, S. M. F., Simões, J. C., Cota, B. B., Rosa, C. A., & Rosa, L. H. (2020). Fungi in glacial ice of Antarctica: Diversity, distribution and bioprospecting of bioactive compounds. *Extremophiles: Life Under Extreme Conditions*, *24*(3), 367–376. Available from https://doi.org/10.1007/s00792-020-01161-5.

Del-Cid, A., Ubilla, P., Ravanal, M. C., Medina, E., Vaca, I., Levicán, G., Eyzaguirre, J., & Chávez, R. (2014). Cold-active xylanase produced by fungi associated with Antarctic marine sponges. *Applied Biochemistry and Biotechnology*, *172*(1), 524–532. Available from https://doi.org/10.1007/s12010-013-0551-1.

Demirjian, D. C., Morís-Varas, F., & Cassidy, C. S. (2001). Enzymes from extremophiles. *Current Opinion in Chemical Biology*, *5*(2), 144–151. Available from https://doi.org/10.1016/S1367-5931(00)00183-6.

Dickinson, I., Goodall-Copestake, W., Thorne, M. A., Schlitt, T., Ávila-Jiménez, M. L., & Pearce, D. A. (2016). Extremophiles in an Antarctic marine ecosystem. *Microorganisms*, *4*(1), 8.

Dimitrova, S., Pavlova, K., Lukanov, L., Korotkova, E., Petrova, E., Zagorchev, P., & Kuncheva, M. (2013). Production of metabolites with antioxidant and emulsifying properties by antarctic strain *Sporobolomyces salmonicolor* AL1. *Applied Biochemistry and Biotechnology*, *169*(1), 301–311. Available from https://doi.org/10.1007/s12010-012-9983-2.

Dimitrova, S., Pavlova, K., Lukanov, L., & Zagorchev, P. (2010). Synthesis of coenzyme Q10 and β-carotene by yeasts isolated from Antarctic soil and lichen in response to ultraviolet and visible radiations. *Applied Biochemistry and Biotechnology*, *162*(3), 795–804. Available from https://doi.org/10.1007/s12010-009-8845-z.

Divya, K., & Padma, P. N. (2015). Use of Plackett-Burman design for rapid screening of diverse raw pectin sources for cold-active polygalacturonase and amylase production by Geotrichum sp. *International Journal of Current Microbiology and Applied Sciences*, *4*, 821–827.

Duarte, A. W. F., dos Santos, J. A., Vianna, M. V., Vieira, J. M. F., Mallagutti, V. H., Inforsato, F. J., Wentzel, L. C. P., Lario, L. D., Rodrigues, A., Pagnocca, F. C., Pessoa, A., & Durães Sette, L. (2018). Cold-adapted enzymes produced by fungi from terrestrial and marine Antarctic environments. *Critical Reviews in Biotechnology*, *38*(4), 600–619. Available from https://doi.org/10.1080/07388551.2017.1379468.

Duarte, A. W. F., Lopes, A. M., Molino, J. V. D., Pessoa, A., & Sette, L. D. (2015). Liquid-liquid extraction of lipase produced by psychrotrophic yeast *Leucosporidium scottii* L117 using aqueous two-phase systems. *Separation and Purification Technology, 156*, 215–225. Available from https://doi.org/10.1016/j.seppur.2015.10.001.

Dumorné, K., Córdova, D. C., Astorga-Eló, M., & Renganathan, P. (2017). Extremozymes: A potential source for industrial applications. *Journal of Microbiology and Biotechnology, 27*(4), 649–659. Available from https://doi.org/10.4014/jmb.1611.11006.

Ekelöf, E. (1908). *Bakteriologische Studien während der schwedischen Südpolar-Expedition 1901–1903*. von Erik Ekelöf.

Feller, G., D'Amico, D., & Gerday, C. (1999). Thermodynamic stability of a cold-active α-amylase from the antarctic bacterium *Alteromonas haloplanctis*. *Biochemistry, 38*(14), 4613–4619. Available from https://doi.org/10.1021/bi982650.

Fenice, M., Selbmann, L., Zucconi, L., & Onofri, S. (1997). Production of extracellular enzymes by Antarctic fungal strains. *Polar Biology, 17*(3), 275–280. Available from https://doi.org/10.1007/s003000050132.

Florczak, T., Daroch, M., Wilkinson, M. C., Białkowska, A., Bates, A. D., Turkiewicz, M., & Iwanejko, L. A. (2013). Purification, characterisation and expression in *Saccharomyces cerevisiae* of LipG7 an enantioselective, cold-adapted lipase from the Antarctic filamentous fungus *Geomyces* sp. P7 with unusual thermostability characteristics. *Enzyme and Microbial Technology, 53*(1), 18–24. Available from https://doi.org/10.1016/j.enzmictec.2013.03.021.

Franzmann, P. D., Liu, Y., Balkwill, D. L., Aldrich, H. C., Conway De Macario, E., & Boone, D. R. (1997). *Methanogenium frigidum* sp. nov., a psychrophilic, H2-using methanogen from Ace Lake, Antarctica. *International Journal of Systematic Bacteriology, 47*(4), 1068–1072. Available from https://doi.org/10.1099/00207713-47-4-1068.

Franzmann, P. D., Stackebrandt, E., Sanderson, K., Volkman, J. K., Cameron, D. E., Stevenson, P. L., Mcmeekin, T. A., & Burton, H. R. (1988). *Halobacterium lacusprofundi* sp. nov., a halophilic bacterium isolated from Deep Lake, Antarctica. *Systematic and Applied Microbiology, 11*(1), 20–27. Available from https://doi.org/10.1016/S0723-2020(88)80044-4.

Fukuda, W., Kimura, T., Araki, S., Miyoshi, Y., Atomi, H., & Imanaka, T. (2013). *Lysobacter oligotrophicus* sp. nov., isolated from an Antarctic freshwater lake in Antarctica. *International Journal of Systematic and Evolutionary Microbiology, 63*(9), 3313–3318. Available from https://doi.org/10.1099/ijs.0.051805-0.

García-Echauri, S. A., Gidekel, M., Gutiérrez-Moraga, A., Santos, L., & de León-Rodríguez, A. (2011). Isolation and phylogenetic classification of culturable psychrophilic prokaryotes from the Collins Glacier in the Antarctica. *Folia Microbiologica, 56*(3), 209–214. Available from https://doi.org/10.1007/s12223-011-0038-9.

Garg, A. P., Roberts, J. C., & McCarthy, A. J. (1998). Bleach boosting effect of cellulase-free xylanase of Streptomyces thermoviolaceus and its comparison with two commercial enzyme preparations on birchwood kraft pulp. *Enzyme and Microbial Technology, 22*(7), 594–598. Available from https://doi.org/10.1016/S0141-0229(97)00250-0.

Garwin, J. L., & Cronan, J. E. (1980). Thermal modulation of fatty acid synthesis in *Escherichia coli* does not involve de novo enzyme synthesis. *Journal of Bacteriology, 141* (3), 1457–1459.

Gavrilescu, M., Demnerová, K., Aamand, J., Agathos, S., & Fava, F. (2015). Emerging pollutants in the environment: Present and future challenges in biomonitoring, ecological risks and bioremediation. *New Biotechnology, 32*(1), 147–156. Available from https://doi.org/10.1016/j.nbt.2014.01.001.

Gazert, H. (1901). *Undersuchungen uber Meeresbakterien und chren Einfl uss auf den Stoffweckselim Meer*. Deutschi Biidpolar Expedition.

Gentile, G., Bonasera, V., Amico, C., Giuliano, L., & Yakimov, M. M. (2003). *Shewanella* sp. GA-22, a psychrophilic hydrocarbonoclastic antarctic bacterium producing polyunsaturated fatty acids. *Journal of Applied Microbiology, 95*(5), 1124–1133. Available from https://doi.org/10.1046/j.1365-2672.2003.02077.x.

Gerday, C., Aittaleb, M., Bentahir, M., Chessa, J. P., Claverie, P., Collins, T., D'Amico, S., Dumont, J., Garsoux, G., Georlette, D., Hoyoux, A., Lonhienne, T., Meuwis, M. A., & Feller, G. (2000). Cold-adapted enzymes: From fundamentals to biotechnology. *Trends in Biotechnology, 18*(3), 103–107. Available from https://doi.org/10.1016/S0167-7799(99)01413-4.

Gerginova, M., Manasiev, J., Yemendzhiev, H., Terziyska, A., Peneva, N., & Alexieva, Z. (2013). Biodegradation of phenol by Antarctic strains of *Aspergillus fumigatus. Zeitschrift Fur Naturforschung - Section C Journal of Biosciences, 68*(9–10), 384–393. Available from https://doi.org/10.1515/znc-2013-9-1006.

Gesheva, V., & Vasileva-Tonkova, E. (2012). Production of enzymes and antimicrobial compounds by halophilic *Antarctic nocardioides* sp. grown on different carbon sources. *World Journal of Microbiology and Biotechnology, 28*(5), 2069–2076. Available from https://doi.org/10.1007/s11274-012-1009-2.

Gilbert, J. A., Davies, P. L., & Laybourn-Parry, J. (2005). A hyperactive, Ca^{2+}-dependent antifreeze protein in an *Antarctic bacterium. FEMS Microbiology Letters, 245*(1), 67–72. Available from https://doi.org/10.1016/j.femsle.2005.02.022.

Godinho, V. M., Gonçalves, V. N., Santiago, I. F., Figueredo, H. M., Vitoreli, G. A., Schaefer, C. E. G. R., Barbosa, E. C., Oliveira, J. G., Alves, T. M. A., Zani, C. L., Junior, P. A. S., Murta, S. M. F., Romanha, A. J., Kroon, E. G., Cantrell, C. L., Wedge, D. E., Duke, S. O., Ali, A., Rosa, C. A., & Rosa, L. H. (2015). Diversity and bioprospection of fungal community present in oligotrophic soil of continental Antarctica. *Extremophiles: Life Under Extreme Conditions, 19*(3), 585–596. Available from https://doi.org/10.1007/s00792-015-0741-6.

Gomes, J., Gomes, I., & Steiner, W. (2000). Thermolabile xylanase of the Antarctic yeast *Cryptococcus adeliae*: Production and properties. *Extremophiles: Life Under Extreme Conditions, 4*(4), 227–235. Available from https://doi.org/10.1007/s007920070024.

Gupta, R., Beg, Q. K., Khan, S., & Chauhan, B. (2002). An overview on fermentation, downstream processing and properties of microbial alkaline proteases. *Applied Microbiology and Biotechnology, 60*(4), 381–395. Available from https://doi.org/10.1007/s00253-002-1142-1.

Hamamoto, T., Kaneda, M., Horikoshi, K., & Kudo, T. (1994). Characterization of a protease from a psychrotroph, *Pseudomonas fluorescens* 114. *Applied and Environmental Microbiology, 60*(10), 3878–3880. Available from https://doi.org/10.1128/aem.60.10.3878-3880.1994.

Hamid, B., Rana, R. S., Chauhan, D., Singh, P., Mohiddin, F. A., Sahay, S., & Abidi, I. (2014). Psychrophilic yeasts and their biotechnological applications—A review. *African Journal of Biotechnology, 22*.

Hassan, N., Rafiq, M., Hayat, M., Shah, A. A., & Hasan, F. (2016). Psychrophilic and psychrotrophic fungi: A comprehensive review. *Reviews in Environmental Science and Biotechnology, 15*(2), 147–172. Available from https://doi.org/10.1007/s11157-016-9395-9.

Herrera, L. M., García-Laviña, C. X., Marizcurrena, J. J., Volonterio, O., de León, R. P., & Castro-Sowinski, S. (2017). Hydrolytic enzyme-producing microbes in the *Antarctic oligochaete* Grania sp. (Annelida). *Polar Biology, 40*(4), 947–953. Available from https://doi.org/10.1007/s00300-016-2012-0.

Hoyoux, A., Jennes, I., Dubois, P., Genicot, S., Dubail, F., François, J. M., Baise, E., Feller, G., & Gerday, C. (2001). Cold-adapted β-galactosidase from the Antarctic psychrophile *Pseudoalteromonas haloplanktis. Applied and Environmental Microbiology, 67*(4), 1529–1535. Available from https://doi.org/10.1128/AEM.67.4.1529-1535.2001.

Hua, M. X., Chi, Z., Liu, G. L., Buzdar, M. A., & Chi, Z. M. (2010). Production of a novel and cold-active killer toxin by *Mrakia frigida* 2E00797 isolated from sea sediment in

Antarctica. *Extremophiles: Life Under Extreme Conditions, 14*(6), 515–521. Available from https://doi.org/10.1007/s00792-010-0331-6.

Hughes, K. A., Bridge, P., & Clark, M. S. (2007). Tolerance of Antarctic soil fungi to hydrocarbons. *Science of the Total Environment, 372*(2–3), 539–548. Available from https://doi.org/10.1016/j.scitotenv.2006.09.016.

Humphry, D. R., George, A., Black, G. W., & Cummings, S. P. (2001). *Flavobacterium frigidarium* sp. nov., an aerobic, psychrophilic, xylanolytic and laminarinolytic bacterium from Antarctica. *International Journal of Systematic and Evolutionary Microbiology, 51*(4), 1235–1243. Available from https://doi.org/10.1099/00207713-51-4-1235.

Huston, A. L. (2008). *Biotechnological Aspects of Cold-Adapted Enzymes.* In: Margesin R., Schinner F., Marx J. C., Gerday C. (eds) *Psychrophiles: from Biodiversity to Biotechnology.* Berlin, Heidelberg: Springer. Available from https://doi.org/10.1007/978-3-540-74335-4_20.

Isabelle, R.-B., Jacqueline, G., Yevgen, Z., Jennifer, R., Martina, S., Corien, B., & Whyte, L. G. (2018). Conserved genomic and amino acid traits of cold adaptation in subzero-growing Arctic permafrost bacteria. *FEMS Microbiology Ecology, 94*(4). Available from https://doi.org/10.1093/femsec/fiy023.

Jansson, J. (2014). The microbial ecology of permafrost. *Nature, 12.*

Ji, M., van Dorst, J., Bissett, A., Brown, M. V., Palmer, A. S., Snape, I., Siciliano, S. D., & Ferrari, B. C. (2016). Microbial diversity at Mitchell Peninsula, Eastern Antarctica: A potential biodiversity "hotspot. *Polar Biology, 39*(2), 237–249. Available from https://doi.org/10.1007/s00300-015-1776-y.

Jones, P. G., & Inouye, M. (1994). The cold-shock response—A hot topic. *Molecular Microbiology, 11*(5), 811–818. Available from https://doi.org/10.1111/j.1365-2958.1994.tb00359.x.

Joseph, B., Ramteke, P. W., & Thomas, G. (2008). Cold active microbial lipases: Some hot issues and recent developments. *Biotechnology Advances, 26*(5), 457–470. Available from https://doi.org/10.1016/j.biotechadv.2008.05.003.

Joseph, B., Ramteke, P. W., Thomas, G., & Shrivastava, N. (2007). Cold-active microbial lipases: A versatile tool for industrial applications. *Biotechnology and Molecular Biology Reviews, 2*, 39–48.

Jung, W., Gwak, Y., Davies, P. L., Kim, H. J., & Jin, E. S. (2014). Isolation and characterization of antifreeze proteins from the Antarctic marine microalga *Pyramimonas gelidicola. Marine Biotechnology, 16*(5), 502–512. Available from https://doi.org/10.1007/s10126-014-9567-y.

Jung, Y. H., Yi, J. Y., Jung, H. J., Lee, Y. K., Lee, H. K., Naicker, M. C., Uh, J. H., Jo, I. S., Jung, E. J., & Im, H. (2010). Overexpression of cold shock protein a of *Psychromonas arctica* KOPRI 22215 confers cold-resistance. *Protein Journal, 29*(2), 136–142. Available from https://doi.org/10.1007/s10930-010-9233-9.

Junge, K., Eicken, H., & Deming, J. W. (2004). Bacterial Activity at −2 to −20°C in Arctic Wintertime Sea Ice. *Applied and Environmental Microbiology, 70*(1), 550–557. Available from https://doi.org/10.1128/AEM.70.1.550-557.2004.

Karasova, P. E. T. R. A., Spiwok, V. O. J. T. Ě. C. H., Mala, S., Kralova, B. L. A. N. K. A., & Russell, N. J. (2002). Beta-galactosidase activity in psychrotrophic microorganisms and their potential use in food industry. *Czech Journal of Food Sciences, 20*(2), 43–47.

Karr, E. A., Ng, J. M., Belchik, S. M., Sattley, W. M., Madigan, M. T., & Achenbach, L. A. (2006). Biodiversity of methanogenic and other Archaea in the permanently frozen Lake Fryxell, Antarctica. *Applied and Environmental Microbiology, 72*(2), 1663–1666. Available from https://doi.org/10.1128/AEM.72.2.1663-1666.2006.

Kerry, E. (1990). Microorganisms colonizing plants and soil subjected to different degrees of human activity, including petroleum contamination, in the Vestfold hills and MacRobertson Land, Antarctica. *Polar Biology, 10*(6), 423–430. Available from https://doi.org/10.1007/BF00233690.

Kim, H. D., Kim, S. M., & Choi, J. I. (2018). Purification, characterization, and cloning of a cold-adapted protease from antarctic *Janthinobacterium lividum*. *Journal of Microbiology and Biotechnology*, *28*(3), 448–453. Available from https://doi.org/10.4014/jmb.1711.11006.

Kimura, T., Fukuda, W., Sanada, T., & Imanaka, T. (2015). Characterization of water-soluble dark-brown pigment from Antarctic bacterium, *Lysobacter oligotrophicus*. *Journal of Bioscience and Bioengineering*, *120*(1), 58–61. Available from https://doi.org/10.1016/j.jbiosc.2014.11.020.

Kol, E., & Flint, E. A. (1968). Algae in green ice from the Balleny Islands, Antarctica. *New Zealand Journal of Botany*, *6*(3), 249–261. Available from https://doi.org/10.1080/0028825X.1968.10428810.

Kostadinova, N., Krumova, E., Tosi, S., Pashova., & Angelova, M. (2009). Isolation and identification of filamentous fungi from island livingston, antarctica. *Biotechnology and Biotechnological Equipment*, *23*, 267–270. Available from https://doi.org/10.1080/13102818.2009.10818416.

Králová, S. (2017). Role of fatty acids in cold adaptation of Antarctic psychrophilic *Flavobacterium* spp. *Systematic and Applied Microbiology*, *40*(6), 329–333. Available from https://doi.org/10.1016/j.syapm.2017.06.001.

Lario, L. D., Chaud, L., Almeida, M. d. G., Converti, A., Durães Sette, L., & Pessoa, A. (2015). Production, purification, and characterization of an extracellular acid protease from the marine Antarctic yeast *Rhodotorula mucilaginosa* L7. *Fungal Biology*, *119*(11), 1129–1136. Available from https://doi.org/10.1016/j.funbio.2015.08.012.

Laybourn-Parry, J. (2009). No place too cold. *Science (New York, N.Y.)*, *324*(5934), 1521–1522.

Learn-Han, L., Yoke-Kqueen, C., Shiran, M. S., Vui-Ling, C. M., Nurul-Syakima, A. M., Son, R., & Andrade, H. M. (2012). Identification of actinomycete communities in Antarctic soil from Barrientos Island using PCR-denaturing gradient gel electrophoresis. *Genetics and Molecular Research: GMR*, *11*(1), 277–291. Available from https://doi.org/10.4238/2012.February.8.3.

Loperena, L., Soria, V., Varela, H., Lupo, S., Bergalli, A., Guigou, M., Pellegrino, A., Bernardo, A., Calviño, A., Rivas, F., & Batista, S. (2012). Extracellular enzymes produced by microorganisms isolated from maritime Antarctica. *World Journal of Microbiology and Biotechnology*, *28*(5), 2249–2256. Available from https://doi.org/10.1007/s11274-012-1032-3.

López-Bueno, A., Tamames, J., Velázquez, D., Moya, A., Quesada, A., & Alcamí, A. (2009). High diversity of the viral community from an Antarctic lake. *Science (New York, N.Y.)*, *326*(5954), 858–861. Available from https://doi.org/10.1126/science.1179287.

Maharana, A. K., & Ray, P. (2015). Optimization and characterization of cold-active endoglucanase produced by *Aspergillus terreus* strain AKM-F3 grown on sugarcane bagasse. *Turkish Journal of Biology*, *39*(1), 175–185. Available from https://doi.org/10.3906/biy-1408-22.

Maida, I., Bosi, E., Fondi, M., Perrin, E., Orlandini, V., Papaleo, M. C., Mengoni, A., de Pascale, D., Tutino, M. L., Michaud, L., Lo Giudice, A., & Fani, R. (2015). Antimicrobial activity of Pseudoalteromonas strains isolated from the Ross Sea (Antarctica) vs cystic fibrosis opportunistic pathogens. *Hydrobiologia*, *761*(1), 443–457. Available from https://doi.org/10.1007/s10750-015-2190-8.

Margesin, R., & Feller, G. (2010). Biotechnological applications of psychrophiles. *Environmental Technology*, *31*(8–9), 835–844. Available from https://doi.org/10.1080/09593331003663328.

Melo, I. S., Zucchi, T. D., Silva, R. E., Vilela, E. S. D., Sáber, M. L., Rosa, L. H., & Pellizari, V. H. (2014). Isolation and characterization of cellulolytic bacteria from the Stain house Lake, Antarctica. *Folia Microbiologica*, *59*(4), 303–306. Available from https://doi.org/10.1007/s12223-013-0295-x.

Mohammed, S., Te'O, J., & Nevalainen, H. (2013). A gene encoding a new cold-active lipase from an Antarctic isolate of *Penicillium expansum*. *Current Genetics*, *59*(3), 129–137. Available from https://doi.org/10.1007/s00294-013-0394-x.

Mojib, N., Farhoomand, A., Andersen, D. T., & Bej, A. K. (2013). UV and cold tolerance of a pigment-producing Antarctic *Janthinobacterium* sp. Ant5–2. *Extremophiles: Life Under Extreme Conditions*, *17*(3), 367–378. Available from https://doi.org/10.1007/s00792-013-0525-9.

Mojib, N., Philpott, R., Huang, J. P., Niederweis, M., & Bej, A. K. (2010). Antimycobacterial activity in vitro of pigments isolated from Antarctic bacteria. *Antonie van Leeuwenhoek, International Journal of General and Molecular Microbiology*, *98*(4), 531–540. Available from https://doi.org/10.1007/s10482-010-9470-0.

Möller, C., & Dreyfuss, M. M. (1996). Microfungi from antarctic lichens, mosses and vascular plants. *Mycologia*, *88*(6), 922–933. Available from https://doi.org/10.2307/3761054.

Mou, S., Xu, D., Ye, N., Zhang, X., Liang, C., Liang, Q., Zheng, Z., Zhuang, Z., & Miao, J. (2012). Rapid estimation of lipid content in an Antarctic ice alga (*Chlamydomonas* sp.) using the lipophilic fluorescent dye BODIPY505/515. *Journal of Applied Phycology*, *24*(5), 1169–1176. Available from https://doi.org/10.1007/s10811-011-9746-4.

Muñoz, P. A., Márquez, S. L., González-Nilo, F. D., Márquez-Miranda, V., & Blamey, J. M. (2017). Structure and application of antifreeze proteins from Antarctic bacteria. *Microbial Cell Factories*, *16*(1). Available from https://doi.org/10.1186/s12934-017-0737-2.

Nadeau, T. L., & Castenholz, R. W. (2000). Characterization of psychrophilic oscillatorians (cyanobacteria) from antarctic meltwater ponds. *Journal of Phycology*, *36*(5), 914–923. Available from https://doi.org/10.1046/j.1529-8817.2000.99201.x.

Nascimento, T. C. E. d. S., Sena, A. R. d., Gomes, J. E. G., Santos, W. L. d., Agamez Montalvo, G. S., Tambourgi, E. B., Medeiros, E. V. d., Sette, L. D., Pessoa Junior, A., & Moreira, K. A. (2015). Extracellular serine proteases by *Acremonium* sp. L1–4B isolated from Antarctica: Overproduction using cactus pear extract with response surface methodology. *Biocatalysis and Agricultural Biotechnology*, *4*(4), 737–744. Available from https://doi.org/10.1016/j.bcab.2015.10.006.

Nicolaus, B., Lama, L., Esposito, E., Bellitti, M. R., Improta, R., Panico, A., & Gambacorta, A. (2000). Extremophiles in Antarctica. *Italian Journal of Zoology*, *67*, 169–174. Available from https://doi.org/10.1080/11250000009356373.

Niederberger, T. D., Sohm, J. A., Gunderson, T. E., Parker, A. E., Tirindelli, J., Capone, D. G., Carpenter, E. J., & Cary, S. C. (2015). Microbial community composition of transiently wetted Antarctic Dry Valley soils. *Frontiers in Microbiology*, *6*. Available from https://doi.org/10.3389/fmicb.2015.00009.

Ozerskaya, S., Kochkina, G., Ivanushkina, N., & Gilichinsky, D. A. (2009). Fungi in permafrost. In R. Margesin (Ed.), *Permafrost soils. Soil Biology* (16). Berlin, Heidelberg: Springer. Available from https://doi.org/10.1007/978-3-540-69371-0_7.

Pan, Y., Tan, G. Y., Convey, P., Pearce, D. A., & Tan, I. K. (2013). Diversity and bioactivity of actinomycetes from Signy Island terrestrial soils, maritime Antarctic. *Advances in Polar Science*, *24*(4), 208–212.

Pandey, K. D., Shukla, S. P., Shukla, P. N., Giri, D. D., Singh, J. S., Singh, P., & Kashyap, A. K. (2004). Cyanobacteria in Antarctica: Ecology, physiology and cold adaptation. *Cellular and Molecular Biology (Noisy-Le-Grand, France)*, *50*(5), 575–584.

Papa, R., Parrilli, E., & Sannia, G. (2009). Engineered marine Antarctic bacterium *Pseudoalteromonas haloplanktis* TAC125: A promising micro-organism for the bioremediation of aromatic compounds. *Journal of Applied Microbiology*, *106*(1), 49–56. Available from https://doi.org/10.1111/j.1365-2672.2008.03971.x.

Papale, M., Lo Giudice, A., Conte, A., Rizzo, C., Rappazzo, A. C., Maimone, G., Caruso, G., La Ferla, R., Azzaro, M., Gugliandolo, C., Paranhos, R., Cabral, A. S., Spica, V. R., & Guglielmin, M. (2019). Microbial assemblages in pressurized antarctic brine pockets

(Tarn flat, northern Victoria Land): A hotspot of biodiversity and activity. *Microorganisms, 7*(9). Available from https://doi.org/10.3390/microorganisms7090333.

Papaleo, M. C., Fondi, M., Maida, I., Perrin, E., Lo Giudice, A., Michaud, L., Mangano, S., Bartolucci, G., Romoli, R., & Fani, R. (2012). Sponge-associated microbial Antarctic communities exhibiting antimicrobial activity against *Burkholderia cepacia* complex bacteria. *Biotechnology Advances, 30*(1), 272–293. Available from https://doi.org/10.1016/j.biotechadv.2011.06.011.

Paudel, B., Bhattarai, H. D., Koh, H. Y., Lee, S. G., Han, S. J., Lee, H. K., Oh, H., Shin, H. W., & Yim, J. H. (2011). Ramalin, a novel nontoxic antioxidant compound from the Antarctic lichen *Ramalina terebrata. Phytomedicine: International Journal of Phytotherapy and Phytopharmacology, 18*(14), 1285–1290. Available from https://doi.org/10.1016/j.phymed.2011.06.007.

Pazgier, M., Turkiewicz, M., Kalinowska, H., & Bielecki, S. (2003). The unique cold-adapted extracellular subtilase from psychrophilic yeast *Leucosporidium antarcticum. Journal of Molecular Catalysis B: Enzymatic, 21*(1–2), 39–42. Available from https://doi.org/10.1016/S1381-1177(02)00134-0.

Pearce, D. A. (2009). Antarctic subglacial lake exploration: A new frontier in microbial ecology. *ISME Journal, 3*(8), 877–880. Available from https://doi.org/10.1038/ismej.2009.53.

Pearce, D. A. (2017). *Extremophiles in Antarctica: Life at low temperatures. Adaption of microbial life to environmental extremes: Novel research results and application* (2nd ed., pp. 99–131). Springer International Publishing. Available from https://doi.org/10.1007/978-3-319-48327-6_5.

Pearce, D. A., Bridge, P. D., Hughes, K. A., Sattler, B., Psenner, R., & Russell, N. J. (2009). Microorganisms in the atmosphere over Antarctica. *FEMS Microbiology Ecology, 69*(2), 143–157. Available from https://doi.org/10.1111/j.1574-6941.2009.00706.x.

Pearce, D. A., Hughes, K. A., Lachlan-Cope, T., Harangozo, S. A., & Jones, A. E. (2010). Biodiversity of air-borne microorganisms at Halley station, Antarctica. *Extremophiles: Life Under Extreme Conditions, 14*(2), 145–159. Available from https://doi.org/10.1007/s00792-009-0293-8.

Pereira, D. C., Giese, E. C., Moretti, G., Perrone, O. M., Boscolo, M., Martins., & B, D. A. (2017). Effect of metal ions, chemical agents and organic compounds on lignocellulolytic enzymes activities. In: M. Senturk (Ed.), *Enzyme Inhibitors and Activators, 29*, 139–164.

Petrescu, I., Lamotte-Brasseur, J., Chessa, J. P., Ntarima, P., Claeyssens, M., Devreese, B., Marino, G., & Gerday, C. (2000). Xylanase from the psychrophilic yeast *Cryptococcus adeliae. Extremophiles: Life Under Extreme Conditions, 4*(3), 137–144. Available from https://doi.org/10.1007/s007920070028.

Pirie, J. H. H. (1904). First Antarctic voyage of the Scotia. *Journal of Bacterial Scottish Geographical Magazine, 20*, 129–132.

Poveda, G., Gil-Durán, C., Vaca, I., Levicán, G., & Chávez, R. (2018). Cold-active pectinolytic activity produced by filamentous fungi associated with Antarctic marine sponges. *Biological Research, 51*(1). Available from https://doi.org/10.1186/s40659-018-0177-4.

Prescott, G. W. (1978). *How to Know the Freshwater Algae, Pictured Key Nature Series, 3rd edn. W. M. C Brown Company Publishers, Dubuque, Iowa.*

Pulicherla, K. K., Ghosh, M., Kumar, P. S., & Rao, K. R. S. S. (2011). Psychrozymes—The next generation industrial enzymes. *Journal of Marine Science Research and Development, 2*.

Quesada, A., & Vincent, W. F. (1997). Strategies of adaptation by antarctic cyanobacteria to ultraviolet radiation. *European Journal of Phycology, 32*(4), 335–342. Available from https://doi.org/10.1080/09670269710001737269.

Ramli, A. N. M., Azhar, M. A., Shamsir, M. S., Rabu, A., Murad, A. M. A., Mahadi, N. M., & Md. Illias, R. (2013). Sequence and structural investigation of a novel psychrophilic α-amylase from Glaciozyma antarctica PI12 for cold-adaptation analysis. *Journal of Molecular Modeling, 19*(8), 3369–3383. Available from https://doi.org/10.1007/s00894-013-1861-5.

Rani, K., Rana, R., & Datt, S. (2015). Review on characteristics and application of amylases. *International Journal of Microbiology and Bioinformatics, 5*, 1–5.

Rao, M. B., Tanksale, A. M., Ghatge, M. S., & Deshpande, V. V. (1998). Molecular and biotechnological aspects of microbial proteases. *Microbiology and Molecular Biology Reviews, 62*(3), 597–635. Available from https://doi.org/10.1128/mmbr.62.3.597-635.1998.

Ray, M. K., Devi, K. U., Kumar, G. S., & Shivaji, S. (1992). Extracellular protease from the antarctic yeast *Candida humicola. Applied and Environmental Microbiology, 58*(6), 1918–1923. Available from https://doi.org/10.1128/aem.58.6.1918-1923.1992.

Säwström, C., Anesio, M. A., Granéli, W., & Laybourn-Parry, J. (2007). Seasonal viral loop dynamics in two large ultraoligotrophic antarctic freshwater lakes. *Microbial Ecology, 53* (1), 1–11. Available from https://doi.org/10.1007/s00248-006-9146-5.

Schleper, C., Jurgens, G., & Jonuscheit, M. (2005). Genomic studies of uncultivated archaea. *Nature Reviews Microbiology, 3*(6), 479–488. Available from https://doi.org/10.1038/nrmicro1159.

Schuchardt, J. P., Huss, M., Stauss-Grabo, M., & Hahn, A. (2010). Significance of long-chain polyunsaturated fatty acids (PUFAs) for the development and behaviour of children. *European Journal of Pediatrics, 169*(2), 149–164. Available from https://doi.org/10.1007/s00431-009-1035-8.

Shahidi, F., & Janak Kamil, Y. V. A. (2001). Enzymes from fish and aquatic invertebrates and their application in the food industry. *Trends in Food Science and Technology, 12*(12), 435–464. Available from https://doi.org/10.1016/S0924-2244(02)00021-3.

Shivaji, S., & Reddy, G. S. N. (2010). *Polar microbiology: The ecology, biodiversity and bioremediation potential of microorganisms in extremely cold environments. edited by Bej A. K., Aislabie J. and Atlas R. M.*, 61–95, CRC Press (Taylor and Francis group).

Shivaji, S., Reddy, G. S. N., & Chattopadhyay, M. K. (2017). Bacterial biodiversity, cold adaptation and biotechnological importance of bacteria occurring in Antarctica. *Proceedings of the Indian National Science Academy, 83*(2), 327–352. Available from https://doi.org/10.16943/ptinsa/2017/48956.

Siddiqui, K. S., & Cavicchioli, R. (2006). Cold-adapted enzymes. *Annual Review of Biochemistry, 75*, 403–433. Available from https://doi.org/10.1146/annurev.biochem.75.103004.142723.

Singh, A. K., Singh, M. P., & Sisinthy, S. (2015). Molecular mechanisms of cold adaptation in bacteria. *Aperito Journal of Cellular and Molecular Biology, 1*(104), 2–9. Available from https://doi.org/10.14437/AJCMB-1-104.

Singh, S. M., Puja, G., & Bhat, G. J. (2006). *Psychrophilic fungi* from Schirmacher Oasis, East Antarctica. *Current Science, 90*(10), 1388–1392. Available from http://www.ias.ac.in/currsci/may252006/1388.pdf.

Soares, F. L., Melo, I. S., Dias, A. C. F., & Andreote, F. D. (2012). Cellulolytic bacteria from soils in harsh environments. *World Journal of Microbiology and Biotechnology, 28*(5), 2195–2203. Available from https://doi.org/10.1007/s11274-012-1025-2.

Song, C., Chi, Z., Li, J., & Wang, X. (2010). β-Galactosidase production by the psychrotolerant yeast *Guehomyces pullulans* 17–1 isolated from sea sediment in Antarctica and lactose hydrolysis. *Bioprocess and Biosystems Engineering, 33*(9), 1025–1031. Available from https://doi.org/10.1007/s00449-010-0427-5.

Srinivasan, M. C., & Rele, M. V. (1995). Cellulase-free xylanases from microorganisms and their application to pulp and paper biotechnology: An overview. *Indian Journal of Microbiology, 35*, 93, 93.

Steven, B., Léveillé, R., Pollard, W. H., & Whyte, L. G. (2006). Microbial ecology and biodiversity in permafrost. *Extremophiles: Life Under Extreme Conditions, 10*(4), 259–267. Available from https://doi.org/10.1007/s00792-006-0506-3.

Sukumaran, R. K., Singhania, R. R., & Pandey, A. (2005). Microbial cellulases-production, applications and challenges. *Journal of Scientific and Industrial Research, 64*(11), 832–844.

Suutari, M., & Laakso, S. (1992). Changes in fatty acid branching and unsaturation of *Streptomyces griseus* and *Brevibacterium fermentans* as a response to growth temperature.

Applied and Environmental Microbiology, 58(7), 2338–2340. Available from https://doi.org/10.1128/aem.58.7.2338-2340.1992.

Szczęsna-Antczak, M., Kamińska, J., Florczak, T., & Turkiewicz, M. (2014). *Cold-active yeast lipases: Recent issues and future prospects. Cold-adapted yeasts* (pp. 353–375). Berlin, Heidelberg: Springer.

Thomas, D. N., & Dieckmann, G. S. (2002). Antarctic sea ice—A habitat for extremophiles. *Science (New York, N.Y.), 295*(5555), 641–644. Available from https://doi.org/10.1126/science.1063391.

Thomas-Hall, S. R., Turchetti, B., Buzzini, P., Branda, E., Boekhout, T., Theelen, B., & Watson, K. (2010). Cold-adapted yeasts from Antarctica and the Italian Alps-description of three novel species: *Mrakia robertii* sp. nov., *Mrakia blollopis* sp. nov. and *Mrakiella niccombsii* sp. nov. *Extremophiles: Life Under Extreme Conditions, 14*(1), 47–59. Available from https://doi.org/10.1007/s00792-009-0286-7.

Tropeano, M., Coria, S., Turjanski, A., Cicero, D., Bercovich, A., Mac Cormack, W., & Vazquez, S. (2012). Culturable heterotrophic bacteria from Potter Cove, Antarctica, and their hydrolytic enzymes production. *Polar Research, 31*(1), 18507. Available from https://doi.org/10.3402/polar.v31i0.18507.

Tsiklinsky, M. (1908). La flore microbienne dans les regions du Pole Sud. *Expedition Antarctique Française, 1903–1905*, 1–33.

Tsuji, M., Yokota, Y., Shimohara, K., Kudoh, S., & Hoshino, T. (2013). An application of wastewater treatment in a cold environment and stable lipase production of Antarctic Basidiomycetous yeast *Mrakia blollopis*. *PLOS ONE, 8*(3). Available from https://doi.org/10.1371/journal.pone.0059376.

Turk, B. (2006). Targeting proteases: Successes, failures and future prospects. *Nature Reviews Drug Discovery, 5*(9), 785–799. Available from https://doi.org/10.1038/nrd2092.

Turkiewicz, M., Kur, J., Białkowska, A., Cieśliński, H., Kalinowska, H., & Bielecki, S. (2003). Antarctic marine bacterium *Pseudoalteromonas* sp. 22b as a source of cold-adapted β-galactosidase. *Biomolecular Engineering, 20*(4–6), 317–324. Available from https://doi.org/10.1016/S1389-0344(03)00039-X.

Van De Voorde, I., Goiris, K., Syryn, E., Van Den Bussche, C., & Aerts, G. (2014). Evaluation of the cold-active *Pseudoalteromonas haloplanktis* β-galactosidase enzyme for lactose hydrolysis in whey permeate as primary step of d-tagatose production. *Process Biochemistry, 49*(12), 2134–2140. Available from https://doi.org/10.1016/j.procbio.2014.09.010.

Vaz, A. B. M., Rosa, L. H., Vieira, M. L. A., de Garcia, V., Brandão, L. R., Teixeira, L. C. R. S., Moliné, M., Libkind, D., van Broock, M., & Rosa, C. A. (2011). The diversity, extracellular enzymatic activities and photoprotective compounds of yeasts isolated in Antarctica. *Brazilian Journal of Microbiology, 42*(3), 937–947. Available from https://doi.org/10.1590/S1517-83822011000300012.

Vollú, R. E., Jurelevicius, D., Ramos, L. R., Peixoto, R. S., Rosado, A. S., & Seldin, L. (2014). Aerobic endospore-forming bacteria isolated from Antarctic soils as producers of bioactive compounds of industrial interest. *Polar Biology, 37*(8), 1121–1131. Available from https://doi.org/10.1007/s00300-014-1505-y.

Violot, S., Aghajari, N., Czjzek, M., Feller, G., Sonan, G. K., Gouet, P., … Receveur-Brechot, V. (2005). Structure of a full length psychrophilic cellulase from Pseudoalteromonas haloplanktis revealed by X-ray diffraction and small angle X-ray scattering. *Journal of molecular biology, 348*(5), 1211–1224.

Wang, M., Tian, J., Xiang, M., & Liu, X. (2017). Living strategy of cold-adapted fungi with the reference to several representative species. *Mycology, 8*(3), 178–188. Available from https://doi.org/10.1080/21501203.2017.1370429.

Wang, Q. F., Hou, Y. H., Xu, Z., Miao, J. L., & Li, G. Y. (2008). Purification and properties of an extracellular cold-active protease from the psychrophilic bacterium Pseudoalteromonas sp. NJ276. *Biochemical Engineer Journal, 38*, 362–368.

Wang, X., Kan, G., Ren, X., Yu, G., Shi, C., Xie, Q., ... Betenbaugh, M. (2018). Molecular cloning and characterization of a novel alpha-amylase from Antarctic sea ice bacterium *Pseudoalteromonas* sp. M175 and Its primary application in detergent. *BioMed Research International, 2018*(12), 1–16.

Wang, Q. F., Miao, J. L., Hou, Y. H., Ding, Y., Wang, G. D., & Li, G. Y. (2005). Purification and characterization of an extracellular cold-active serine protease from the psychrophilic bacterium *Colwellia* sp. NJ341. *Biotechnology Letters, 27*(16), 1195–1198. Available from https://doi.org/10.1007/s10529-005-0016-x.

Wi, A. R., Jeon, S. J., Kim, S., Park, H. J., Kim, D., Han, S. J., ... Kim, H. W. (2014). Characterization and a point mutational approach of a psychrophilic lipase from an arctic bacterium. *Bacillus pumilus. Biotechnology letters, 36*(6), 1295–1302.

Wille, M., Aban, M., Wang, J., Moore, N., Shan, S., Marshall, J., González-Acuña, D., Vijaykrishna, D., Butler, J., Wang, J., Hall, R. J., Williams, D. T., & Hurt, A. C. (2019). Antarctic penguins as reservoirs of diversity for avian avulaviruses. *Journal of Virology, 93*(11). Available from https://doi.org/10.1128/JVI.00271-19.

Wille, M., Harvey, E., Shi, M., Gonzalez-Acuña, D., Holmes, E. C., & Hurt, A. C. (2020). Sustained RNA virome diversity in Antarctic penguins and their ticks. *ISME Journal, 14*(7), 1768–1782. Available from https://doi.org/10.1038/s41396-020-0643-1.

Williams, T. J., Burg, D. W., Raftery, M. J., Poljak, A., Guilhaus, M., Pilak, O., & Cavicchioli, R. (2010). Global proteomic analysis of the insoluble, soluble, and supernatant fractions of the psychrophilic archaeon Methanococcoides burtonii part I: The effect of growth temperature. *Journal of Proteome Research, 9*(2), 640–652. Available from https://doi.org/10.1021/pr900509n.

Wynn-Williams, D. D. (1996). Response of pioneer soil microalgal colonists to environmental change in Antarctica. *Microbial Ecology, 31*(2), 177–188. Available from https://doi.org/10.1007/BF00167863.

Xiao, N., Suzuki, K., Nishimiya, Y., Kondo, H., Miura, A., Tsuda, S., & Hoshino, T. (2010). Comparison of functional properties of two fungal antifreeze proteins from *Antarctomyces psychrotrophicus* and *Typhula ishikariensis. FEBS Journal, 277*(2), 394–403. Available from https://doi.org/10.1111/j.1742-4658.2009.07490.x.

Yamashita, Y., Nakamura, N., Omiya, K., Nishikawa, J., Kawahara, H., & Obata, H. (2002). Identification of an antifreeze lipoprotein from *Moraxella* sp. of antarctic origin. *Bioscience, Biotechnology, and Biochemistry, 66*(2), 239–247. Available from https://doi.org/10.1271/bbb.66.239.

Yang, J. K., Liu, L. Y., Dai, J. H., & Li, Q. (2013). De novo design and synthesis of *Candida antarctica* lipase B gene and α-factor leads to high-level expression in *Pichia pastoris. PLOS ONE, 8*(1). Available from https://doi.org/10.1371/journal.pone.0053939.

Zawar-Reza, P., Argüello-Astorga, G. R., Kraberger, S., Julian, L., Stainton, D., Broady, P. A., & Varsani, A. (2014). Diverse small circular single-stranded DNA viruses identified in a freshwater pond on the McMurdo Ice Shelf (Antarctica). *Infection, Genetics and Evolution, 26*, 132–138. Available from https://doi.org/10.1016/j.meegid.2014.05.018.

Zhang, X., Hua, M., Song, C., & Chi, Z. (2012). Occurrence and diversity of marine yeasts in Antarctica environments. *Journal of Ocean University of China, 11*(1), 70–74. Available from https://doi.org/10.1007/s11802-012-1820-2.

CHAPTER

13

Isolation methods for evaluation of extremophilic microbial diversity from Antarctica region

Janki Ruparelia, Aniruddh Rabari, Nishra Joshi and Chaitanya Kumar Jha

Microbiology Department, Gujarat Arts and Science College, Ahmedabad, India

13.1 Introduction

In Antarctica, the landforms are a complex montage of extremophilic habitats of single-celled members of the domain Eukarya, Bacteria, and Archaea. Over the last eras, scientists have been fascinated by role-playing extremophilic microorganisms that inhabit in Antarctica region. Microbiological investigations of Antarctic ice at various profundities may convey significant correlations, by way of their purview records of contagious cells and ancient environment (Antony et al., 2012). The microorganisms existing in Antarctica's biodiversity show the foremost role in nitrogen, carbon, phosphorus, and sulfur cycles (Ogaki et al., 2020). For the study of extremophilic diversity from Antarctica hotspot, proper cultivation techniques are required for their further studies and characterization. For the study of extremophilic diversity, certain isolation methods are used which are divided into the main two groups, namely, isolation methods for cultivable and noncultivable microorganisms.

The cells that form a colony on specific nutrient media are known as "cultivable organisms." The cells which are metabolically or physiologically active are known as "viable cells." The cells which are metabolically or physiologically active but cannot be cultivated on a particular medium are known as (VBNC) "viable but nonculturable cells" (Bogosian & Bourneuf, 2001).

Microbial Diversity and Ecology in Hotspots
DOI: https://doi.org/10.1016/B978-0-323-90148-2.00008-0

13.2 Diversity of extremophilic organisms in Antarctic hotspot

Recent endeavors to discover marine microbial diversity and the universal extreme microbiome have indicated a large proportion of previously unknown diversity (Iain et al., 2016). Antarctica atmospheres endure a tremendous microbial diversity, especially of extremophiles. The topography of Antarctica provides for glacial ice, seawater, snow, lake ice, soil, and sediments varying temperature normal to ultra-cold and salinity fresh to saturations disparity has also been found in pressure and pH as shown in Fig. 13.1. These extremophiles are able to grow only in a constricted sort of low temperature (stenothermal) which are usually known as "obligate psychrophiles." Additional group is designated as facultative psychrophiles or psychrotrophic organisms which are able to cultivate a broader range of temperatures (Loperena et al., 2012) as given in Table 13.1.

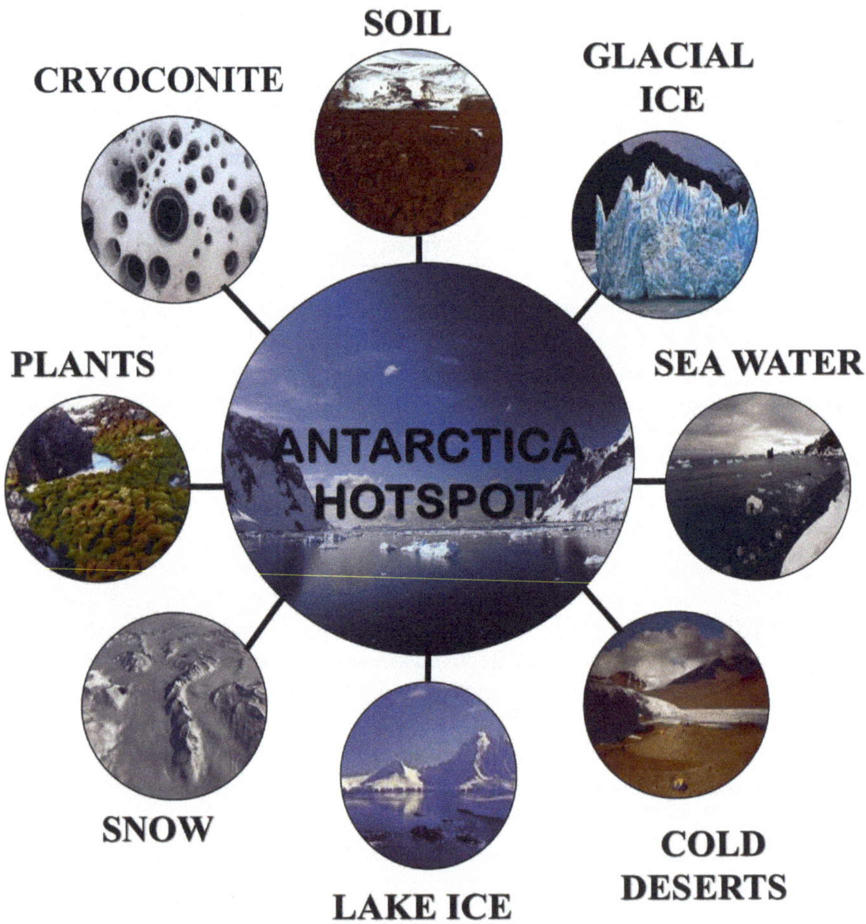

FIGURE 13.1 The complex topography of Antarctic hotspot.

TABLE 13.1 Extremophiles-their types, nature, and identification method.

Name of the strain	Location	Method of identification	Extreme nature of microorganism	Eukaryote/ Prokaryote/ Archae	Reference
Alicyclobacillus acidocaldarius subsp. *rittmannii*	Terranova Bay Station in Antarctica	16S rDNA sequencing	Thermophile (63°C)	Bacteria	Nicolaus et al. (1998)
Bacillus thermoantarctius (DSM 9572)	Near the crater of Mount Melbourne, Antarctica	16S rDNA sequencing	Thermophile (63°C)	Bacteria	Nicolaus et al. (1996)
Methanogenium fngidum	Ace lake in the Vestfold Hills of Antarctica	16S rRNA sequencing	Psychrophilic methanogen (15°C)	Archae	Franzmann et al. (1997)
Clostridium schirmacherense	Lake sediment of Schirmacher Oasis, Antarctica	16S rRNA sequencing	Psychrophilic anaerobe (5°C –10°C)	Bacteria	Alam et al. (2006)
Halobacterium lacusprofundi	Monomictic lake in the Vestfold Hills, Antarctica	16S rRNA sequencing	Halophile (3.5 mol/L Na^+) and Psychrophile (4°C)	Bacteria	Franzmann et al. (1988)
Nesterenkonia aurantiaca	Cape King, Antarctica	16S rRNA sequencing	Haloalkaliphilic (pH 9.0 and 2% NaCl)	Actinobacteria	Finore et al. (2016)
Halomonas alkaliantarctica	Cape Russell in Antarctica	16S rRNA sequencing + high-pressure liquid chromatography	Alkali tolerant and halotolerant (pH 9.0 and 22.2% NaCl)	Bacteria	Poli et al. (2007)
Halohasta litchfieldiae	Deep Lake, Antarctica	Cultural method and 16S rRNA sequencing	Halophile (180 g/L NaCl)	Archae	Mou et al. (2012)
Cellulophaga algicola	Eastern Antarctic Coastal Zone	16S rRNA sequencing	Psychrophilic algae (2°C–10°C)	Algae	Bowman (2000)
Penicillium italicum	Archipelago South Shetland Islands, West Antarctica	Cultural method using Cooke Rose Bengal Agar medium	Psychrophilic fungi (10°C)	Fungi	Kostadinova et al. (2009)

(*Continued*)

TABLE 13.1 (Continued)

Name of the strain	Location	Method of identification	Extreme nature of microorganism	Eukaryote/ Prokaryote/ Archae	Reference
Monodictys austrina Tubaki	Archipelago South Shetland Islands, West Antarctica	Cultural method using Cooke Rose Bengal Agar medium	Psychrophilic fungi (10°C)	Fungi	Kostadinova et al. (2009)
Colwellia rossensis	Ice interface of fast sea ice, McMurdo South Antarctica	16S rDNA sequencing	Psychrophile (10°C–12°C)	Bacteria	Bowman et al. (1998)
Cryptococcus nyarrowii	Lichen Valley Vestfold Hills, Davis Base, Antarctica	26S rDNA sequencing	Psychrophilic fungi (15°C)	Yeast (Fungi)	Thomas-Hall et al. (2002)
Colwellia demingiae ACAM 459	Prydz Bay Coast, Antarctica	16S rDNA sequencing	Psychrophile (10°C–12°C)	Bacteria	Bowman et al. (1998)
Glaciozyma martinii	Lake Fletcher and Lichen Valley, Vestfold Hills Area of Davis Base, Antarctica	26S rDNA sequencing	Psychrophilic fungi (10°C)	Yeast (Fungi)	Turchetti et al. (2011)
Hymenobacter artigasi	King George Island, Maritime Antarctica	16S rRNA sequencing	Psychrophile (10°C)	Bacteria	Roldán et al. (2020)
Chlorella sp. strain B1	Ross Sea, South of Bratina Island Antarctica	16S rRNA sequencing and 18S rRNA sequencing	Psychrophilic algae (10°C)	Algae	Morgan-Kiss et al. (2008)
Glaciozyma watsonii	Lake Fletcher and Lichen Valley, Vestfold Hills Area of Davis Base, Antarctica	26S rDNA sequencing	Psychrophilic fungi (10°C)	Yeast (Fungi)	Turchetti et al. (2011)
Psychrosinus fermentans	McMurdo Dry Valleys, Antarctica	16S rRNA sequencing	Psychrophile (14°C)	Bacteria	Sattley et al. (2008)

The phylogenetic tree constructed on 16S rRNA illustrate that all organisms have a common ancestor as these extremophiles are divided into three groups, Eucarya, Bacteria, and Archaea (Imanaka, 2002). This region is found to be a commendable source of extremophiles which shows the compound topography of this region (Fig. 13.1) where extremophiles show diversity within adaption of temperature (Thermophiles, Psychrophiles), pressure (Piezophiles), and salt concentrations (Halophiles) (Table 13.1). It also provides information on eukaryotes, prokaryotes, and archaea identification methods using 16S rRNA sequencing, high-pressure liquid chromatography, 26S rDNA, 18S rRNA sequencing, and use of different culture medium (Table 13.1). In a current survey, it is found that (Pellizzari et al., 2016) 104 macroalgal taxa (28 Phaeophyceae, 24 Chlorophyta, and 52 Rhodophyta) in the South Shetland archipelago, demonstrated about assortment was cumulative for the reason that additional effectual random sample techniques or modifications in a particular region that might be a result of regular scattering which goes along with meteoceanographic variations. Environmental diversity catalogs revealed about the fungal gathering are assorted and ironic. The genera *Cladophialophora, Cladosporium,* and *Penicillium* displayed extensive distribution with rocks on several islets (Alves et al., 2019).

13.3 Isolation and identification methodology

To identify and study a particular organism from a mixed group of population proper cultivation and isolation methods are indispensable. These are divided into isolation methods for cultivable and noncultivable microorganisms. The isolation and identification methods for culturable extremophiles is shown in Fig. 13.2.

Collection of sample: As discussed earlier in this chapter, Antarctic hotspot has complex topography as described in Fig. 13.1. Samples can be collected from either of any region, namely, cryoconite, glacial ice, seawater, snow, lake ice, soil, and sediments. All samples should be first collected then selected for further screening and air plants should be detached, followed by washing, and then it should be well maintained in formalin followed by seawater dilution up to 4% in the workroom for further cultivation and microscopic evaluation (Furbino et al., 2018). The sample collection and preserving techniques may vary according to environmental habitats and substance availability.

Methods for sampling: A sample is a subsection of entities from abundant inhabitants. Probability sampling methods include simple random sampling, systematic sampling, stratified sampling, cluster sampling, multistage sampling, and area sampling.

1. *Simple random sampling*: Indiscriminately selected specimen of the population is known as simple random sampling where every

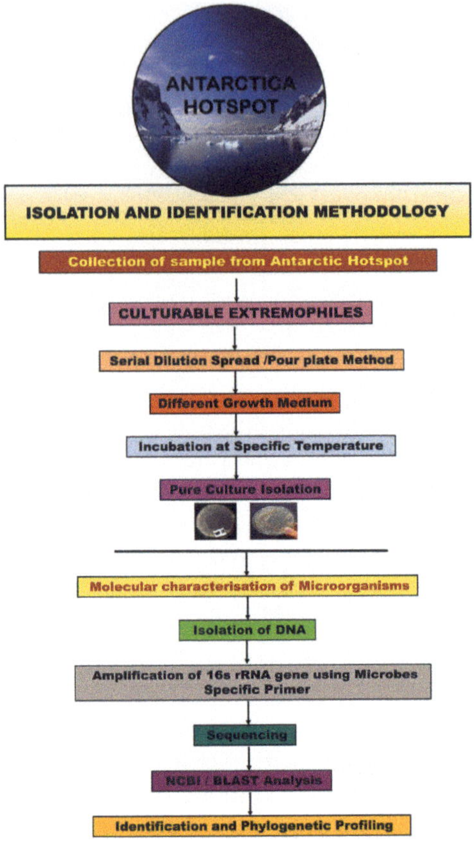

FIGURE 13.2 Isolation and identification method for culturable extremophiles.

member of the population has the same probability of selection. This technique is the supreme forthright of all the probability sampling methods; meanwhile, it only includes a sole random collection and needs little advanced information about the population. Because of randomization, studies involving these samples should have high inside and peripheral plausibility. (Sundar & Richard, 2004).

2. *Systemic sampling*: This method involves the first sampling collection randomly and then the following methods are adopted in an intermittent process. The systematic sampling method selects every kth item for sampling and k is dictated by diving the number of items in the sample frame from the desired sample size. Here, random sampling is done for the initial sample and then every kth item on the list is selected, where k is dictated by dividing a total number of items in the whole population with the preferred sample size (Acharya et al., 2013). For example, if we take $N = 100$ and $n = 25$, so,

$k = N/n$, that is, 4, which develops the sampling interval. At this instant, if we select a random number from 1 to 4. Assume it is "2," so the number "2" applicant is our first matter. Then we go on adding "4" to this number. Our following matters would be 6, 10, 14, 18, and 22 so on till we complete the mandatory sample size of 25.

3. *Stratified sampling*: This sampling type includes a proper division of the population into subdivisions and then performing random sampling from every subdivision. A subdivision is a normal set of matters. This sampling method is preferred when there is a vast difference exists within the population (Hamed, 2016).
4. *Cluster sampling*: Cluster sampling is chiefly used when the population is scattered as clusters. Consequently, random sampling will be performed from all these clusters which collectively used in the final sample (Wilson, 2014). Steps to perform cluster sampling could involve these criteria: Select cluster groups for the given population, for example, type of education, income etc. ... label every cluster, perform random sampling from each cluster (Hamed, 2016).
5. *Multistage sampling*: In this type of sampling, cluster sampling is performed at multiple stages. This sampling method is used when the population to be studied is immensely large and it is impossible to make clusters directly without making the smaller subcluster in every cluster. This type of downscaling makes clusters smaller at every stage and random sampling shall be performed at the smallest stage of the clusters which is also called the "Final sampling unit." (Menon et al., 2004).
6. *Area sampling*: It is a kind of design sampling that compacts with the sector of environment that signifies bunches of units that concentrated on terrestrial locations (Ilker, 2017).

13.3.1 Isolation methodology for culturable extremophiles

13.3.1.1 *Culturable extremophiles*

Culturable extremophiles use numerous synthetic cultures for their growth and metabolism. To isolate and identify specific organisms from a mixed population, a serial dilution of the sample is a preliminary technique. The serial dilution technique is divided into two methods for screening purpose, namely, serial dilution spread plate technique and serial dilution pour plate method.

13.3.1.2 *Serial dilution spread/pour plate method*

Serial dilution technique: Serial dilution is the incremental dilution technique for a particular component, that is, water or salt in the

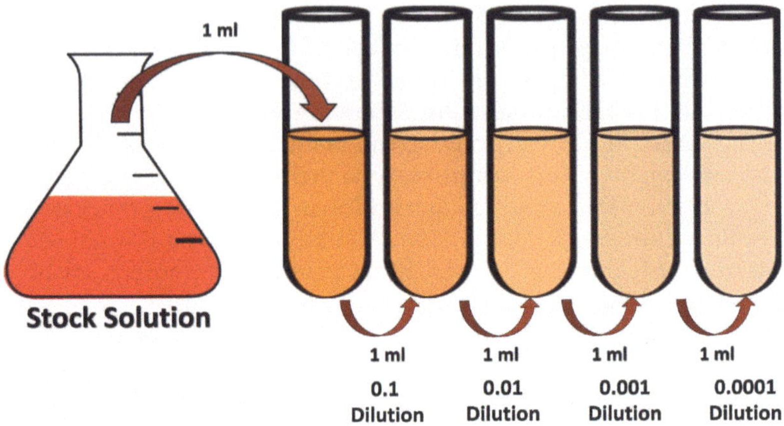

FIGURE 13.3 Serial dilution method.

solution. Ordinarily, dilution factors remain constant, and dilution increments logarithmically. The serial dilution method is shown in Fig. 13.3. A 10-fold serial dilution can be 1, 0.1, 0.01, 0.001 M, and so on (Fig. 13.3). It is the most direct culturable method for the isolation of extremophilic organisms. This method is also known as culture-most probable number assay (SDC-MPN, also called the extinction dilution method) (Throndsen & Kristiansen, 1991). The methodology is created on a bacterial assay established several ages ago (Cullen & MacIntyre, 2016). The principal aim of this method is to determine the probable number of organisms, that is, bacteria, viruses etc. in an unknown sample by counting colonies grown from the ultimate dilution of the sample and then multiplying it with the dilution factor we have imparted (Ben-David & Davidson, 2014).

Spread plate technique: This methodology involves a technique to plate an accurate amount of a liquid sample comprising microorganisms onto the specific culture medium and spread it thoroughly using a sterile glass spreader so that the organisms could be spread thoroughly and it becomes easy for isolation. An efficacious spread plate would be having a countable number of isolated bacterial colonies thoroughly disseminated on the plate. The plate is then kept inverted and incubated at 37°C for 24–48 hours.

Pour plate method: The pour plate method is used for counting the number of colony-forming bacteria present in a liquid sample. In this method, the inoculum (1.0 mL) from a broth is placed in the center of the sterile Petri plate using a sterile pipette. Molten cooled agar (15.0 mL) is then poured into the Petri plate containing the inoculum and mixed well. After the solidification of the agar, the plate is inverted and incubated at 37°C for 24–48 hours. Microorganisms will grow both on the surface and within the medium. Colonies that grow within the medium generally are small in size and maybe confluent; the few that grow on the agar surface are of the same size and appearance as those on a streak plate. Each colony that represents the "colony-forming unit" (CFU) is counted using a magnifying colony counter. The number of microorganisms present in the particular test sample is determined using the formula, CFU/mL = CFU × dilution factor × 1/aliquot (Goldman & Green, 2015). For precise counts, the optimum count should be within the range of 30–300 colonies per plate. To ensure a countable plate, a series of dilutions should be plated.

For example, cultivable bacteria found in the Antarctic region Comprised adherents of the *Firmicutes, Actinobacteria, Proteobacteria, Cyanobacteria,* and *Deinococcus thermus* has been reported by Antibus et al. (2012) in which the techniques used were heterotrophic plate counts, autotrophic and heterotrophic enrichment cultures studies from 50 Antarctic rock samples, 386 fungal isolates were obtained and identified as 33 taxes of 20 genera. The genera *Cladophialophora, Cladosporium, Cyphellophora, Eichleriella, Paracladophialophora,* and *Penicillium* displayed not only bacteria but the cultivable fungus was also isolated by these techniques. Fungi CFUs were used to define the density of fungus (Alves et al., 2019).

Incubation at specific temperature: Incubation temperature is one of the most significant steps for cultivation. If an appropriate temperature is not given for incubation, it might have a probability of mislaying the culture. Where straight pouring/plating of the igneous frost samples on solidified agar media cannot give rise to colonies on any media despite approximately half-year of incubation at both 4°C ± 0.1°C and 20°C ± 0.5°C. Nevertheless, in the liquescent medium, progressive growth can be obtained within 6–29 days of cultivation at 20°C ± 0.5°C (Antony et al., 2012). This shows that organisms prefer various temperatures (Table 13.2).

Identification of pure culture: Identification procedure of pure culture is followed by molecular characterization which uses isolation of DNA and sequencing techniques described in the 13.3.2.3 of the chapter.

13.3.2 Isolation methodology for unculturable extremophiles

Unculturable extremophiles: The organisms which do not grow under laboratory conditions are known to be uncultivable organisms. Specific physical and chemical parameters (environmental) which are firm or unmanageable to replicate in the workroom are necessary for many species. These species are known as "unculturable species." The different approaches for the characterization and identification of unculturable species are represented in Fig. 13.4. Identification of such extremophiles is based on metagenomics, which is followed by isolation of Community DNA, metagenomics library, 16S rRNA/18S rRNA, activity-based sequencing, clone showing activity, and study of expression function.

TABLE 13.2 Culture medium with extremophilic microorganisms and their types.

Culturable microorganisms	Culture medium	Type of extremophilic	References
Alicyclobacillus sp.	Yeast extract	Thermophilic	Nicolaus et al. (2001)
Planococcus psychrophilus	Peptone-yeast extract medium	Psychrophilic	Reddy et al. (2002)
Glaciozyma antarctica PI12	Nitrogen-free Davis medium	Psychrophilic yeast	Firdaus-Raih et al. (2018)
Microbacterium sp. LEMMJ01	Lb medium	Unknown	Reis-Mansur et al. (2019)
Penicillium solitum	Marine agar	Psychrotolerant	Gonçalves et al. (2013)
Halomonas glaciei	ZoBell 2216 Agar	Halotolerant Psychrophile	Celussi et al. (2008)
Leucosporidiella fragaria	Dichloran Rose Bengala Chloramphenicol Agar (DRBCA)	Psychrophile (yeast)	Matrinez et al. (2016)
Antarctomyces pellizariae	Sabouraud Dextrose Agar (SDA)	Psychrophilic	de-Menezes et al. (2017)
Sporosarcina aquimarina	Luria–Bertani (LB) medium	Psychrophilic	Santos et al. (2015)
Halobacterium lacusprofundi	Artificial Deep Lake Vitamin Agar	Halophilic	Franzmann et al. (1988)
Marinobacter guineae	Marine Agar	Halophilic (moderate)	Montes et al. (2008)
Hymenobacter amundsenii	R2A (Oxoid) Agar	Psychrophilic	Sedláček et al. (2019)

13.3.2.1 Metagenome

Metagenomics word was firstly given by Jo Handelsman, Jon Clardy, Robert M. Goodman, Sean F. Brady, and others which was first time published in the year 1998 (Handelsman et al., 1998). Metagenomics, environmental genomics, economics, or community genomics can also be referred to as the broad field, Initial sequencing of environmental gene copied exact genes frequently the 16S rRNA gene to create a form of variety in a likely sample, whereas conventional microbial technology, genome sequencing of microbes and genomics depends on cultured clones. Such studies found that the enormous mainstream of microbial biodiversity was overlooked by methodology focused on cultivation (Handelsman et al., 1998).

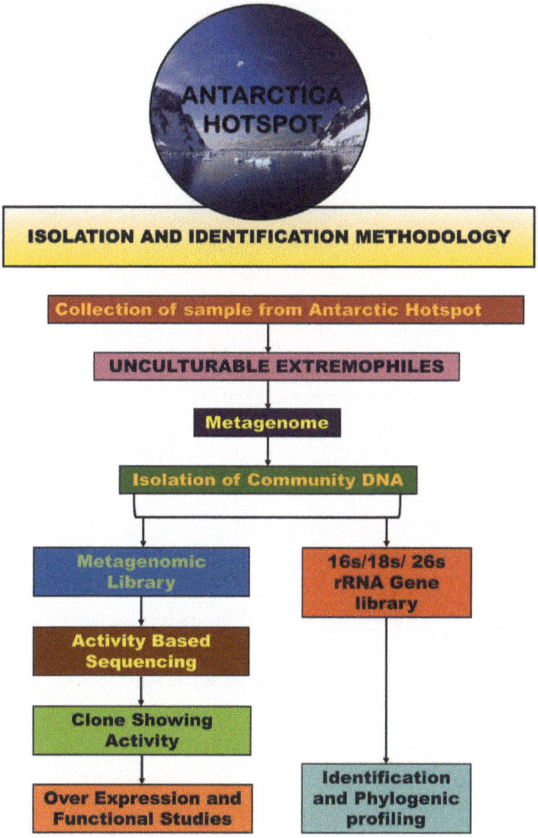

FIGURE 13.4 Characterization and identification for identification for unculturable extremophiles.

Metagenomics is an enormously powerful omics technique used for investigation of nonculturable microbes and arduous mainly on having a subterranean appreciative of worldwide microbial ecosystem in various niches of the ecological unit (Handelsman, 2004). Against the context of the general conjecture that in any given area, more than 99% of bacteria are incompetent of growing when traditional methodologies are used. Metagenomic approaches have materialized as an attractive solution to allow the microbial genomes present in these environments to be studied (Amann et al., 1995).

13.3.2.2 Isolation of community DNA

Nucleic acid extraction is the key of every evaluation in molecular biology and is therefore regarded as a core part. In 1869, the Swiss scientist Friedrich Miescher performed the first crude extraction of DNA, when he was studying proteins from leukocytes, he unintentionally purified DNA from the nucleus and noticed that the assets of this constituent was profoundly diverse from proteins, therefore devised the word "nucleic" (Dahm, 2005).

The key difficulties in extracting genetic material from environmental samples include the quality and quantity of retrieval of the intact material for further study. Therefore it depends on the intent of the analysis to choose the optimal method for nucleic acid extraction of a sample (Dias et al., 2014). It can be taken directly or indirectly to extract the genetic material from microorganisms found in a given marine sample. The isolation and selection of cells from the matrix and its subsequent lysis and purification of genetic material consists of indirect DNA extraction in comparison to direct extraction techniques which involve prior Cell lysing in the model conditions and the consequent collection and purification of genetic material. But the drawback is this technique typically has lower yields (Leff et al., 1995; Gabor et al., 2003).

Direct DNA isolation within the sample matrix is based on cell lysis and subsequent separation of DNA from the matrix and cell debris, whereas the indirect approach includes the removal of cells from the environmental material prior to the lytic release of DNA (Gabor et al., 2003; Holben et al., 1988; Ogram et al., 1987) has been found that this genetic material (environmental sample) can consist of up to 93% eukaryotic DNA, which also prohibits the use of this material for bacterial gene screening, whereas indirect techniques have also been considered as an alternative for metagenomic research, with a focus on preserving microbial diversity, gene integrity and prokaryotic material purity in relation to eukaryotic DNA content.

The extraction of genomic DNA is shown in. The key steps behind DNA extraction are (1) breakdown of the cytoplasmic and nuclear membranes; (2) isolation and purification of DNA from other cell lysates elements, such as lipids, proteins, and other nucleic acids; and (3) DNA concentration and purification (Ali et al., 2017) (Fig. 13.5). There are several DNA extraction techniques available, so choose an effective DNA extraction tool, depending on what is the determination of your study.

13.3.2.3 16S rRNA, 18S rRNA, and 26S rRNA

16S ribosomal RNA (rRNA) sequencing is a technique of amplicon sequencing used to classify and compare bacterial species existing in a known sample. The gene of 16S rRNA sequencing is used to research phylogeny and taxonomy of samples from diverse microbiomes or ecosystems that are difficult or impossible to research. 18S rRNA sequencing for eukaryotic phylogeny investigation of both rare and abundant fungal and micro eukaryotic organisms from limited amounts of DNA in complex populations, 18S rRNA sequencing that captures the

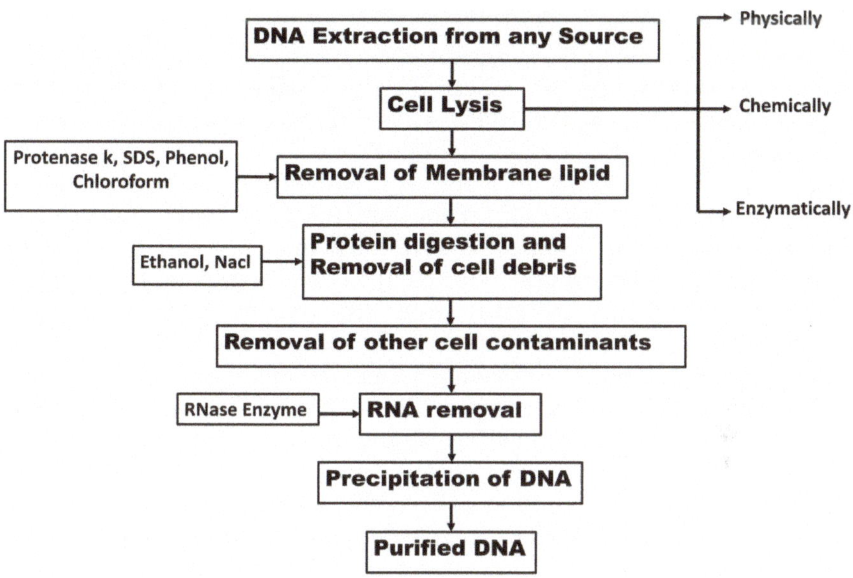

FIGURE 13.5 Extraction of genomic DNA.

ITS1and/or ITS2 hypervariable Internal Transcribed Spacer (ITS) regions are recommended (PerkinElmer).

Generally 16S and 18S rRNA sequencing governing the steps like extraction of DNA from the marine sample, polymerase chain reaction requires for making copies of extracted DNA, Agarose gel electrophoresis for separating DNA on the basis of their molecular charge, Elution of DNA describes the definite bands extraction of DNA from agarose gel from which they are separated by means of using electrophoresis, Radiolabeling technique has the ability to label nucleic acids, Autoradiography a system used to distinguish radioactive compounds through the utility of photographic emulsion is known as Autoradiography. These are the basic methods for 16S rRNA and 18S rRNA sequencing (Gartner, 2020).

According to Christner et al. (2003) organisms were isolated from an Antarctic cryoconite hole. The sequences acquired discovered the existence of eight bacterial ancestries, namely, *Acidobacterium, Actinobacteria, Cyanobacteria, Cytophagales, Gemmimonas, Planctomycetes, Proteobacteria, Verrucomicrobia*, and *metazoan, truffle, ciliate*, and *green algal* Eukarya. The *D1/D2* domain is part of the ribosomal RNA (Fell et al., 2000) nuclear large subunit and is therefore located as ITS in the same ribosomal tandem repeat cluster of the gene. *D1/D2* (or ITS) maybe used alone with respect to the molecular identification of basidiomycetous yeasts (Scorzetti et al., 2002). However (Fell et al., 2000) the combined analysis of the *D1/D2* and ITS regions.

13.3.2.4 Identification and phylogenetic analysis

A diagram presenting the progression interrelations of a cluster of species resulting on or after a mutual inherited type is called the phylogenetic tree. The phylogenetic tree is also called a Dendrogram. The distance between two groups explains the amount of connection; for example, nearby correlated groups are positioned on branches neighboring to each other. Species evolved from it are positioned at the ends of phylogenic tree branches. Though speculative, phylogenetic trees provide an expedient tool for studying phylogenetic relationships (Sapp, 2009). The corresponding sequences for the definite strains are taken from its complete genome sequence. Genome sequencing is done by innumerable sequencing methods and after that sequence is submitted to National Center for Biotechnology Information (NCBI). This is followed by constructing a FASTA format of the partial genome sequence. The obtained sequences are independently Basic Local Alignment Search Tool (BLAST) against GenBank in order to identify the homologs with maximum sequence distinctiveness for the phylogenetic analysis which further goes for multiple sequence alignment and generating of a phylogenetic tree with the assistance of any bioinformatics program software such as "MEGA" and treasure the relationship in the percentage between microorganisms (Hall, 2013; Winsor et al., 2010; Vásquez-Ponce et al., 2018), that is, *Constrictibacter antarcticus* was isolated from the White rock Skallen region, Antarctica (Yamada et al., 2011). *Amycolatopsis antarctica* was isolated from the South Shetland Islands, Antarctic Peninsula (Wang et al., 2018), *Paenibacillus wynnii* from Mars Oasis on Alexander Island, Antarctica (Rodríguez-Díaz et al., 2005) and *Aliterella antarctica* sp. nov. from Admiralty Bay, King George Island, Antarctica (Rigonato et al., 2016). These all are novel sp. identified on the basis of phylogenetic tree analysis.

13.3.2.5 Metagenomics library

Metagenomics elaborate the cloning of gene taken from environmental sample to understand different kinds of genes and library sequence (Amann et al., 1995; Stein et al., 1996). Steps for constructing a metagenomic library are described as follows:

1. Isolation of metagenomic DNA from marine sample.
2. 16S rDNA phylogenetic analyses.
3. Construction of a metagenomic library (Fig. 13.6). PCR amplification and sequence-based screening—are used for distinguishing gene targets from conserved regions of recognized genes. Screening based on functions shepherded expanding robotic systems that search for a well-defined phenotype (Acernese et al., 2014).

Samples could be collected in superior quantities of at least 200 L of a CTD fitted with a Niskin 24- and 10-L rosette bottle. The isolation of

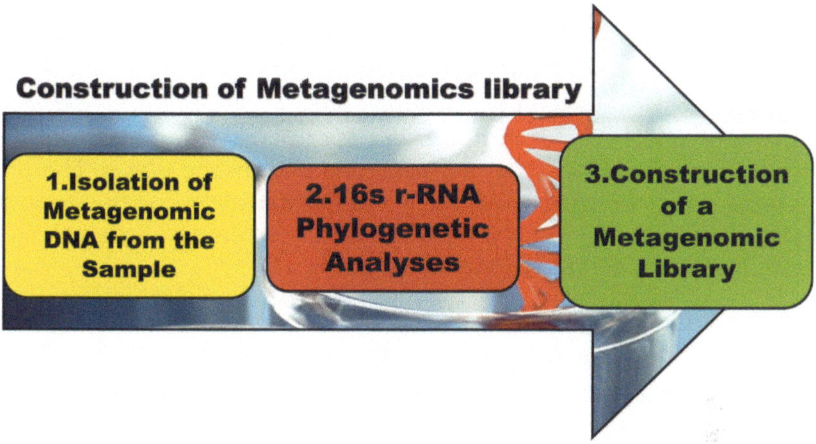

FIGURE 13.6 Steps involved in the construction of metagenomic library.

DNA can be performed through direct lysis of microorganisms with help of several methods, for example, beads in a cell disruptor or by vortexing, frosting, and defrosting; oven/dry heating; and grinding with liquid nitrogen (Dias et al., 2014; Gabor et al., 2003). Perform 16S rDNA sequencing. The copies of isolated metagenomic DNA requisite can be prepared through PCR amplification with bacterial-specific primer. Purified metagenomic DNA is transferred into the vector-like Fosmid and bacterial artificial chromosome (BAC). BAC vectors were established to make replica up to 40 kb and ~120 kb of large genomic DNA fragments with the help of the single-copy F-factor replicon, these vectors replicate and display excellent constancy when transporting a huge amount of DNA (Lane et al., 1991). Therefore Fosmid and BAC are very important tools for constructing a metagenomic library. Hundreds of replica metagenomic DNA stowed on the plates in the library (Weiland et al., 2010; Wild et al., 1996)

13.3.2.6 *Activity-based sequencing*

Two methods commonly used in metagenomic approaches to test and classify novel genes from environmental DNA libraries are, namely, sequence, and activity-based screening. Function-based screening is an easy way of acquiring genes with desired functions (Uchiyama & Miyazaki, 2009). The likelihood (hit rate = positive numbers divided by the total number of screened clones) of finding an assured gene dependent on many elements that are inextricably associated with one another, the system of the host-vector, the target gene size, profusion in the metagenome basis, the techniques of the assay, and the efficacy of a surrogate host's heterologous expression of the gene (Simon et al., 2009; Uchiyama & Miyazaki, 2009).

First, take any environmental sample for activity-based sequencing and perform metagenomics on it, then distinguish the gene and transfer it with the aid of vectors such as a plasmid, cosmid, and fosmid into an appropriate host. Successful transformed gene's activity of the particular enzyme is typically tested with nutrient agar plates complemented with substrates by creating a library of metagenomics on the plates. The positive clones could be distinguished by visual detection for the presence of a clear/halo zone or color and that kind of screening needs no different equipment (Uchiyama & Miyazaki, 2009).

13.3.2.7 Clone showing activity

The study of genetic material picked up sprightly from environmental sample is known as metagenomics. Side-steps involves restriction by making the complete DNA easily accessible from environmental samples into a host cell by cloning and/or sequencing, accompanied by either sequence or activity-driven screening (Handelsman, 2004). Activity-based sequencing process for particular genes transferred to the host and halo zone activity regulation and stored as a metagenomic library, whereas in clone activity, further steps are included: (1) screening for the desired gene: clones were initially reactivated in agar plates containing desired substrate (if we are interested in cellulase enzyme, the substrate would be cellulose) (Ruparelia et al., 2020); (2) sequence analysis of specific gene; (3) sequence alignment homology modeling; (4) subcloning of clones having gene of interest; (4) clones should be stored as the library or use as commercially (Araujo et al., 2020).

13.3.2.8 Overexpression and functional studies

Overexpression studies were started in the middle era and their device is used as detecting tools in the genetic engineering (Beggs, 1978). This kind of study is used to discriminate the apparatus responsible for overexpression. And the constituents are used to control the functionality loss for the desired gene phenotype. For example, deletion of *WOR1* blocks white-opaque switching in some species of *Candida*, while overexpression starts interchanging (Zordan et al., 2006). The explanation is that overexpression results in an unbound and/or hyperactive protein. This hypermorphic effect is revealing of true stimulatory role in the mechanism, either with the expression of a limiting factor or transforming protein which is essential for that mechanism or by overexpression stabilizing of an inhibitor (Prelich, 2012).

13.4 Future prospects

The use of combined techniques for the cultivation of extremophiles has proven successful in many. For cases, for example, (Nichols et al., 2008) cultivated an innovative *Psychrobacter* strain by the solicitation of applicable ecological environments for composed growth with the practice of aid strains and familiar media. Enzymes from these microorganisms possess significant prospective for high-tech expenditures as they are well known to have innovative activities at lower temperatures compared to mesophilic and thermophilic surroundings (Loperena et al., 2012). Extremophilic organisms are known to have several enzymes as enzymes obtained from that microorganisms deal with multipurpose tools in ecological improvement with several ranges of applications in the industry. They also show significant environmentally friendly stability due to their biodegradation capability, precise steadiness under thrilling conditions, enhanced use of natural resources, and reduced quantities of leftover products.

13.5 Concluding remarks

Antarctica comprises miscellaneous primeval environments conquered by extremophile microbial assemblies. Extremophilic organisms were isolated from Antarctic hotspot signifying that organisms have the potential to withstand the viable condition if appropriate cultivation methods are applied. Despite the environmental eccentric prominence of microorganisms from Antarctica showed many of them can symbolize fascinating sources of bioproducts with budding industrial applications (Furbino et al., 2018). Eventually, upcoming shifts in isolation and identification methodology succeeding to environmental conditions favoring more pluralistic assemblies can have a quirky influence on the diversity of microbes with various functionalities in several regions of Antarctica. However, this study that such methodology, targeting to the study novel organisms might deliver intermittent acumen and understanding of microbial function in the Antarctica hotspot.

Acknowledgment

We gratefully acknowledge support from the Microbiology Department, Gujarat Arts and Science College, Ahmedabad.

Conflict of interest

The authors declare that there is no conflict of interest with regards to the chapter published and authorship.

References

Acernese, F. A., Agathos, M., Agatsuma, K., Aisa, D., Allemandou, N., Allocca, A., & Meidam, J. (2014). Advanced Virgo: a second-generation interferometric gravitational wave detector. *Classical and Quantum Gravity, 32*(2), 024001.

Acharya, A. S., Prakash, A., Saxena, P., & Nigam, A. (2013). Sampling: Why and how of it. *Indian Journal of Medical Specialties*, *4*(2), 330–333.

Alam, S. I., Dixit, A., Reddy, G. S. N., Dube, S., Palit, M., Shivaji, S., & Singh, L. (2006). Clostridium schirmacherense sp. nov., an obligately anaerobic, proteolytic, psychrophilic bacterium isolated from lake sediment of Schirmacher Oasis, Antarctica. *International Journal of Systematic and Evolutionary Microbiology*, *56*(4), 715–720.

Ali, N., Rampazzo, R. D. C. P., Costa, A. D. T., & Krieger, M. A. (2017). Current nucleic acid extraction methods and their implications to point-of-care diagnostics. *BioMed Research International*, 2017. Available from https://doi.org/10.1155/2017/9306564.

Alves, I. M., Gonçalves, V. N., Oliveira, F. S., Schaefer, C. E., Rosa, C. A., & Rosa, L. H. (2019). The diversity, distribution, and pathogenic potential of cultivable fungi present in rocks from the South Shetlands archipelago, Maritime Antarctica. *Extremophiles*, *23*(3), 327–336. Available from https://doi.org/10.1007/s00792-019-01086-8.

Amann, R. I., Ludwig, W., & Schleifer, K. H. (1995). Phylogenetic identification and in situ detection of individual microbial cells without cultivation. *Microbiological Reviews*, *59*(1), 143–169. Available from https://doi.org/10.1128/mmbr.59.1.143-169.1995.

Antibus, D. E., Leff, L. G., Hall, B. L., Baeseman, J. L., & Blackwood, C. B. (2012). Cultivable bacteria from ancient algal mats from the McMurdo Dry Valleys, Antarctica. *Extremophiles: Life Under Extreme Conditions*, *16*(1), 105–114. Available from https://doi.org/10.1007/s00792-011-0410-3.

Antony, R., Krishnan, K. P., Laluraj, C. M., Thamban, M., Dhakephalkar, P. K., Engineer, A. S., & Shivaji, S. (2012). Diversity and physiology of culturable bacteria associated with a coastal Antarctic ice core. *Microbiological Research*, *167*(6), 372–380. Available from https://doi.org/10.1016/j.micres.2012.03.003.

Araujo, F. J., Hissa, D. C., Silva, G. O., Antunes, A. S. L. M., Nogueira, V. L. R., Gonçalves, L. R. B., & Melo, V. M. M. (2020). A novel bacterial carboxylesterase identified in a metagenome derived-clone from Brazilian mangrove sediments. *Molecular Biology Reports*, *47*(5), 3919–3928. Available from https://doi.org/10.1007/s11033-020-05484-6.

Beggs, J. D. (1978). Transformation of yeast by a replicating hybrid plasmid. *Nature*, *275*(5676), 104–109. Available from https://doi.org/10.1038/275104a0.

Beier-Sexton, M., Driscoll, T. P., Azad, A. F., Gillespie, J. J., Goldman, E., & Green, L. H. (2015). Practical handbook of microbiology.

Ben-David, A., & Davidson, C. E. (2014). Estimation method for serial dilution experiments. *Journal of Microbiological Methods*, *107*, 214–221. Available from https://doi.org/10.1016/j.mimet.2014.08.023.

Bogosian, G., & Bourneuf, E. V. (2001). A matter of bacterial life and death. *EMBO Reports*, *2*(9), 770–774. Available from https://doi.org/10.1093/embo-reports/kve182.

Bowman, J. P. (2000). Description of Cellulophaga algicola sp. nov., isolated from the surfaces of Antarctic algae, and reclassification of Cytophaga uliginosa (ZoBell and Upham 1944) Reichenbach 1989 as Cellulophaga uliginosa comb. nov. *International Journal of Systematic and Evolutionary Microbiology*, *50*(5), 1861–1868.

Bowman, J. P., Gosink, J. J., McCammon, S. A., Lewis, T. E., Nichols, D. S., Nichols, P. D., Skerratt, J. H., Staley, J. T., & McMeekin, T. A. (1998). Colwellia demingiae sp. nov., Colwellia hornerae sp. nov., Colwellia rossensis sp. nov. and Colwellia psychrotropica sp. nov.: psychrophilic Antarctic species with the ability to synthesize docosahexaenoic acid (22: ω63). *International Journal of Systematic and Evolutionary Microbiology*, *48*(4), 1171–1180.

Celussi, M., Balestra, C., Fabbro, C., Crevatin, E., Cataletto, B., Fonda Umani, S., & Del Negro, P. (2008). Organic-matter degradative potential of Halomonas glaciei isolated from frazil ice in the Ross Sea (Antarctica). *FEMS Microbiology Ecology*, *65*(3), 504–512.

Christner, B. C., Kvitko, B. H., & Reeve, J. N. (2003). Molecular identification of Bacteria and Eukarya inhabiting an Antarctic cryoconite hole. *Extremophiles: Life Under Extreme Conditions*, *7*(3), 177–183. Available from https://doi.org/10.1007/s00792-002-0309-0.

Cullen, J. J., & MacIntyre, H. L. (2016). On the use of the serial dilution culture method to enumerate viable phytoplankton in natural communities of plankton subjected to ballast water treatment. *Journal of Applied Phycology*, *28*(1), 279–298. Available from https://doi.org/10.1007/s10811-015-0601-x.

Dahm, R. (2005). Friedrich Miescher and the discovery of DNA. *Developmental Biology*, *278* (2), 274–288. Available from https://doi.org/10.1016/j.ydbio.2004.11.028.

de Menezes, G. C., Godinho, V. M., Porto, B. A., Gonçalves, V. N., & Rosa, L. H. (2017). Antarctomyces pellizariae sp. nov., a new, endemic, blue, snow resident psychrophilic ascomycete fungus from Antarctica. *Extremophiles*, *21*(2), 259–269.

Dias, R. S., Silva, L. C. F., Eller, M. R., Oliveira, V. M., De Paula, S. O., & Silva, C. C. (2014). *Metagenomics: Library construction and screening methods. Metagenomics: Methods, applications and perspectives* (pp. 45–65). Nova Science Publishers, Inc. Available from https://www.novapublishers.com/catalog/product_info.php?products_id = 50916.

Fell, J. W., Boekhout, T., Fonseca, A., Scorzetti, G., & Statzell-Tallman, A. (2000). Biodiversity and systematics of basidiomycetous yeasts as determined by large-subunit rDNA D1/D2 domain sequence analysis. *International Journal of Systematic and Evolutionary Microbiology*, *50*(3), 1351–1371. Available from https://doi.org/10.1099/00207713-50-3-1351.

Finore, I., Orlando, P., Di Donato, P., Leone, L., Nicolaus, B., & Poli, A. (2016). Nesterenkonia aurantiaca sp. nov., an alkaliphilic actinobacterium isolated from Antarctica. *International Journal of Systematic and Evolutionary Microbiology*, *66*(3), 1554–1560.

Firdaus-Raih, M., Hashim, N. H. F., Bharudin, I., Abu Bakar, M. F., Huang, K. K., Alias, H., Lee, B. K. B., Mat Isa, M. N., Mat-Sharani, S., & Sulaiman, S. (2018). , The Glaciozyma antarctica genome reveals an array of systems that provide sustained responses towards temperature variations in a persistently cold habitat. *PLoS One*, *13* (1), e0189947.

Franzmann, P., Stackebrandt, E., Sanderson, K., Volkman, J., Cameron, D., Stevenson, P., McMeekin, T., & Burton, H. (1988). Halobacterium lacusprofundi sp. nov., a halophilic bacterium isolated from Deep Lake, Antarctica. *Systematic and applied microbiology*, *11* (1), 20–27.

Franzmann, P. D., Liu, Y., Balkwill, D. L., Aldrich, H. C., De Macario, E. C., & Boone, D. R. (1997). Methanogenium frigidum sp. nov., a psychrophilic, H2-using methanogen from Ace Lake, Antarctica. *International Journal of Systematic and Evolutionary Microbiology*, *47* (4), 1068–1072.

Furbino, L. E., Pellizzari, F. M., Neto, P. C., Rosa, C. A., & Rosa, L. H. (2018). Isolation of fungi associated with macroalgae from maritime Antarctica and their production of agarolytic and carrageenolytic activities. *Polar Biology*, *41*(3), 527–535. Available from https://doi.org/10.1007/s00300-017-2213-1.

Gabor, E. M., De Vries, E. J., & Janssen, D. B. (2003). Efficient recovery of environmental DNA for expression cloning by indirect extraction methods. *FEMS Microbiology Ecology*, *44*(2), 153–163. Available from https://doi.org/10.1016/S0168-6496(02)00462-2.

Gartner, L. P. (2020). *Textbook of histology*. E-Book.

Gonçalves, V. N., Campos, L. S., Melo, I. S., Pellizari, V. H., Rosa, C. A., & Rosa, L. H. (2013). Penicillium solitum: a mesophilic, psychrotolerant fungus present in marine sediments from Antarctica. *Polar Biology*, *36*(12), 1823–1831.

Hall, B. G. (2013). Building phylogenetic trees from molecular data with MEGA. *Molecular Biology and Evolution*, *30*(5), 1229–1235. Available from https://doi.org/10.1093/molbev/mst012.

Hamed, T. (2016). Sampling methods in research methodology: How to choose a sampling technique for research. *SSRN Electronic Journal, 5*(2), 18–27. Available from https://doi.org/10.2139/ssrn.3205035.

Handelsman, J. (2004). Metagenomics: Application of genomics to uncultured microorganisms. *Microbiology and Molecular Biology Reviews, 68*(4), 669–685. Available from https://doi.org/10.1128/MMBR.68.4.669-685.2004.

Handelsman, J., Rondon, M. R., Brady, S. F., Clardy, J., & Goodman, R. M. (1998). Molecular biological access to the chemistry of unknown soil microbes: A new frontier for natural products. *Chemistry and Biology, 5*(10), R245–R249. Available from https://doi.org/10.1016/S1074-5521(98)90108-9.

Holben, W. E., Jansson, J. K., Chelm, B. K., & Tiedje, J. M. (1988). DNA probe method for the detection of specific microorganisms in the soil bacterial community. *Applied and Environmental Microbiology, 54*(3), 703–711. Available from https://doi.org/10.1128/aem.54.3.703-711.1988.

Iain, D., William, G. -C., Michael, T., Thomas, S., Maria, Á. -J., & David, P. (2016). Extremophiles in an Antarctic marine ecosystem. *Microorganisms,* 8. Available from https://doi.org/10.3390/microorganisms4010008.

Ilker, E. (2017). Sampling and sampling methods. *Biometrics & Biostatistics International Journal, 5*(6), 215–217. Available from https://doi.org/10.15406/bbij.2017.05.00149.

Imanaka, T. (2002). *Isolation and application of extremophiles—Monograph*. Seibutsu-Kogaku Kaishi.

Kostadinova, N., Krumova, E., Tosi, S., Pashova., & Angelova, M. (2009). Isolation and identification of filamentous fungi from island Livingston, Antarctica. *Biotechnology & Biotechnological Equipment, 23*(supl), 267–270.

Lane, B. G., Bernier, F., Dratewka-Kos, E., Shafai, R., Kennedy, T. D., Pyne, C., Munro, J. R., Vaughan, T., Walters, D., & Altomare, F. (1991). Homologies between members of the germin gene family in hexaploid wheat and similarities between these wheat germins and certain *Physarum spherulins. Journal of Biological Chemistry, 266*(16), 10461–10469.

Leff, L. G., Dana, J. R., McArthur, L. V., & Shimkets, L. J. (1995). Comparison of methods of DNA extraction from stream sediments. *Applied and Environmental Microbiology, 61*(3), 1141–1143. Available from https://doi.org/10.1128/aem.61.3.1141-1143.1995.

Loperena, L., Soria, V., Varela, H., Lupo, S., Bergalli, A., Guigou, M., Pellegrino, A., Bernardo, A., Calviño, A., Rivas, F., & Batista, S. (2012). Extracellular enzymes produced by microorganisms isolated from maritime Antarctica. *World Journal of Microbiology and Biotechnology, 28*(5), 2249–2256. Available from https://doi.org/10.1007/s11274-012-1032-3.

Martinez, A., Cavello, I., Garmendia, G., Rufo, C., Cavalitto, S., & Vero, S. (2016). Yeasts from sub-Antarctic region: biodiversity, enzymatic activities and their potential as oleaginous microorganisms. *Extremophiles, 20*(5), 759–769.

Menon, J., Pease, D., & Rees, R. (2004). *U.S. Patent Application No. 10/323,113.*

Montes, M. J., Bozal, N., & Mercade, E. (2008). Marinobacter guineae sp. nov., a novel moderately halophilic bacterium from an Antarctic environment. *International Journal of Systematic and Evolutionary Microbiology, 58*(6), 1346–1349.

Morgan-Kiss, R. M., Ivanov, A. G., Modla, S., Czymmek, K., Hüner, N. P., Priscu, J. C., Lisle, J. T., & Hanson, T. E. (2008). Identity and physiology of a new psychrophilic eukaryotic green alga, Chlorella sp., strain BI, isolated from a transitory pond near Bratina Island, Antarctica. *Extremophiles, 12*(5), 701–711.

Mou, Y. Z., Qiu, X. X., Zhao, M. L., Cui, H. L., Oh, D., & Dyall-Smith, M. L. (2012). Halohasta litorea gen. nov. sp. nov., and Halohasta litchfieldiae sp. nov., isolated from the Daliang aquaculture farm, China and from Deep Lake, Antarctica, respectively. *Extremophiles, 16*(6), 895–901.

Nicolaus, B., Improta, R., Manca, M. C., Lama, L., Esposito, E., & Gambacorta, A. (1998). Alicyclobacilli from an unexplored geothermal soil in Antarctica: Mount Rittmann. *Polar Biology, 19*(2), 133–141.

Nicolaus, B., Lama, L., Esposito, E., Manca, M. C., Gambacorta, A., & di Prisco, G. (1996). "Bacillus thermoantarcticus" sp. nov., from Mount Melbourne, Antarctica: a novel thermophilic species. *Polar Biology, 16*(2), 101–104.

Nicolaus, B., Manca, M. C., Lama, L., Esposito, E., & Gambacorta, A. (2001). Lipid modulation by environmental stresses in two models of extremophiles isolated from Antarctica. *Polar Biology, 24*(1), 1–8.

Nichols, D., Lewis, K., Orjala, J., Mo, S., Ortenberg, R., O'Connor, P., Zhao, C., Vouros, P., Kaeberlein, T., & Epstein, S. S. (2008). Short peptide induces an uncultivable microorganism to grow in vitro. *Applied and Environmental Microbiology, 74*(15), 4889–4897. Available from https://doi.org/10.1128/AEM.00393-08.

Ogaki, M. B., Coelho, L. C., Vieira, R., Neto, A. A., Zani, C. L., Alves, T. M. A., Junior, P. A. S., Murta, S. M. F., Barbosa, E. C., Oliveira, J. G., Ceravolo, I. P., Pereira, P. O., Cota, B. B., Viana, R. O., Alves, V. S., & Rosa, L. H. (2020). Cultivable fungi present in deep-sea sediments of Antarctica: Taxonomy, diversity, and bioprospecting of bioactive compounds. *Extremophiles: Life Under Extreme Conditions, 24*(2), 227–238. Available from https://doi.org/10.1007/s00792-019-01148-x.

Ogram, A., Sayler, G. S., & Barkay, T. (1987). The extraction and purification of microbial DNA from sediments. *Journal of Microbiological Methods, 7*(2–3), 57–66. Available from https://doi.org/10.1016/0167-7012(87)90025-X.

Pellizzari, F., Velez-Rubio, G., Silva, M., & Carranza, A. (2016). The seaweeds Myriogloea major Asensi (Chordariacea, Phaeophyceae) and Gayralia oxysperma (Kützing) K. L. Vinogradova ex Scagel et al. (Ulvophyceae, Chlorophyta): A case of range extension in the Southwestern Atlantic Ocean? Marine Biodiversity. *In and 18S Library Preparation Kits for Metagenomics Studies, 46*. Available from https://doi.org/10.1007/s12526-016-0447-8, perkinelmer. 16S.

Poli, A., Esposito, E., Orlando, P., Lama, L., Giordano, A., de Appolonia, F., Nicolaus, B., & Gambacorta, A. (2007). *Halomonas alkaliantarctica sp. nov., isolated from saline lake Cape Russell in Antarctica, an alkalophilic moderately halophilic, exopolysaccharide-producing bacterium, 30*(1), 31–38.

Prelich, G. (2012). Gene overexpression: Uses, mechanisms, and interpretation. *Genetics, 190*(3), 841–854. Available from https://doi.org/10.1534/genetics.111.136911.

Reddy, G., Prakash, J., Vairamani, M., Prabhakar, S., & Matsumoto, G. (2002). and Shivaji, S., Planococcus antarcticus and Planococcus psychrophilus spp. nov. isolated from cyanobacterial mat samples collected from ponds in Antarctica. *Extremophiles, 6*(3), 253–261.

Reis-Mansur, M. C. P., Cardoso-Rurr, J. S., Silva, J. V. M. A., de Souza, G. R., da Silva Cardoso, V., Mansoldo, F. R. P., Pinheiro, Y., Schultz, J., Lopez Balottin, L. B., da Silva, A. J. R., Lage, C., dos Santos, E. P., Rosado, A. S., da Silva., & Vermelho, A. B. (2019). Carotenoids from UV-resistant Antarctic Microbacterium sp. *LEMMJ01. Scientific reports, 9*(1), 1–14.

Rigonato, J., Gama, W. A., Alvarenga, D. O., Branco, L. H. Z., Brandini, F. P., Genuário, D. B., & Fiore, M. F. (2016). Aliterella atlantica gen. nov., sp. nov., and *Aliterella antarctica* sp. nov., novel members of coccoid Cyanobacteria. *International Journal of Systematic and Evolutionary Microbiology, 66*(8), 2853–2861. Available from https://doi.org/10.1099/ijsem.0.001066.

Rodríguez-Díaz, M., Lebbe, L., Rodelas, B., Heyrman, J., De Vos, P., & Logan, N. A. (2005). *Paenibacillus wynnii* sp. nov., a novel species harbouring the nifH gene, isolated from Alexander Island, Antarctica. *International Journal of Systematic and Evolutionary Microbiology, 55*(5), 2093–2099. Available from https://doi.org/10.1099/ijs.0.63395-0.

Roldán, D. M., Kyrpides, N., Woyke, T., Shapiro, N., Whitman, W. B., Králová, S., Sedláček, I., Busse, H.-J., & Menes, R. J. (2020). Hymenobacter artigasi sp. nov., isolated from air sampling in maritime Antarctica. *International Journal of Systematic and Evolutionary Microbiology, 70*(9), 4935–4941.

Ruparelia, J., Saraf, M., & Jha, D. C. (2020). Screening and optimization for cellulase production by soil bacterial isolates JRC1 AND JRC2. *Bioscience Biotechnology Research Communications, 13*, 159–164.

Santos, A. F., Pires, F., Jesus, H. E., Santos, A. L., Peixoto, R., Rosado, A. S., D'Avila-Levy, C. M., & Branquinha, M. H. (2015). Detection of proteases from Sporosarcina aquimarina and Algoriphagus antarcticus isolated from Antarctic soil. *Anais da Academia Brasileira de Ciências, 87*(1), 109–119.

Sattley, W. M., Jung, D. O., & Madigan, M. T. (2008). Psychrosinus fermentans gen. nov., sp. nov., a lactate-fermenting bacterium from near-freezing oxycline waters of a meromictic Antarctic lake. *FEMS Microbiology Letters, 287*(1), 121–127.

Sapp, J. (2009). *The new foundations of evolution: on the tree of life*. Oxford University Press.

Sedláček, I., Pantůček, R., Králová, S., Mašlaňová, I., Holochová, P., Staňková, E., Vrbovská, V., Švec, P., & Busse, H.-J. (2019). Hymenobacter amundsenii sp. nov. resistant to ultraviolet radiation, isolated from regoliths in Antarctica. *Systematic and Applied Microbiology, 42*(3), 284–290.

Scorzetti, G., Fell, J. W., Fonseca, A., & Statzell-Tallman, A. (2002). Systematics of basidiomycetous yeasts: A comparison of large subunit D1/D2 and internal transcribed spacer rDNA regions. *FEMS Yeast Research, 2*(4), 495–517. Available from https://doi.org/10.1016/S1567-1356(02)00128-9.

Simon, C., Herath, J., Rockstroh, S., & Daniel, R. (2009). Rapid identification of genes encoding DNA polymerases by function-based screening of metagenomic libraries derived from glacial ice. *Applied and Environmental Microbiology, 75*(9), 2964–2968. Available from https://doi.org/10.1128/AEM.02644-08.

Stein, J. L., Marsh, T. L., Wu, K. Y., Shizuya, H., & Delong, E. F. (1996). Characterization of uncultivated prokaryotes: Isolation and analysis of a 40-kilobase-pair genome fragment from a planktonic marine archaeon. *Journal of Bacteriology, 178*(3), 591–599. Available from https://doi.org/10.1128/jb.178.3.591-599.1996.

Sundar, P., & Richard, R. (2004). *An introduction to biostatistics, a manual for students in health sciences, Eastern Economy edition*. Prentice-Hall of India Plc.

Thomas-Hall, S., & Watson, K. (2002). Cryptococcus nyarrowii sp. nov., a basidiomycetous yeast from Antarctica. *International Journal of Systematic and Evolutionary Microbiology, 52* (3), 1033–1038.

Throndsen, J., & Kristiansen, S. (1991). Micromonas pusilla (Prasinophyceae) as part of pico-and nanoplankton communities of the Barents Sea. *Polar Research, 10*(1), 201–208. Available from https://doi.org/10.1111/j.1751-8369.1991.tb00646.x.

Turchetti, B., Hall, S. R. T., Connell, L. B., Branda, E., Buzzini, P., Theelen, B., Müller, W. H., & Boekhout, T. (2011). Psychrophilic yeasts from Antarctica and European glaciers: description of Glaciozyma gen. nov., Glaciozyma martinii sp. nov. and Glaciozyma watsonii sp. nov. *Extremophiles, 15*(5), 573.

Uchiyama, T., & Miyazaki, K. (2009). Functional metagenomics for enzyme discovery: Challenges to efficient screening. *Current Opinion in Biotechnology, 20*(6), 616–622. Available from https://doi.org/10.1016/j.copbio.2009.09.010.

Vásquez-Ponce, F., Higuera-Llantén, S., Pavlov, M. S., Marshall, S. H., & Olivares-Pacheco, J. (2018). Phylogenetic MLSA and phenotypic analysis identification of three probable novel Pseudomonas species isolated on King George Island, South Shetland, Antarctica. *Brazilian Journal of Microbiology, 49*(4), 695–702. Available from https://doi.org/10.1016/j.bjm.2018.02.005.

Wang, J., Leiva, S., Huang, J., & Huang, Y. (2018). *Amycolatopsis antarctica* sp. nov., isolated from the surface of an Antarctic brown macroalga. *International Journal of Systematic and Evolutionary Microbiology, 68*(7), 2348–2356. Available from https://doi.org/10.1099/ijsem.0.002844.

Weiland, N., Löscher, C., Metzger, R., & Schmitz, R. (2010). Construction and screening of marine metagenomic libraries. *Methods in Molecular Biology (Clifton, N.J.), 668*, 51–65. Available from https://doi.org/10.1007/978-1-60761-823-2_3.

Wild, J., Hradečná, Z., Pósfai, G., & Szybalski, W. (1996). A broad-host-range in vivo pop-out and amplification system for generating large quantities of 50- to 100-kb genomic fragments for direct DNA sequencing. *Gene, 179*(1), 181–188. Available from https://doi.org/10.1016/S0378-1119(96)00487-8.

Wilson, J. (2014). *Essentials of business research: A guide to doing your research project*. Sage.

Winsor, W. G., Lam, D. K., Fleming, L., Lo, R., Whiteside, M. D., Yu, N. Y., Hancock, R. E. W., & Brinkman, F. S. L. (2010). Pseudomonas genome database: Improved comparative analysis and population genomics capability for *Pseudomonas genomes*. *Nucleic Acids Research, 39*, D596–D600. Available from https://doi.org/10.1093/nar/gkq869.

Yamada, K., Fukuda, W., Kondo, Y., Miyoshi, Y., Atomi, H., & Imanaka, T. (2011). *Constrictibacter antarcticus* gen. nov., sp. nov., a cryptoendolithic micro-organism from Antarctic white rock. *International Journal of Systematic and Evolutionary Microbiology, 61* (8), 1973–1980. Available from https://doi.org/10.1099/ijs.0.026625-0.

Zordan, R. E., Galgoczy, D. J., & Johnson, A. D. (2006). Epigenetic properties of white–opaque switching in Candida albicans are based on a self-sustaining transcriptional feedback loop. *Proceedings of the National Academy of Sciences, 103*(34), 12807–12812. Available from https://doi.org/10.1073/pnas.0605138103.

CHAPTER

14

Recent advances in microbial databases with special reference to kinetoplastids

Divya Narayanan Prakash, Chinmayee Bar Routaray, Renuka Bhor and Kalpana Pai

Centre for Advanced Studies, Department of Zoology, Savitribai Phule Pune University, Pune, India

14.1 Introduction

Microorganisms comprise about 60% of the earth's total biomass and constitute an important part of earth's evolution and environment biosphere (Staley et al., 1997). Around 3 billion years ago, the first living forms which existed on earth were the microbes and in due course, adaptations to grow in extreme environmental conditions like temperature, pH, salt concentrations and oxygen levels led to their extraordinary physiological, genetic and metabolic diversity compared to plants and animals making them unique and robust to extreme ecological niches (Fraser et al., 2000). Out of the estimated trillion prokaryotic microbial species, less than 1% have been identified until now (Lewin et al., 2018); while, only around 10–15 million species of eukaryotes (0.2%) are known (Hinchliff et al., 2015). In-depth studies of their genes, proteins, transcriptomes and metabolites is essential in order to understand their ecology and diversity, which would help in strategizing novel therapeutics, metabolites, genetic modifications, synthetic biology, etc. Further, advances in the field of genome sequencing technology, proteomics, transcriptomics, informatics, automation and artificial

Microbial Diversity and Ecology in Hotspots
DOI: https://doi.org/10.1016/B978-0-323-90148-2.00014-6

intelligence have assisted in improved knowledge, utilization, sustenance, and biodiversity conservation (Lewin et al., 2018). With the advent of innovative technologies to discover new species, cellular, metabolic and subcellular processes, the colossal data generated by genome/proteome sequencing due to high-throughput and low-cost DNA sequencing technologies has drastically increased over the last decade. However, the data generated would be vital and informative only after its analysis, curation, and archival. This enormous data generated around the world is increasing at an exponential rate and its interpretation, storage and retrieval is extremely difficult. This has ultimately led to a surge in the number of biological databases. Currently, there are around 1613 biological databases which can be publicly accessed online (Rigden & Fernández, 2019) making this count of publicly accessible databases conservative. However, it is time-consuming and daunting to navigate into these databases for the subject of specific interest. Most of the biological databases contain similar information, but no two databases are alike.

Data retrieval, analytical tools, visualization, automated integration of data from various databases, and web application programming interfaces are some of the major objectives of biological databases (Zou et al., 2015). Therefore biological databases develop essential tools in bioinformatics to analyze and integrate gigantic volumes of omics data submitted by various researchers around the world. These data become relevant and indispensable to other scientists who carry out wet-lab scientific experiments for design, comparison, analysis and interpretation of their data. Biological databases have an array of applications, encompass different types of heterogenous data and are curated by different methods at diverse levels making them applicable to the user using several distinct criteria. Various biological databases provide paramount information on whole genomes, genes and their functions, protein expression data, protein sequences and its 3D-structures, functional metabolites, metabolic reactions, nomenclature, literature, protein–protein/gene interactions and cellular pathways (Zhulin, 2015). Therefore biological databases can be classified based on their content and applications. The discovery of several useful genes, proteins and unique metabolic pathways will continue as a result of sequencing and annotation of several previously uncharacterized genomes and proteomes. This will provide vital clues for clinical diagnosis, therapeutics, synthetic biology methods for bioactive compounds production, sustainable agriculture and feedstock, biotechnological applications, etc.

There are 36 biodiversity hotspots around the world according to the Critical Ecosystem Partnership Fund. Among these, the Eastern Himalayas, Indo-Burma and Western Ghats located in India offer diversity and uniqueness in the inherent flora and fauna (Bhattacharjee & Joshi, 2014). The microbial ecosystem is the most sensitive ecosystem to even the slightest change in the microenvironment which occurs mainly due to anthropogenic activities leading to imbalance, invasion of different species, endangerment and ultimately loss of valuable and novel microbial species. The microbial diversity is represented by the variability among prokaryotes (bacteria and archaea), eukaryotes (protozoa, fungi, and algae) along with acellular viruses in the natural world. The abundant information available in various microbial databases on the microbial diversity and ecology has enhanced our understanding of microbes with modern biology advancements in medicine, biotechnology and agriculture.

In-depth study of their habitats and diversity for the exploration of novel genes/gene products and monitoring the microbial pattern to study the environmental changes can be correlated to understand the evolutionary processes and interactions among the microbes (Satyanarayana et al., 2019). Prokaryotic and eukaryotic microbes especially those isolated from biodiversity hotspots reveal unique cellular adaptations making them interesting to study. The first prokaryotic genome of *Haemophilus influenzae* was completely sequenced in the year 1995 (Fleischmann et al., 1995) while, the first eukaryotic genome of *Saccharomyces cerevisiae* was available in the year 1996. Around 85,000 prokaryotic and eukaryotic genomes have been sequenced from 1995 to 2018 (https://www.labiotech.eu/in-depth/genome-sequencing-review-projects/). The biological data generated from prokaryotic microbes are considerably more than the eukaryotic microbial data. Prokaryotic microbes have gained importance due to their predominance and earlier existence compared to the eukaryotic microbes. It has been recently documented that bacterial genetic variation is predominant in temperate ecosystems compared to fungi and other eukaryotes. This implies that fluctuations in the genetic composition of microbes are highly correlated with environmental factors than with geographical locations (Bahram et al., 2018). Among the eukaryotic microbes, protists especially kinetoplastids have gained attention due to their complex, unknown and intriguing interactions with the mammalian immune system (Stuart et al., 2008). Kinetoplastids are unicellular flagellated eukaryotic protists having an organelle with a substantial mass of DNA (kDNA) called kinetoplast. This class of organisms include the pernicious *Trypanosoma* and *Leishmania* parasites responsible for the fatal diseases such as African sleeping sickness, Chagas disease, and leishmaniasis in humans. These diseases are recognized as important neglected tropical diseases by World Health Organization (WHO) which mostly affects the population with low socioeconomic status.

Trypanosoma brucei causes human African trypanosomiasis (HAT) or sleeping sickness which is a severe meningoencephalitic disease causing acute lymphatic infection (Almeida et al., 2018). Similarly, *Trypanosoma cruzi* causes Chagas disease or American sleeping sickness affecting nearly 6–7 million people worldwide (Sutherland et al., 2015). In order to establish successful infection, the microbial parasite avoids host killing and evades the host immune system (Ponte-Sucre, 2016).

Leishmania spp. also belongs to the same family of Trypanosomatidae and consists of 20 protozoan species. It causes the tropical neglected disease leishmaniasis in humans (mammalian host) and is transmitted by the bite of phlebotomine sandflies which is the secondary or intermediate host (Maslov et al., 2001). The clinical expressions of leishmaniasis include cutaneous leishmaniasis (CL), mucocutaneous leishmaniasis (MCL), diffuse cutaneous leishmaniasis (DCL) or the fatal visceral leishmaniasis (VL) also known as Kala-azar (Burza et al., 2018; Rodrigues et al., 2016). An estimated 1–1.6 million new cases every year, incidence of CL is reported while VL accounts for 90,000 new cases per year with approximately 10% mortality rate (Alvar et al., 2012; Burza et al., 2018). There has been significant research in leishmaniasis in the last two decades but information regarding the diversity, evolution, and genetics of *Leishmania donovani* species is still inadequate and intriguing.

The complexity of HAT, Chagas disease, and leishmaniasis along with host–pathogen interactions, underlying mechanisms is not yet decoded. A deeper understanding of these aspects at the genome, molecular and cellular levels will provide vital clues for the development of therapeutics, vaccines, diagnostics, and other tools required against these fatal diseases. Recent advances in the field of "Omics" have played a vital part in better comprehension of these deadly diseases. In this chapter, we highlight the recent advances in microbial databases with emphasis on kinetoplasts parasitic microbes especially *Trypanosoma* and *Leishmania* as it is beyond the scope to cover of all other microbes from the enormous number of available microbial databases.

14.2 Classification of biological databases

According to the available information, data submission and tools developed for data analysis, the biological databases can be classified as Global resources/Comprehensive databases and Community/Specialized databases.

14.3 Global resources/comprehensive databases

The global or comprehensive databases include GenBank (http://www.ncbi.nlm.nih.gov), European Molecular Biology Laboratory (EMBL) (https://www.ebi.ac.uk), and DNA Data Bank of Japan (DDBJ) (http://www.ddbj.nig.ac.jp) which curates and archives different types of updated raw data from several laboratories across the world. These three global databases which were established as the International Nucleotide Sequence Database Collaboration in 1988 function to collect and disseminate sequence information on DNA, protein and RNA molecules. The National Center for Biotechnology Information (NCBI) at the National Institutes of Health was created in 1988 to develop robust information system technologies for molecular and cellular biology. NCBI provides GenBank which is primarily a nucleic acid sequence database and receives data from other global resource databases like DDBJ and the European Nucleotide Archive (ENA). In addition to the direct submission of raw data files from the scientific community to GenBank, NCBI also provides other biological data, analyses tools, retrieval and resources of other microbial and biological data.

As of October 2020, GenBank consists of numerous prokaryotic genomes (2,82,609); microbial eukaryotes: fungi (6884) and protists (943) making it highly comprehensive and resourceful. There is a surge in the genome sequencing of kinetoplastid microbes especially *Trypanosoma* spp. and *Leishmania* spp. during the past 5 years which is evident in Table 14.1 indicating analysis and interpretation of their gene/protein sequences for therapeutic applications and in-depth understanding of their pathogenesis. A more specific search submitted to NCBI reveals 44 and 57 genome assemblies of kinetoplastids like *Trypanosoma* (Taxonomy ID: 5690) and *Leishmania* (Taxonomy ID: 5658) respectively, available along with other gene, protein, PubChem, literature, etc., databases as seen in Table 14.2.

The proteins sequence deposition in NCBI belonging to *Trypanosoma* spp. and *Leishmania* spp. are also sourced from databases like Protein Data Bank (PDB), RefSeq, UniPortKB/SwissPort and updated regularly. NCBI database is one of the most widely and frequently used global biological database due to the innumerable submission, analytical and retrieval tools/softwares available. All the data relevant to DNA, RNA, proteins, etc., is available at http://www.ncbi.nlm.nih.gov, the NCBI home page. Since 2019, the Webin submission interfaces extends this support to several thousand active raw data submission from several countries across the world resulting in 419,490 submissions which includes 5700 studies, 493,000 runs, 620,000 samples and 197,000 assemblies (Amid et al., 2020). The total number of sequence submissions of *Trypanosoma* is 256,040 and 53 assembly sets while, those of *Leishmania* include 72,514 sequences and 82 assembly sets.

The ENA (https://www.ebi.ac.uk/ena) is Europe's open and free global resource database (since its existence for the past 37 years) like the GenBank for DNA and RNA data. The three main databases namely, Sequence Read Archive (SRA), Trace Archive, and EMBL-Bank are an integral part of the EMBL (R. et al., 1995). The wide spectrum of ENA extends from raw reads obtained from experimental data to the assembled and annotated sequences. Therefore, it offers exhaustive support for the

TABLE 14.1 Genome sequences of *Trypanosoma* and *Leishmania* from various projects submitted to NCBI-GenBank during the period 2015–20.

Organism name	BioSample	Assembly	Size (Mb)	GC%	CDS	Release date	rRNA
Leishmania peruviana	SAMEA2767110	GCA_001403695.1	33.8902	53.202	0	2015-09-17T00:00:00Z	0
Leishmania tarentolae	SAMD00191432	GCA_009731335.1	35.4165	57.4	8703	2019-11-27T21:02:00Z	0
Trypanosoma congolense	SAMN07142734	GCA_002287245.1	41.2334	47.4	0	2017-09-06T00:00:00Z	0
Leishmania aethiopica	SAMN09984807	GCA_003992445.1	33.6484	60.4	0	2019-01-04T00:00:00Z	0
Trypanosoma rangeli	SAMN04566061	GCA_003719475.1	21.1573	51.9	10104	2018-11-08T00:00:00Z	3
Leishmania amazonensis	SAMN10955936	GCA_005317125.1	32.1565	59.497	0	2019-05-14T00:00:00Z	0
Leishmania chagasi	SAMN10964361	GCA_014466935.1	31.925	59.558	0	2020-09-03T00:00:00Z	0
Trypanosomatidae sp. JR-2017a	SAMN07508845	GCA_003671325.1	54.075	0	0	2018-10-22T00:00:00Z	0
Trypanosoma conorhini	SAMN04565988	GCA_003719485.1	21.3342	57.2	10148	2018-11-08T00:00:00Z	4
Leishmania tropica	SAMN09984808	GCA_014139745.1	32.7007	59.662	0	2020-08-10T00:00:00Z	0
Leishmania chagasi	SAMN10964360	GCA_014466975.1	31.9246	59.556	0	2020-09-03T00:00:00Z	0
Leishmania infantum	SAMN08583618	GCA_003020905.1	32.0091	59.562	0	2018-03-27T00:00:00Z	0
Leishmania guyanensis	SAMEA5854544	GCA_902369315.1	0.01962	20.5	0	2019-10-03T00:00:00Z	0
Leishmania guyanensis	SAMN09763187	GCA_003664525.1	33.816	58	0	2018-10-16T00:00:00Z	0
Leishmania mexicana	SAMN09984810	GCA_003992435.1	32.0572	59.8	0	2019-01-04T00:00:00Z	0
Leishmania sp. AIIMS/LM/SS/PKDL/LD-974	SAMN03457191	GCA_000981925.2	27.8483	55.8	0	2015-04-29T00:00:00Z	0
Leishmania adleri	SAMEA5854542	GCA_902369305.1	0.01921	21.8	0	2019-10-03T00:00:00Z	0
Trypanosomatidae sp. JR-2017a	SAMN07508846	GCA_003671345.1	54.718	0	0	2018-10-22T00:00:00Z	0
Leishmania tropica	SAMN13269096	GCA_011316065.1	30.8702	59.9	0	2020-03-16T00:00:00Z	0
Trypanosoma cruzi	SAMN10335319	GCA_013358655.1	40.8013	51.2	0	2020-06-18T00:00:00Z	46
Trypanosoma theileri	SAMN03296061	GCA_002087225.1	29.8222	41.4	11312	2017-04-12T00:00:00Z	0
Leishmania lainsoni	SAMN10094664	GCA_003664395.1	34.152	57.8	0	2018-10-16T00:00:00Z	0
Leishmania major strain Friedlin	SAMEA5948836	GCA_902498725.1	0.00273	43	0	2019-10-03T00:00:00Z	0

TABLE 14.2 Data availability for kinetoplastids like *Trypanosoma* and *Leishmania* at NCBI and EBI-EMBL enhancing the research carried out in NTDs.

NCBI		*Trypanosoma*	*Leishmania*
Genes	Gene GEO	88,778	57,475
	DataSets GEO	1713	1060
	Profiles	122,470	54,678
	HomoloGene	0	4
	PopSet	549	528
Proteins	Conserved domains	96	68
	Identical protein	234,576	85,754
	Groups protein	483,458	1,491,368
	Protein clusters	8351; 465	7647; 128
	Sparcle structure	1987	1105
PubChem	BioAssays	12,191	8930
	Compounds	0	1
	Pathways	243	205
	Substances	14	12
EBI-EMBL	Protein sequences	535,921	234,942
	Gene expression	8883	461

sequence data operations around the world and distinct thematic integration of collaborative data for expressed sequence tag (EST), whole genome sequencing, patents and other high-throughput sequences by assembling the genomic sequences and annotation of the sequence fragments to whole DNA/RNA replicons (Leinonen et al., 2010). The extensive protein sequences of *Trypanosoma* and *Leishmania* available at EBI-EMBL database are 535,921 and 234,942, respectively. Similarly, gene expression data for *Trypanosoma* and *Leishmania* are 8883 and 461, respectively while, protein expression data available for *Trypanosoma* and *Leishmania* are 111 and 39, respectively.

ENA is now a part of the Elixir infrastructure and offers a platform for data submission, annotation, curation as well as 17,380 exhaustive tools for analysis and interpretation of Transcriptomics, gene transcripts, nucleic acid reads (https://elixir-europe.org/platforms/tools). The Elixir Deposition Databases for Biomolecular Data in ENA ensures long-term preservation and accessibility of molecular data after deposition of experimental data. It consists of several important databases for deposition of biomolecular data like ArrayExpress, BioSamples, EMDB, IntAct, MetaboLights, etc. The integration of Elixir Bio tools, OpenEBench, and BioContainers Flow and Galaxy provides a platform for the identification, deployment and benchmark software tools.

The DDBJ was established in 1984 with operation at the Research Organization of Information and System National Institute of Genetics (NIG) in Mishima, Japan with endorsement of the Ministry of Education, Culture, Sports, Science, and Technology, Japan. It has been a public database of nucleotide sequences which are collected and annotated as a database service along with international collaboration with NCBI and ENA within the INSDC framework (Mashima et al., 2017). DDBJ, NCBI and ENA/EBI centers virtually offer the same data at any given time since DDBJ Center shares the released data with these databases on a daily basis and all the three centers are updated. DDBJ offer various software tools developed by DDBJ for data analyses through web services or NIG Supercomputer (Ogasawara et al., 2020). Presently, there are 274,691 annotated/assembled data sequence entries for *Trypanosoma* and 95,640 for *Leishmania*. This excludes the Sequence Read Archive (SRA) entries in DDBJ. More information can be accessed at http://www.ddbj.nig.ac.jp.

The genetic information is complemented with information on proteins as they are the products of gene expression affecting the phenotype and cellular functions. With the advancement in MALDI, X-ray crystallography, NMR and other proteomics techniques, there has been a steady increase in protein data deposition. These data are accommodated in various global resource biological databases specific for proteins like PIR, SIB, PDB, UniProt, etc.

The Protein Information Resource (PIR) was established by the National Biomedical Research Foundation in 1984. PIR is an integrated public biological database to maintain genomic, proteomic and systems biology research (Wu et al., 2002). In 2002, PIR along with NCBI and EBI formed the UniProt, a single protein database web resource by integrating the PIR-PSD, TrEMBL, and Swiss-Prot databases. The iProClass resource database in PIR provides significant information for UniProtKB and unique NCBI Entrez protein sequences in UniParc, linking about 175 biological databases which includes databases for protein functions, pathways, interactions, protein families, structures, genes and genomes, ontologies and literature. It combines various methods like data warehouse and hypertext navigation for integration of data which provides complete protein information in detail which aid in prediction and functional characterization of hypothetical or uncharacterized proteins and various protein groups. iProLINK resource in PIR provides data mining tools for in-depth protein studies and literature. iPTMnet is a resource which connects multiple bioinformatics tools and text mining resources, visualization tools, databases leading to a cutting-edge research database for exploring and understanding the posttranslational modifications in proteins.

The world's principal and complete catalog for protein information is the Universal Protein Resource (UniProt) which is a central repository of protein sequences and their functions. The UniProt consortium members include EMBL-EBI, SIB, and PIR leading to a comprehensive protein knowledge database having protein data from external sources like EMBL/GenBank/DDBJ, VEGA, Ensembl, PDB, RefSeq, etc. The UniParc is a database resource for sequence archive containing old, new, and obsolete sequences. The UniRef database is the UniProt Reference Clusters resource, UniMES contains metagenomic and environmental sequences. UniMES clusters has Cluster 90 and 100. UniProt provides proteomes with completely sequenced genomes of species and the UniProt databases is the UniProt Knowledgebase (UniProtKB) forming the core data of UniProt. Currently, there are 17 proteome data of *Trypanosoma*; out of which, 10 are Reference proteomes (https://www.uniprot.org/proteomes/?query = trypanosoma&sort = score). There are 15 proteome datasets for *Leishmania* out of which only Reference proteomes for *L. major*, *L. infantum*, and *L. braziliensis* are available. Detailed reference proteomes for *L. donovani*, *L. tropica*, *L. mexicana* are still missing. The supporting data in UniProt consists of many databases for taxonomy, literature, cross-references, annotation systems, human disease proteins-controlled vocabulary, etc., which makes UniProt an exhaustive and complete resource for protein studies. Some of the recent protein sequence and expression data available in UniProt/EBI with respect to *Trypanosoma* and *Leishmania* are indicated in Tables 14.3 and 14.4.

Many other resourceful databases include SIB-ExPAsy which is extensible and integrative providing access to over 160 databases and software tools (Artimo et al., 2012). BioStudies which provides the latest data using comprehensive literature search from other databases and many others specific for the user's requirement and data analyses are shown in Table 14.5.

BioCyc (https://biocyc.org/) is a comprehensive database resource by curation of publications, databases, metabolic pathways, regulatory networks, GO annotations providing reference to the genomes and metabolic pathways of an array of sequenced organisms. It is a collection of Pathway/Genome Databases (PGDBs) that predict the metabolic pathways of organisms whose sequences are available and predict genes for enzymes missing in the metabolic pathways and predict operons. It also provides software tools for searching and visualization of various data, data analysis of "omics" raw files, comparison of different metabolic

TABLE 14.3 Protein sequence data availability of *Trypanosoma* spp. and *Leishmania* spp. in UniProt since 2019.

UniProt ID	Organism	Protein	Gene Ontology (GO)—molecular function	GO—cellular component	Function	Domain	URL
UniProtKB - Q27052 (FCA1_TRYRA)	*Trypanosoma rangeli*	Flagellar calcium-binding protein	Calcium ion binding	Motile cilium	Involved in the rapid motility of trypanosomes, flagellar structure as well as in calcium metabolism. Present in either GDP-bound inactive form or the calcium/GTP-bound active form.	4 EF-hand domains, 3 of which functional calcium-binding sites	https://www.uniprot.org/uniprot/Q27052 (last modified: June 17, 2020)
UniProtKB - O15885 (CLP_TRYCR)	*Trypanosoma cruzi*	Heat shock protein 100	ATP binding	—	ATP binding, chaperone, nucleotide binding	Belongs to ClpA/ClpB family	https://www.uniprot.org/uniprot/O15885 (last modified: August 12, 2020)
UniProtKB - O15886 (PHLC_TRYCR)	*Trypanosoma cruzi*	Variant-surface-glycoprotein (VSG) phospholipase C	Glycosylphosphatidylinositol diacylglycerol -lyase activity Phosphoric diester hydrolase activity	Membrane	By hydrolysis of the attached glycolipid, releases soluble variant-surface-glycoprotein containing phosphoinositol from the cell of *Trypanosoma brucei* after cell lysis. VSG lipase play a role in parasite differentiation or antigenic variation.	PI-PLC X-box	https://www.uniprot.org/uniprot/O15886 (last modified: October 7, 2020)
UniProtKB - P28593 (TYTR_TRYCR)	*Trypanosoma cruzi*	Trypanothione reductase	Flavin adenine dinucleotide binding Trypanothione-disulfide reductase activity	Cytoplasm	Trypanothione is a parasite analog of glutathione.	Belongs to the class-I pyridine nucleotide-disulfide oxidoreductase family	https://www.uniprot.org/uniprot/P28593 (last modified: August 12, 2020)
UniProtKB - P09194 (PHLC_TRYBB)	*Trypanosoma brucei brucei*	Variant-surface-glycoprotein phospholipase C	Glycosylphosphatidylinositol diacylglycerol-lyase activity Phospholipase C activity	Plasma membrane	By hydrolysis of the attached glycolipid, releases soluble variant-surface-glycoprotein containing phosphoinositol from the cell of *Trypanosoma brucei* after cell lysis. VSG lipase play a role in parasite differentiation or antigenic variation.	PI-PLC X-box	https://www.uniprot.org/uniprot/P09194 (last modified: October 7, 2020)
UniProtKB - O96507 (KARG_TRYCR)	*Trypanosoma cruzi*	Arginine kinase	Arginine kinase activity ATP binding Creatine kinase activity	—	Catalytic activity Cofactor	Phosphagen kinase *N*-terminal, phosphagen kinase *C*-terminal	https://www.uniprot.org/uniprot/O96507 (last modified: August 12, 2020)

UniProtKB - P18061 (CALM_TRYCR)	*Trypanosoma cruzi*	Calmodulin	Calcium ion binding	—	Calmodulin mediates the control of a large number of enzymes, ion channels and other proteins by Ca^{2+}.	EF-hand 1,2,3,4	https://www.uniprot.org/uniprot/P18061 (last modified: August 12, 2020)
UniProtKB - P21899 (RL40_TRYBB)	*Trypanosoma brucei brucei*	Ubiquitin-60S ribosomal protein L40	Structural constituent of ribosome	Cytoplasm Nucleus Ribosome	Ubiquitin and 60S ribosomal protein L40.	Ubiquitin-like	https://www.uniprot.org/uniprot/P21899 (last modified: June 17, 2020)
UniProtKB - O61078 (TOP2_LEICH)	*Leishmania chagasi*	DNA topoisomerase 2	ATP binding, DNA binding, DNA topoisomerase type II, metal ion binding	Nucleus	Control of topological states of DNA by transient breakage and subsequent rejoining of DNA strands. Topoisomerase II makes double-strand breaks.	Toprim	https://www.uniprot.org/uniprot/O61078 (last modified: June 17, 2020)
UniProtKB - Q9NGA0 (DCAM_LEIIN)	*Leishmania infantum*	*S*-adenosylmethionine decarboxylase proenzyme	Adenosylmethionine decarboxylase activity	—	Catalytic activity, cofactor	Belongs to the eukaryotic AdoMetDC family	https://www.uniprot.org/uniprot/Q9NGA0 (last modified: August 12, 2020)
UniProtKB - P21620 (IMDH_LEIDO)	*Leishmania donovani*	Inosine-5′-monophosphate dehydrogenase	IMP dehydrogenase activity, metal ion binding, nucleotide binding	Glycosome	Catalytic activity, cofactor, activity regulation, kinetics, pathways: Xanthosine monophosphate (XMP) biosynthesis via de novo pathways	CBS 1 and CBS 2	https://www.uniprot.org/uniprot/P21620 (last modified: October 7, 2020)
UniProtKB - P45520 (ACT_LEIMA)	*Leishmania major*	Actin	ATP binding	Cytoplasm, cytoskeleton	Actins are highly conserved proteins that are involved in various types of cell motility and are ubiquitously expressed in all eukaryotic cells.	Belongs to actin family	https://www.uniprot.org/uniprot/P45520 (last modified: October 7, 2020)
UniProtKB - P21148 (TBB_LEIME)	*Leishmania mexicana*	Tubulin beta-chain	GTPase activity, GTP binding, structural constituent of cytoskeleton	Cytoplasm, microtubule	Tubulin is the major constituent of microtubules. It binds two moles of GTP, one at an exchangeable site on the beta-chain and one at a nonexchangeable site on the alpha-chain.	Belongs to tubulin family	https://www.uniprot.org/uniprot/P21148 (last modified: August 12, 2020)

(*Continued*)

TABLE 14.3 (Continued)

UniProt ID	Organism	Protein	Gene Ontology (GO)—molecular function	GO—cellular component	Function	Domain	URL
UniProtKB - Q8MNZ1 (GP63_LEITR)	*Leishmania tropica*	Leishmanolysin	Metal ion binding, metalloendopeptidase activity	Integral component of membrane	Role during the infection of macrophages in the mammalian host, catalytic activity, cofactor.	Belongs to peptidase M8 family	https://www.uniprot.org/uniprot/Q8MNZ1 (last modified: August 12, 2020)
UniProtKB - P27116 (DCOR_LEIDO)	*Leishmania donovani*	Ornithine decarboxylase	Ornithine decarboxylase activity	—	Catalyzes the first and rate-limiting step of polyamine biosynthesis that converts ornithine into putrescine, which is the precursor for the polyamines, spermidine and spermine.	Pyridoxal phosphate binding and substrate binding. Belongs to the Orn/Lys/Arg decarboxylase class-II family.	https://www.uniprot.org/uniprot/P27116 (last modified: October 7, 2020)

TABLE 14.4 Protein expression data of *Trypanosoma* spp. and *Leishmania* spp. available at EBI-EMBL since 2015.

Project ID	Organism	Protein	Title	Description	URL
Project PXD019032	*Leishmania mexicana*	Cytoskeletal protein	An important and essential cytoskeletal protein complex for the division of intracellular amastigotes of *Leishmania mexicana* (LAND-483 and LAND-536).	Cytoskeletal protein known as "KHARON" is an important flagellar trafficking of the glucose transporter GT1 and for cytokinesis. Essential for the survival of amastigote inside of mammalian macrophages.	https://www.ebi.ac.uk/pride/archive/projects/PXD019032 (publication pate: August 17, 2020)
Project PXD014077	*Leishmania mexicana*	CFAP44	Proteomic dissection of CFAP44-deficient axonemes.	Deletion of gene CFAP44 caused a reduction in promastigote motility.	https://www.ebi.ac.uk/pride/archive/projects/PXD014077 (publication date: January 20, 2020)
Project PXD000724	*Leishmania donovani*	Glycosomes	Proteomic analysis of the glycosomes from *Leishmania donovani*.	Glycosomes are known to harbor most peroxisomal enzymes and they possess glycolytic enzymes.	https://www.ebi.ac.uk/pride/archive/projects/PXD000724 (publication date: March 25, 2015)
Project PXD002465	*Leishmania major*	Noncoding RNAs	A novel protein responsible for the coding potential of long intergenic noncoding RNAs (lincRNAs) in the kinetoplastid parasite *Leishmania major*.	Identification of 10 novel protein-coding genes based on peptide data and 100 proteins coding genes based on homology classified as ncRNA genes.	https://www.ebi.ac.uk/pride/archive/projects/PXD002465 (publication date: May 11, 2016)
Project PXD011057	*Leishmania mexicana*	Flagellar proteome	*Leishmania mexicana* promastigote flagellar proteome.	Flagellar protein used for motility, attachment to the sand fly vector, sensory functions.	https://www.ebi.ac.uk/pride/archive/projects/PXD011057 (publication date: December 14, 2018)
Project PXD014580	*Leishmania mexicana*	De novo protein synthesis of newly synthesized proteins (NSPs)	Quantitative proteomics profiling of NSPs in *Leishmania mexicana* parasites during starvation.	De novo protein synthesis in the starving *Leishmania mexicana* with the help of BONCAT technique. 166 NSPs identified in the parasite and quantified the relative changes in abundance of these proteins as a function of duration of starvation.	https://www.ebi.ac.uk/pride/archive/projects/PXD014580 (publication date: December 3, 2019)
Project PXD002766	*Leishmania* sp.	Arginine	Arginine sensing in *Leishmania*.	Lack of arginine environment causes rapid upregulation in amino acid transporter (AAP3) expression as well as several other transporters. Arginine deprivation activates parasites during macrophage infection.	https://www.ebi.ac.uk/pride/archive/projects/PXD002766 (publication date: February 17, 2016)

(*Continued*)

TABLE 14.4 (Continued)

Project ID	Organism	Protein	Title	Description	URL
Project PXD005907	*Trypanosoma brucei*	Bloodstream form proteins	Identification of *Trypanosoma brucei* bloodstream form proteins by mass spectrometry. These proteins interact with Ins(1,4,5)P3 and Ins(1,3,4,5)P4.	Identification of parasites bloodstream form proteins by mass spectrometry that interacts with Ins(1,4,5)P3 and Ins(1,3,4,5)P4.	https://www.ebi.ac.uk/pride/archive/projects/PXD005907 (publication date: October 26, 2018)
Project PXD010285	*Trypanosoma cruzi*	*O*-GlcNAcylated protein	Identification of *O*-GlcNAcylated proteins in the protozoan parasite *Trypanosoma cruzi*.	*O*-GlcNAcylation, is highly conserved monosaccharide posttranslational modifications found on serine and threonine residues of proteins mainly from nucleus, cytoplasm and mitochondria.	https://www.ebi.ac.uk/pride/archive/projects/PXD010285 (publication date: April 18, 2019)
Project PXD009073	*Trypanosoma brucei*	Protein kinase A (PKA)-dependent proteome	Label-free quantitative analysis of PKA-dependent proteome changes in *Trypanosoma brucei*.	PKA is a second messenger cyclic adenosine monophosphate (cAMP), highly conserved among eukaryotes. PKA holoenzymes in *Trypanosoma* are unresponsive to cAMP in vitro and in vivo. This literature describes the PKA-inducible proteome of the bloodstream stage of *Trypanosoma brucei*.	https://www.ebi.ac.uk/pride/archive/projects/PXD009073 (publication date: February 25, 2019)
Project PXD000051	*Trypanosoma brucei*	Flagellar protein	Proteomic analysis of intact flagella of *Trypanosoma brucei* procyclic cells. This identified several novel flagellar proteins with unique sublocalization and dynamics.	Identification of intact flagella of procyclic *Trypanosoma brucei* cells by liquid chromatography with tandem mass spectrometry (LC-MS/MS) after a trypsin shaving.	https://www.ebi.ac.uk/pride/archive/projects/PXD000051 (publication date: December 17, 2018)
Project PXD002970	*Trypanosoma cruzi*	*N*-myristoylome	The *N*-myristoylome of *Trypanosoma cruzi*.	Identification of *N*-myristoylated proteins in *Trypanosoma cruzi* using a chemically tractable myristic acid analog, azidomyristate, further labeled proteins was carried out prior to identification by LC-Ms/MS.	https://www.ebi.ac.uk/pride/archive/projects/PXD008741 (publication date: March 21, 2018)
Project PXD008741	*Trypanosoma brucei*	Proteins of cell cycle	Proteomic analysis of the cell cycle of procylic form *Trypanosoma brucei*.	5325 proteins were quantified by using tenplex tandem mass tag technology. 384 proteins were classified as cell cycle regulated and subdivided into nine distinct clusters of temporal regulation. 40 novel and essential cell cycle regulated proteins that could be considered as future drug targets.	https://www.ebi.ac.uk/pride/archive/projects/PXD002970 (publication date: August 8, 2016)
Project PXD001617	*Trypanosoma vivax*	Gene expression	Global gene expression profiling through the complete life cycle of *Trypanosoma vivax*.	Literature describes *Trypanosoma vivax* differs significantly from other African trypanosomes in the developmentally regulated proteins it expresses on its cell surface and in the structure of host–parasite interface. These provide targets for the therapy.	https://www.ebi.ac.uk/pride/archive/projects/PXD001617 (publication date: July 20, 2015)
Project PXD019073	*Trypanosoma brucei*	RNA-binding protein	A novel essential RNA-binding protein complex in *Trypanosoma brucei*.	ZC3H5 is an important cytoplasmic protein with a single Cx7Cx5Cx3H zinc finger domain in trypanosome. RNA immunoprecipitation indicated that ZC3H5 is preferentially associated with poorly translated, low-stability mRNAs which were enriched in the motif (U/A)UAG(U/A) at the 5′-untranslated regions and coding regions.	https://www.ebi.ac.uk/pride/archive/projects/PXD019073 (publication date: September 23, 2020)

TABLE 14.5 SIB-ExPAsy/UniProt data resources and software tools for data analyses.

SIB resources	Functions
Bgee	Knowledgebase for gene expression data including transcriptomics after curation with a software tool.
Tromer	The transcriptome analyzer provides tools for the determination and documentation of all the transcribed parts of a genome-like expressed sequence tags and other sequences of mRNA.
ShoRah	Software for the analysis of deep-sequencing (next-generation sequencing) data and reconstruction of haplotypes in genetically heterogeneous samples.
mOTUs	When metagenomic data is used, it uses phylogenetic marker gene-based operational taxonomic units (mOTUs) to identify and resolve microorganisms at the species level.
miRmap	Provides complete and exhaustive prediction of microRNA target repression score.
STRING	Knowledge base and analytical software tool for known and predicted protein–protein interactions which includes physical and functional parameters.
World-Two-Dimensional Polyacrylamide Gel Electrophoresis (2D-PAGE)	Public repository for gel-based proteomics data published in the literature.
Make2D-DB	Environment to create, convert, publish, interconnect two-dimensional polyacrylamide gel electrophoresis (2D-PAGE) databases offering updation of data from numerous external data resources.
PRIDE	The world's largest data repository for mass spectrometry-based proteomics data is the Proteomics Identifications database.
MetaNetX	Allows to access, analyze genome-scale metabolic networks as well as biochemical pathways. Models can be interactively compared, analyzed, manipulated, and exported.

pathways and comparison of diverse genomics. BioCyc also consists of MetaCyc, HumanCyc, BsubCyc, Pathway tools, etc.

14.4 Community/specialized databases

14.4.1 Microbial Genome Database for Comparative Analysis

For comparative genomics of microbes like bacteria, archaea and unicellular eukaryotes (protozoa and fungi) based on comprehensive orthology analysis, the Microbial Genome Database for Comparative Analysis (MBGD; http://mbgd.genome.ad.jp/) was developed. This database contains 6318 complete genomes (Uchiyama et al., 2019). Previously, for both closely and distantly related organisms, MBGD provided two types of ortholog tables. The main drawback of this was that only the conserved genes in their respective genomes were present in the standard ortholog table. To overcome this, ortholog tables were constructed in a stepwise procedure hierarchically in a bottom to top manner. This enables the entire gene repertoire to be stored in MBGD in this standard ortholog table. Along with this key feature, several functionalities have been enhanced which includes rapid and flexible keyword searching in addition to profile-based sequence searching. The Resource Description Framework and SPAR Query Language (SPARQL) interface represents the core data in MBGD for integrative database searching. Microbial Genome Database Resource Description Framework (MBGDRDF) has been utilized under the MicrobeDB.jp project (http://microbedb.jp) for a collaborative development of an integrative database of microbes.

14.4.2 Fusion Database

Microbes are the most diverse group and are classified based on phylogenetic relationships. The basis of this variability is genetic mostly due to the interactions of various environmental factors and the reasons for the influence of environment on the functionality of microbes is not completely understood. A database which connects the functionality of microbes with the environmental factors would greatly aid in the identification of functional similarity of microbes which may not be phylogenetically related. However, the increase in microbial genome sequence has greatly facilitated the availability of molecular functions of the microbes based on the gene/protein sequence available in various databases. Due to lack of experimental evidence, the functional annotation of proteins is limited which can be enhanced using the high-throughput softwares and databases which would ultimately help in identification of novel proteins functionality with therapeutic implications (Zhulin, 2015).

A method was developed based on comparison of the functional similarity of microbes to the environment on the basis of the proteins which was identified by translating it from the available sequenced genomes. The database namely Fusion Database (fusionDB; http://services.bromberglab.org/fusiondb/) uses the functional data in order to represent 1374 taxonomically distinct bacteria which are annotated on the basis of the available metadata, habitat, temperature preferences and oxygen consumption. The microbes in the fusionDB database are encoded as a set of the function-specific to each microbe. The proteome represents this functionality and different microbes are further interlinked to each other by their common function. FusionDB user can search a combination of organisms by name and metadata in the database. The web interface also has other options in which it allows the mapping of novel microbial genome to all the available reference bacterial functional spectrum, with functional interactive networks which are similar with shared functionality. This unique database not only provides a platform for comparing microbes but also allows identification of environment-specific functional similarity among the microbes and potential horizontal gene transfer events.

14.4.3 Marine Metagenomics Portal

Marine Metagenomics Portal (MMP; https://mmp.sfb.uit.no/) is maintained by elixir Norway and it provides free, highly curated diverse genomic and metagenomics data of the marine community (Klemetsen et al., 2018). It offers several services like MarRef. MarRef is a manually curated marine microbial genome database containing 970 complete genomes of prokaryotes in the current version. While MAR (MARine) databases contains 13,237 partly curated noncomplete marine microbial genomes with each entry containing 120 metadata fields having information of the environment from where the sampling was done or information on the host information, phenotype of the organism, its taxonomy, pathogenicity, assembly and abundantly annotated, manually curated contextual sequences. *META*-pipe is another service which consists of the complete workflow for the analysis of marine metagenomic data. It provides assembly of high-throughput sequence data, predicted genes and their functional annotation, and taxonomic profiling. SalDB is another manually curated database for salmonide-associated genomes available in this marine DB.

14.4.4 North-East India Microbial database

One of the megadiversity hotspots in the world is located in the North-East region of Eastern Himalayan region in India. This is a rich source for the bioprospection of novel organisms and their metabolites. The exploration of this rich microbial diversity from diverse ecological niches could lead to the biomining of biotechnologically significant strains. The unique feature of this region is the diverse habitats comprising gradients in altitudes, temperature and soil. North-East India Microbial database (NEMiD) (http://mblabnehu.info/nemid) in its current version offers information on microbial diversity from more than 140 sampling site studies, 229 microbial isolates and their 229 16S rRNA partial sequences. This is a unique platform which offers location of the soil microbes of the Eastern Himalayan range along with compilation of the novel properties of microbes dealing with microbial diversity, information on the microbial characteristics and their molecular data. NEMiD currently has a compilation of 229 entries and database-related research publications of bacteria, fungi and actinomycetes. Cross-linking of new information or integration of newer databases is possible in NEMiD to impart a more extensive study of microbial diversity. NEMiD is a community repertoire for scientists and data analysts to understand the pattern of microbial diversity in the North-Eastern India region along with the identification of several microbes. There is future scope to add information on other microbes from diverse environments other than soil or from different sampling locations along with addition of geoinformatics-based diversity analyses and FTP pages for downloading data. NEMiD allows easy access to microbial diversity data to open new horizons for microbial databases in basic and advanced research.

14.4.5 Biomes of Australian Soil Environments

Biomes of Australian Soil Environments (BASE; http://www.Bioplatforms.Com/soil-biodiversity/) is an open framework database which was developed as a comprehensive national microbial dataset of the microbial biodiversity in the entire Australia region extending into Antarctica which includes the soil, vegetation and climate variations open access data (Bissett et al., 2016). BASE is a platform to investigate the evolution of the diverse Australian soil microorganisms, patterns of the microbial communities across different ecological niches and the environmental contributing factors that drive their changes, effects on genes, their functions, species, or communities driven by land management which can be utilized in the indication of mineral deposits and restoration of degraded ecosystems. It is the first expansive database of microbial diversity associated with metadata collected from 900 exclusive and exhaustive environmental gradients found in the Australian continent. With the number of increasing microbes characterized, the BASE database is robust with upgradation and integration of microbial diversity and function. BASE currently offers amplicon sequences from many prokaryotes like archaea and bacteria, specific eukaryotes like fungi and associated contextual data of diverse regions with biological samples and their vegetation and land-use types. This will soon include metagenomics data. The amplicon data available in this database is provided as raw sequence in the FASTQ format as well as the analyzed OUT table formats. All these data can be accessed via the project's data portal to retrieve data samples of our interest. All the processed data can be visually intersected and examined with other diverse environmental data available. This can be done by using the tools developed by the "Atlas of Living Australia." In the future, exploration of microbial, marine and plant biodiversity will be made possible as BASE will expand and link to other global databases. This will ensure that the key interactions between soil and the microbial hotpots in the soil gradients are not overlooked to ensure sustenance of the soil and natural ecosystems.

14.4.6 LeishCyc and TrypanoCyc databases

Databases which are specific to apicomplexan protist parasites like *Plasmodium falciparum* (PlasmoCyc) and *Toxoplasma gondii* (ToxoCyc) and the kinetoplastid parasites, *Trypanosoma brucei* (TrypanoCyc) and

Leishmania major (LeishCyc) have been developed by the BioCyc database (Saunders et al., 2012). Based on the BioCyc ontology which was initially based on the genomic data of *L. major* strain Friedlin (Doyle et al., 2009), the PGDB for *L. major* is LeishCyc. This database consists of 8520 total gene sequences and 8286 protein sequences. Around 141 pathways integrates the biochemical reaction network and metabolic pathways with the genome and ortholog data availability. All the collective and vital information on transcriptomics, metabolomics, proteomics obtained by experiments, genetic analyses of specific cellular and biochemical pathways is provided by LeishCyc. In addition, metabolic pathways generated by genome wide annotations is also an integral part of this database (Chavali et al., 2008). The detailed information can be accessed at https://www.biocyc.org/LEISH/organism-summary?object = LEISH. The metabolic pathways can be visualized and studied by highlighting the pathway of our interest (β-oxidation pathway) as shown in Fig. 14.1.

TrypanoCyc, maintained at the French National Institute of Agricultural Research collects information on the *T. brucei* metabolism and makes it publicly available. TrypanoCyc is a community database which provides annotated metabolic and cellular pathways in addition to the genome information of *T. brucei*, the causative agent of African Trypanosomiasis (Fig. 14.2). TrypanoCyc allows annotation efforts and uses a collaborative web platform (TrypAnnot) for individual research interests. A developmental stage is added and a separate compartment information on enzyme activity aids in improving the automatic metabolic network reconstruction in TrypanoCyc. These metadata on the

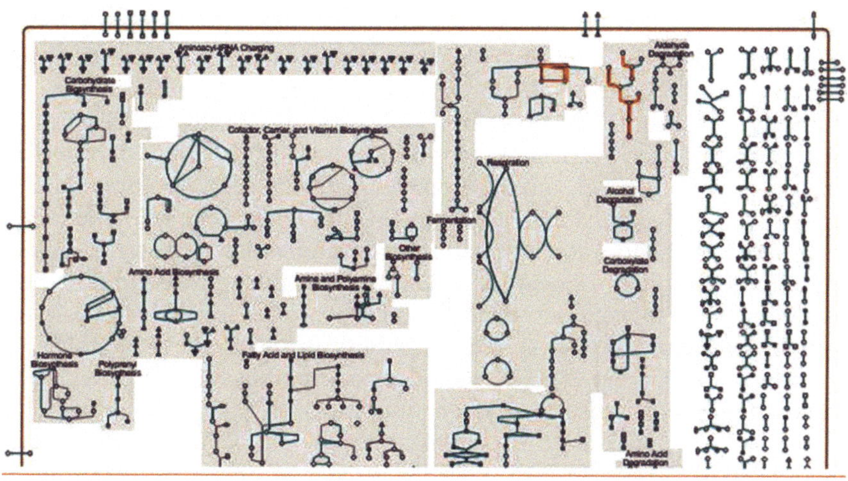

FIGURE 14.1 Metabolic pathways in *Leishmania major* can be visualized by highlighting the pathway/enzymes of our interest (e.g., β-oxidation pathway) in LeishCyc Database.

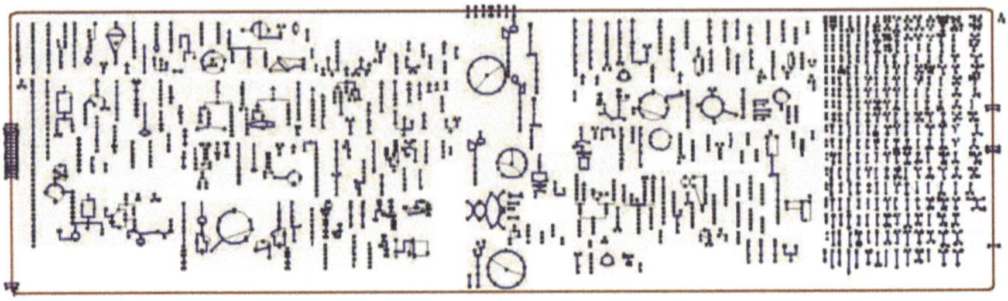

FIGURE 14.2 TrypanoCyc offers features to visualize dead-end metabolites, choke-point reactions, pathway route search, metabolite translational services for *Trypanosoma* spp.

network will help generate customized metabolic networks and help in understanding the parasite metabolism better. In addition, comparative analysis for omics data, organism comparative data, dead-end metabolites, choke-point reactions, pathway route search, metabolite translational service are some of the highlight features available at TrypanoCyc (Chukualim et al., 2008). All these information and tools can be accessed at http://vm-trypanocyc.toulouse.inra.fr/.

14.4.7 VEuPath Database

Eukaryotic Pathogen, Vector and Host Informatics Resource (VEuPathDB; https://veupathdb.org/veupathdb/app) database is funded by the US National Institute of Allergy and Infectious Diseases (NIAID), supported by the Wellcome Trust (United Kingdom) and is one of the important Bioinformatics Resource Centers (BRCs). The resources are obtained from support initially provided for the Plasmodium Genome Database by the Burroughs Wellcome Fund and a research grant from NIAID (Warrenfeltz et al., 2018). In 2004, the BRC program was initiated in order to provide public access to computational platforms and analysis tools which enable the collection, integration and mining of information at the genome level and other large-scale datasets pertaining to pathogens causative of various infectious disease and its management. This also includes their interaction with mammalian hosts and invertebrate vectors of disease. The important focus of the VEuPathDB, the new release in August 2020 is on eukaryotic pathogens and invertebrate vectors of infectious diseases which also includes prior data from BRCs which were dedicated only to parasitic species. The VEuPathDB beta sites includes—ToxoDB, PlasmaDB, GiardiaDB, AmoebaDB, MicrosporidiaDB, CrytoDB, TritrypDB, PiroplasmaDB, PlasmoDB, FungiDB, TrichDB, and also HostDB and vector species (VectorBase). In VEupathDB, all sites on eukaryotic pathogens are in the beta version with basic differences in data content in the respective databases. All relevant information like in-house genome analyses, proteomics, RNA sequences, ChIP sequences, copy number variation, genetic variation and SNP calling, protein array data and metabolic pathways are available for every database included in VEuPathDB. The websites in beta are still under development and is open to user feedback. It includes features linked to social media sites and YouTube channels with access to all the tutorials.

14.4.8 TritrypDB

In VEupathDB, all sites on eukaryotic pathogens including TritrypDB for trypanosomatids are in beta with some basic difference in data content (Warrenfeltz et al., 2018). All relevant information like in-house genome analyses, proteomics, RNA sequences, ChIP sequences, copy number variation, genetic variation and SNP calling, protein array data, and metabolic pathways are available in Tritryp database. The left panel has features for various categories as genes and also includes search filters. All the tools that are linked to the site which includes BLAST, Companion, EuPaGDT, etc. and provides workspace to analyze individual data through Galaxy. It also provides access to the data available in VEuPathDB resource and the available data can be downloaded. The legacy of the TritrypDB site will be curtailed soon once the new beta site is completed. Its data access policy allows all data on these websites freely for public use, through the contributions of many researchers involved in generating genome sequences, functional genomics datasets and additional information. The data available is updated from ongoing research and may not be always accurate. When using data obtained from VEuPathDB, it is important to cite the original publications and contributors. The most recent list of VEuPathDB publications are provided by using the PubMed filter.

14.4.9 Leishmania Exclusive Protein database

A protein sequence repository specific to five *Leishmania* spp. called the Leishmania Exclusive Protein database (Leish-ExP; http://www.hpppi.iicb.res.in/Leish-ex) was developed in CGI PERL with backend retrieval of all data (Das et al., 2020). It comprises two major parts. The first part constitutes all the major species-specific proteins and their sequences. These include the structural and functional features of proteins. Specific proteins and their functions at the genus level constitute the second part of the database. For the construction of this database, all the sequences from *L. braziliensis, L. major, L. donovani, L. infantum, and L. mexicana* were compared against all known protein sequences at the NCBI nonredundant database in addition to species-level comparison. Leish-ExP database contains exclusive proteins which are found only in the *Leishmania* genus. Therefore, these proteins are termed as *Leishmania* genus specific proteins. The proteins exclusively present in specific *Leishmania* species like *L. major* are called species-specific proteins. In addition, using various in silico methods, various genes and proteins available in this database are characterized which contributes to additional information on sequence, function, structure and localization. These have been systemically categorized and available at http://www.hpppi.iicb.res.in/Leish-ex. This community database constitutes the gene/proteins only relevant from five different *Leishmania* species. This database is remarkably similar to the LeishCyc database which gives details on the metabolic pathways and metabolite relationship of all *L. major* genes.

L. major and *S. mansoni* database (LmSmdB) is another integrated database for gene regulatory and metabolic network analyses in *L. major* while, LeishDB is an exclusive database for storing coding genes and noncoding RNAs data related only to *L. braziliensis*. Leish-Exp database is integrated with Leishmania transcriptomics and gene expression studies and information. This is the only database available which identifies, categorizes, and characterizes genes/proteins of *Leishmania* at the genus and individual species as s7een in Fig. 14.3. Various analysis tools like

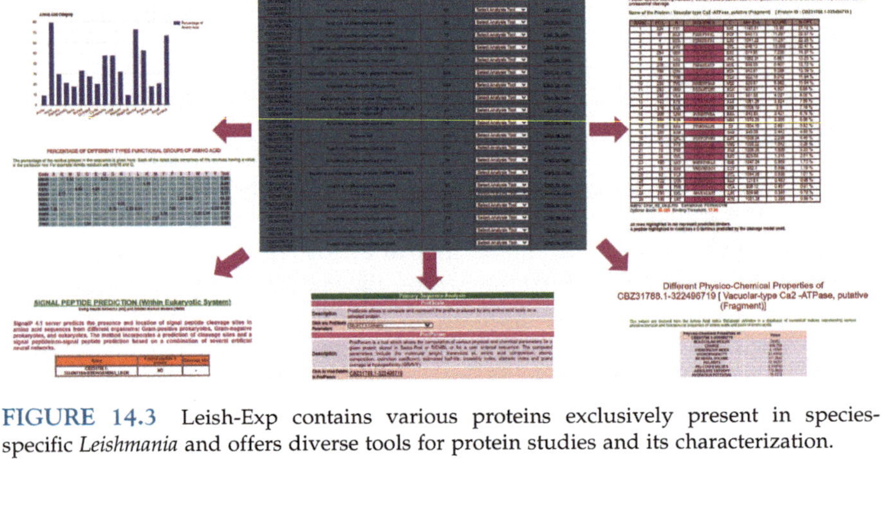

FIGURE 14.3 Leish-Exp contains various proteins exclusively present in species-specific *Leishmania* and offers diverse tools for protein studies and its characterization.

primary gene/protein sequence analysis, prediction of various signal peptides, subcellular localization prediction, prediction of phosphorylation sites and secondary structure prediction are available for species-specific proteins in the database. Thus Leish-Exp database is the only community database which provides genetic and proteomic information relevant for identification of therapeutic targets, drug design for leishmaniasis.

14.4.10 Leishmania Database

Predictive algorithms and updated databases were recently developed using *L. braziliensis* information on the protein-coding genes and noncoding RNAs. A total of 11,491 open-reading frames are available and prediction of 5263 (45.80%) annotated proteins was possible which was made available in public repositories after the reannotation process. The LeishDB (http://www.leishdb.com) represents an increase in 26% of the number of coding gene predictions when compared to other biological databases. Along with this, 11,243 putative noncoding RNAs from different classes were identified after the annotation process. Using RNA-seq public data, validation of both coding and noncoding RNA predictions for their expression was carried out. In this open-source microbial database which is available at: http://www.leishdb.com, different regulatory ncRNAs in *L. braziliensis* is comprehensively and systematically identified and functionally annotated. The process of annotation identified 48.93% (5623 out of 11,491) predictions associated with proteins which includes hypothetical proteins available in other public databases. The gene predictions covered 60.74% (4957 out of 8161) which is available in NCBI and 79.61% (6771 out of 8505) is found in TriTrypDB. An increase of other previously unidentified 6304 predictions is also seen. Exclusion of hypothetical proteins and prediction of proteins annotated with an associated function led to the identification of a total of 5254 coding genes. This incredibly represents around 5254 out of 5623 of annotated predictions which is close to 94.43%. This database also stores a wide array of annotated data from several other organisms like chromosomes, genes, gene ontology, proteins, ncRNAs, literature, genomic sequences, gene coordinates, etc.

14.4.11 Trypanosoma database

This trypanosomes database system (http://www.tanpaku.org/tdb/) provides a complete and comprehensive information of cellular enzymes which are involved in the pyrimidine biosynthesis pathway and various redox reactions in *Trypanosoma* spp. This exhaustive database also provides several search functions such as keyword search, sequence search and motif search for browsing of data. This database also provides several prediction tools for protein structures and protein functions making it valuable to study the cellular enzyme system in *Trypanosoma* spp.

14.5 Future perspectives

Increase in data submission and availability of biological data warrants for better analyses and interpretation for discoveries in biology. With the development of Omics, the integration of data for deeper understanding would provide vital clues for biomining novel microbes as well as their products for downstream applications. Diversity and ecological studies of an organism along with in-depth cellular and molecular analyses will aid in devising better therapeutic strategies against pathogenic microbes like *Trypanosoma* and *Leishmania*.

14.6 Conclusions

Advancement in technologies for critical knowledge of the molecular and cellular processes of microbes will lead to crucial information on strategies devised by various pathogens in escaping the host immune system, host–pathogen interactions and targets for chemotherapeutic drugs and vaccines. The current available data are only the inception for new scientific breakthroughs for drugs and vaccine development against microbial pathogens like kinetoplastids.

References

Almeida, F. A., Kamila, G.-P., Pinheiro, N. M., Ketiuce, Z., Lara, F., Ludimila, P., Danielle, O. N., Luciana, C., & Alexandre, M. (2018). Human kinetoplastid protozoan infections: Where are we going next? *Frontiers in Immunology, 9*, 1493. Available from https://doi.org/10.3389/fimmu.2018.01493.

Alvar, J., Vélez, I. D., Bern, C., Herrero, M., Desjeux, P., Cano, J., Jannin, J., & de Boer, M. (2012). Leishmaniasis worldwide and global estimates of its incidence. *PLoS One, 7*(5). Available from https://doi.org/10.1371/journal.pone.0035671.

Amid, C., Alako, B. T. F., Balavenkataraman Kadhirvelu, V., Burdett, T., Burgin, J., Fan, J., Harrison, P. W., Holt, S., Hussein, A., Ivanov, E., Jayathilaka, S., Kay, S., Keane, T., Leinonen, R., Liu, X., Martinez-Villacorta, J., Milano, A., Pakseresht, A., Rahman, N., ... Cochrane, G. (2020). The European nucleotide archive in 2019. *Nucleic Acids Research, 48*(1), D70–D76. Available from https://doi.org/10.1093/nar/gkz1063.

Artimo, P., Jonnalagedda, M., Arnold, K., Baratin, D., Csardi, G., De Castro, E., Duvaud, S., Flegel, V., Fortier, A., Gasteiger, E., Grosdidier, A., Hernandez, C., Ioannidis, V., Kuznetsov, D., Liechti, R., Moretti, S., Mostaguir, K., Redaschi, N., Rossier, G., ... Stockinger, H. (2012). ExPASy: SIB bioinformatics resource portal. *Nucleic Acids Research, 40*(1), W597–W603. Available from https://doi.org/10.1093/nar/gks400.

Bahram, M., Hildebrand, F., Forslund, S. K., Anderson, J. L., Soudzilovskaia, N. A., Bodegom, P. M., Bengtsson-Palme, J., Anslan, S., Coelho, L. P., Harend, H., Huerta-

Cepas, J., Medema, M. H., Maltz, M. R., Mundra, S., Olsson, P. A., Pent, M., Põlme, S., Sunagawa, S., Ryberg, M., ... Bork, P. (2018). Structure and function of the global top-soil microbiome. *Nature, 560*(7717), 233–237. Available from https://doi.org/10.1038/s41586-018-0386-6.

Bhattacharjee, K., & Joshi, S. R. (2014). NEMiD: A web-based curated microbial diversity database with geo-based plotting. *PLoS One, 9*(4). Available from https://doi.org/10.1371/journal.pone.0094088.

Bissett, A., Fitzgerald, A., Meintjes, T., Mele, P. M., Reith, F., Dennis, P. G., Breed, M. F., Brown, B., Brown, M. V., Brugger, J., Byrne, M., Caddy-Retalic, S., Carmody, B., Coates, D. J., Correa, C., Ferrari, B. C., Gupta, V. V. S. R., Hamonts, K., Haslem, A., ... Young, A. (2016). Introducing BASE: The biomes of Australian soil environments soil microbial diversity database. *GigaScience, 5*(1). Available from https://doi.org/10.1186/s13742-016-0126-5.

Burza, S., Croft, S. L., & Boelaert, M. (2018). Leishmaniasis. *The Lancet, 18*. Available from https://doi.org/10.1016/S0140-6736, 31204–2.

Chavali, A. K., Whittemore, J. D., Eddy, J. A., Williams, K. T., & Papin, J. A. (2008). Systems analysis of metabolism in the pathogenic trypanosomatid *Leishmania major*. *Molecular Systems Biology, 4*, 177. Available from https://doi.org/10.1038/msb.2008.15.

Chukualim, B., Peters, N., Fowler, C. H., & Berriman, M. (2008). TrypanoCyc—A metabolic pathway database for *Trypanosoma brucei*. *BMC Bioinformatics*, S10. Available from https://doi.org/10.1186/1471-2105-9-S10-P5.

Das, A., Biswas, N., & Chakrabarti, S. *Leish-ExP: A database of exclusive proteins from Leishmania bioRxiv*, 2020. https://doi.org/10.1101/2020.05.04.076851

Doyle, M. A., MacRae, J. I., De Souza, D. P., Saunders, E. C., McConville, M. J., & Likić, V. A. (2009). LeishCyc: A biochemical pathways database for *Leishmania major*. *BMC Systems Biology, 3*. Available from https://doi.org/10.1186/1752-0509-3-57.

Fraser, C., Eisen, J. & Salzberg, S., Microbial genome sequencing. Nature 406, 799–803 (2000).

Fleischmann, R. D., Adams, M. D., White, O., Clayton, R. A., Kirkness, E. F., Kerlavage, A. R., Bult, C. J., Tomb, J. F., Dougherty, B. A., Merrick, J. M., et al. (1995). Whole-genome random sequencing and assembly of Haemophilus influenzae Rd. *Science*, 496–512. Available from https://doi.org/10.1126/science.7542800.

Hinchliff, C. E., Smith, S. A., Allman, J. F., Burleigh, J. G., Chaudhary, R., Coghill, L. M., Crandall, K. A., Deng, J., Drew, B. T., Gazis, R., Gude, K., Hibbett, D. S., Katz, L. A., Dail Laughinghouse, H., McTavish, E. J., Midford, P. E., Owen, C. L., Ree, R. H., Rees, J. A., ... Cranston, K. A. (2015). Synthesis of phylogeny and taxonomy into a comprehensive tree of life. *Proceedings of the National Academy of Sciences of the United States of America, 112*(41), 12764–12769. Available from https://doi.org/10.1073/pnas.1423041112.

Klemetsen, T., Raknes, I. A., Fu, J., Agafonov, A., Balasundaram, S. V., Tartari, G., Robertsen, E., & Willassen, N. P. (2018). The MAR databases: Development and implementation of databases specific for marine metagenomics. *Nucleic Acids Research, 46*(1), D693–D699. Available from https://doi.org/10.1093/nar/gkx1036.

Leinonen, R., Akhtar, R., Birney, E., Bower, L., Cerdeno-Tárraga, A., Cheng, Y., Cleland, I., Faruque, N., Goodgame, N., Gibson, R., Hoad, G., Jang, M., Pakseresht, N., Plaister, S., Radhakrishnan, R., Reddy, K., Sobhany, S., Hoopen, P. T., Vaughan, R., ... Cochrane, G. (2010). The European Nucleotide Archive. *Nucleic Acids Research*, D28–D31. Available from https://doi.org/10.1093/nar/gkq967.

Lewin, H. A., Robinson, G. E., Kress, W. J., Baker, W. J., Coddington, J., Crandall, K. A., Durbin, R., Edwards, S. V., Forest, F., Gilbert, M. T. P., Goldstein, M. M., Grigoriev, I. V., Hackett, K. J., Haussler, D., Jarvis, E. D., Johnson, W. E., Patrinos, A., Richards, S., Castilla-Rubio, J. C., ... Zhang, G. (2018). Earth BioGenome Project: Sequencing life for the future of life. *Proceedings of the National Academy of Sciences of the United States of America, 115*(17), 4325–4333. Available from https://doi.org/10.1073/pnas.1720115115.

Mashima J., Kodama Y., Fujisawa T., et al., DNA Data Bank of Japan. Nucleic Acids Research, 45 (D1), 2017, D25–D31. Available from https://doi.org/10.1093/nar/gkw1001.

Maslov, D. A., Podlipaev, S. A., & Lukeš, J. (2001). Phylogeny of the kinetoplastida: Taxonomic problems and insights into the evolution of parasitism. *Memorias Do Instituto Oswaldo Cruz, 96*(3), 397–402. Available from https://doi.org/10.1590/S0074-02762001000300021.

Ogasawara, O., Kodama, Y., Mashima, J., Kosuge, T., & Fujisawa, T. (2020). DDBJ Database updates and computational infrastructure enhancement. *Nucleic acids research, 48*(D1), D45–D50. Available from https://doi.org/10.1093/nar/gkz982.

Ponte-Sucre, A. (2016). An overview of trypanosoma brucei infections: An intense host-parasite interaction. *Frontiers in Microbiology, 7*. Available from https://doi.org/10.3389/fmicb.2016.02126.

Rigden, D. J., & Fernández, X. M. (2019). The 26th annual Nucleic Acids Research database issue and Molecular Biology Database Collection. *Nucleic Acids Research, 47*(1), D1–D7. Available from https://doi.org/10.1093/nar/gky1267.

Rodrigues, V., Cordeiro-Da-Silva, A., Laforge, M., Silvestre, R., & Estaquier, J. (2016). Regulation of immunity during visceral Leishmania infection. *Parasites and Vectors, 9*(1). Available from https://doi.org/10.1186/s13071-016-1412-x.

Satyanarayana, T., Das, S. K., & Johri, B. N. (2019). *Microbial diversity in ecosystem sustainability and biotechnological applications: Volume 2. Soil and agroecosystems* (pp. 1–615). Springer Singapore. Available from https://doi.org/10.1007/978-981-13-8487-5.

Saunders, E. C., MacRae, J. I., Naderer, T., Ng, M., McConville, M. J., & Likić, V. A. (2012). LeishCyc: A guide to building a metabolic pathway database and visualization of metabolomic data. *Methods in Molecular Biology, 881*, 505–529. Available from https://doi.org/10.1007/978-1-61779-827-6_17.

Staley, J.J., et al., The Microbial World. The Foundation of the Biosphere, 1997, American Society for Microbiology, Washington DC.

Stuart, K., Brun, R., Croft, S., Fairlamb, A., Gürtler, R. E., McKerrow, J., Reed, S., & Tarleton, R. (2008). Kinetoplastids: Related protozoan pathogens, different diseases. *Journal of Clinical Investigation, 118*(4), 1301–1310. Available from https://doi.org/10.1172/JCI33945.

Sutherland, C. S., Yukich, J., Goeree, R., & Tediosi, F. (2015). A literature review of economic evaluations for a neglected tropical disease: Human African Trypanosomiasis ("Sleeping Sickness"). *PLoS Neglected Tropical Diseases, 9*(2). Available from https://doi.org/10.1371/journal.pntd.0003397.

Uchiyama, I., Mihara, M., Nishide, H., Chiba, H., & Kato, M. (2019). MBGD update 2018: Microbial genome database based on hierarchical orthology relations covering closely related and distantly related comparisons. *Nucleic Acids Research, 47*(1), D382–D389. Available from https://doi.org/10.1093/nar/gky1054.

Warrenfeltz, S., Basenko, E. Y., Crouch, K., Harb, O. S., Kissinger, J. C., Roos, D. S., Shanmugasundram, A., & Silva-Franco, F. (2018). *EuPathDB: The eukaryotic pathogen genomics database resource. Methods in molecular biology* (Vol. 1757, pp. 69–113). Humana Press Inc. Available from https://doi.org/10.1007/978-1-4939-7737-6_5.

Wu, C. H., Huang, H., Arminski, L., Castro-Alvear, J., Chen, Y., Hu, Z. Z., Ledley, R. S., Lewis, K. C., Mewes, H. W., Orcutt, B. C., Suzek, B. E., Tsugita, A., Vinayaka, C. R., Yeh, L. S. L., Zhang, J., & Barker, W. C. (2002). The protein information resource: An integrated public resource of functional annotation of proteins. *Nucleic Acids Research, 30*(1), 35–37. Available from https://doi.org/10.1093/nar/30.1.35.

Zhulin, I. B. (2015). Databases for microbiologists. *Journal of Bacteriology, 197*(15), 2458–2467. Available from https://doi.org/10.1128/JB.00330-15.

Zou, D., Ma, L., Yu, J., & Zhang, Z. (2015). Biological databases for human research. *Genomics, Proteomics and Bioinformatics, 13*(1), 55–63. Available from https://doi.org/10.1016/j.gpb.2015.01.006.

CHAPTER

15

Advances in sequencing technology, databases, and analyses tools for the assessment of microbial diversity

Shaima Rifaie, Vikas Patil and Kamlesh Jangid

National Centre for Microbial Resource, National Centre for Cell Science, SP Pune University Campus, Pune, India

15.1 Introduction

rRNA gene sequencing has become a standard practice for studying microbial diversity in an ecosystem. It has been in use since the inception of the idea of bacterial identification and classification into species (Lemos et al., 2011). What started with wandering spot analysis (Yarus & Barrell, 1971) was replaced by Sanger's di-deoxy sequencing in 1977 (Luckey et al., 1990). Sanger's method has undergone many changes and substitutions since then. Today, the method is automated and has eased out the hustle in the field of science. However, it is now overpowered by next-generation sequencing (NGS) technologies (Moorthie et al., 2011). NGS platforms deliver high-throughput data yielding millions of sequence reads per sample. However, in most cases, the reads are shorter than the classical Sanger's di-deoxy method. Hence, the complexity of the pipelines processing this enormous data has also increased. Analysis of the raw data generated by NGS is a five-step process involving: (1) quality assessment of reads; (2) alignment of quality reads to a reference database; (3) clustering of sequencing variants; (4) identification of the variants using homology search; (5) data visualization

DOI: https://doi.org/10.1016/B978-0-323-90148-2.00003-1

(Wadapurkar & Vyas, 2018). Regardless of the change in technologies and analyses pipeline, the identification of microbes continues to rely upon rRNA gene sequences as the most used taxonomic marker.

Likewise, databases have also undergone significant modification and have started to display a bigger picture to the researchers. The rate of discovering novel taxa has significantly improved due to the availability of comprehensive 16S rRNA gene databases that cover nearly all species that have been cultured in the laboratory. However, the phylogenetic information in the 16S rRNA gene has also been a limiting step in the identification and classification for certain taxa that show a high degree of relatedness or high variability in certain others, which does not translate into strains being separate species or same species, respectively. With a shift from one-gene based on phylogenetic analysis to utilizing the whole-genome sequences, the phylogenomic investigation has revealed several taxa that require revision. At present, primary databases, such as NCBI contain more than 70,000 genome sequences. However, the metadata associated with these genomes, especially their taxonomic names, need urgent curation. This requirement made way for the development of taxonomically integrated databases for Bacteria and Archaea, such as EzBioCloud (Chun et al., 2007). Today, there are several well-curated databases that are in use for the identification of microbes.

Fueled with the advances in technology and availability of diverse databases, the curiosity of researchers now extends to investigate the reasons behind the existence, functional role and interactions between members of the microbial communities found in an ecosystem. Originally, most analyses were limited to calculating the extent of diversity. Today, these are used for beta-diversity comparisons between communities to assess their relative diversity. The three diversity indices most commonly used are richness, Simpson index and Shannon–Weaver index. These indices allow the statistical comparison between multiple communities but they cannot be compared between them (Bent & Forney, 2008). This limited the scope of information and the basis of the relationship. Thus the need of Operational Taxonomic Units (OTUs) was realized as it gave a better understanding of the functioning of an ecosystem (Md & Bin, 2013). Over a period, the analyses started including algorithms for comparing gene sequence libraries and an explanation for structural and compositional differences. One of the first algorithms introduced for such analysis was LIBSHUFF (Singleton et al., 2001). Soon, AMOVA, -LIBSHUFF, DOTUR, UniFrac, SONS, TreeClimber, LibraryCompare, Metastats, and many others were developed. All these have been discussed in detail (Schloss, 2008). More recently, an advanced algorithm called *K*-Shuff was developed to simultaneously compare the structure and composition of microbial communities (Jangid et al., 2016). *K*-shuff also introduced two new diversity indices: the Intra-*k*-function and the Cross-*k*-function which expand our capacity to statistically understand the observed differences between the compared microbial communities. While some of these use the entire sequence data, some are based on a clustering approach to not only determine the extent of diversity, but also the phylogenetic makeup of the diversity in a given ecosystem. Hence, users rely on multiple algorithms to make sense of the data and to arrive at meaningful outcomes.

In this chapter, the readers will be able to gage the advancements in the field of sequencing technology, databases, and analyses pipelines. While the three developed simultaneously and were in fact, interdependent (Fig. 15.1), we have segregated them for ease of understanding. In addition, a pipeline that is used currently for microbial diversity studies is suggested for the readers.

15.2 Advances in sequencing technology

In the late 19th century, bacterial cultures were defined based on phenotypic characteristics. However, because of the limited number of characteristics, soon physiological, chemotaxonomic and biochemical properties of the bacteria were included into the bacterial identification and classification. The biggest shift came in 1970s with the introduction

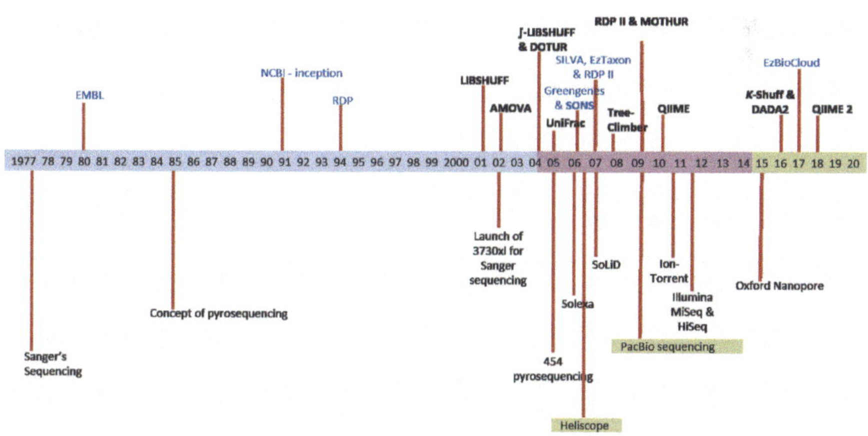

FIGURE 15.1 Overview of the advancement of technology, databases, and algorithms. The number of years has been divided depending on the generation of the sequencing technology. Blue bar indicates first-generation, Violet bar indicates second-generation, and green bar indicates third-generation sequencers. Databases are highlighted in the blue font color and algorithms and pipelines are expressed in bold. Source: *Modified from Bruijns, B., Tiggelaar, R., & Gardeniers, H. (2018). Massively parallel sequencing techniques for forensics: A review.* Electrophoresis, *39(21), 2642–2654.*

of DNA sequencing, inclusiveness of numerical taxonomy and introduction of polyphasic taxonomy (Mahato et al., 2017). Sanger's di-deoxy sequencing method had blown the revolutionary trumpet in sequencing rRNA gene. It exploited the fact that on incorporation of di-deoxy nucleotides, elongation is terminated (Sanger et al., 1977). For nearly three decades Sanger's sequencing method was performed using autoradiography and therefore only experts could perform it. Sometime in the late 1900s, the sequencing process was automated and involved fluorescence dyes which made it easier to perform and more accessible (Luckey et al., 1990).

Even though Sanger's di-dioxy sequencing method is still considered as a gold standard for DNA sequencing, it had many limitations (Table 15.1). These include (1) the requirement of a polymer matrix as a sieving agent; (2) limited throughput as only a few samples could be analyzed; (3) manual sample preparation; (4) requirement of cloning; (5) cost of sequencing; (6) inefficiency in detection of the clinically relevant low-level mutant or allele; (7) difficulty in achieving a throughput for analyzing complex diploid genomes; (8) an uncertain de novo genome assembly. While some of these have been overcome today, these however paved way for the development of high-throughput sequencing technologies (Fakruddin et al., 2013) in the early 2000s.

The first alternative to Sanger's sequencing method with a shift to de novo DNA sequencing was the pyrosequencing method. Originally, the concept of performing sequencing by synthesis was framed in the year 1985 by Nyrén and Lundin (1985) and later on developed by Ronaghi et al. (1998) who first described the real-time pyrophosphate based sequencing method as a viable option. But it was only in the year 2005 that Rothberg and colleagues from 454 Life Sciences demonstrated the parallel pyrosequencing approach (Margulies et al., 2005). The method involves the incorporation of nucleotides by the polymerase during DNA synthesis and the release of pyrophosphate (PPi) and photons, which are proportional to the number of nucleotides being incorporated (Stranneheim & Lundeberg, 2012). Initially, the read length was 100–150 bp with over 200,000 reads and a cumulative output of 20 Mbp per run. Until late 2009, Roche upgraded the system where read length was upgraded to 700 bp, accuracy to 99.9% and output of 14 Gbp per run (Liu et al., 2012). With this shift, pyrosequencing offered an unparalleled throughput at the time.

One of the widely used application of pyrosequencing was single-nucleotide polymorphism (SNP) detection. Before pyrosequencing concept was introduced, conventional Sanger's sequencing method was used. Often this was time-consuming, laborious and required labeling. To increase the understanding of SNPs, use of other techniques like Allele-Specific Oligonucleotides (ASO) hybridization, padlock probes or

TABLE 15.1 Overview of DNA sequencing technologies with their advantages and disadvantages.

Year of conceptualization	Sequencer	Mechanism	Read length	Accuracy	Time/run	Sequencing cost per Gb	Advantages	Disadvantages	References
1977	Sanger	Di-deoxy chain termination	400–900 bp	99.999%	3 h	2400$ (/million bases)	High-quality, long read length	High cost, low throughput	Bruijns et al. (2018); Pettersson et al. (2009)
2005	Roche 454	Pyrosequencing	700 bp	99.9%	24 h	10$ (/million bases)	Good read length	Error rate with polybase more than six, low throughput	Bruijns et al. (2018); Pettersson et al. (2009)
2006	Solexa	Sequencing by reversible terminators	100–300 bp	98%	3–10 days	0.07$ (/million bases)	Highest throughput and compatible with almost all the applications	Challenge in loading samples, major error of substitution and required complex sequence	Bruijns et al. (2018); Pettersson et al. (2009)
2007	SOLiD	Ligation and two-base coding	50 + 35 bp or 50 + 50 bp	99.94% of raw data	7 days for SE and 14 days for PE	0.13$ (/million bases)	Accuracy	Short read assembly	Bruijns et al. (2018); Pettersson et al. (2009)
2010	Ion torrent	Detection of ion released during nucleotide addition	200–400 bp	99.7%	2 h	1000$ (318 chip)	No requirement for optical scanning and fluorescent nucleotides, fast run times and Broad range of applications	High error rates in homo-polymers	Bruijns et al. (2018); Quail et al. (2012); Pettersson et al. (2009)
2010	PacBio	Single-molecule real time	8000–20,000 bp	99.7%	0.5–6 h	2000$	Extremely long reads of 20 kb and even longer	High cost, high overall error rates (~14%), lowest throughput of all platforms and limited range of applications	Bruijns et al. (2018); Quail et al. (2012); Pettersson et al. (2009)
2014	Nanopore	Monitors changes to an electrical current as nucleic acids pass	9545–200,000 bp	99.8%	48 h	Approx. 35$	Portability, easy accessibility and de novo genome assemblies	Slower DNA translocation from microseconds per base to milliseconds	Bruijns et al. (2018); Quail et al. (2012); Pettersson et al. (2009)

allelic discrimination during PCR was implemented (Ahmadian et al., 2000). Thus when pyrosequencing made its way, researcher moved toward it because of sequencing by synthesis approach and reliability on luminometric detection. Pyrosequencing was also able to analyze the loss of heterozygosity in the mixed samples of normal and tumor cells. Pyrosequencing was a rapid, reliable technique and could handle many samples altogether, but the only limiting factor was its long template preparation procedure. While pyrosequencing garnered huge support and was extensively used during its time, newer methods of sequencing with higher output soon picked up.

In 2006, Solexa sequencing became the second most massively parallel sequencing method. The original idea of the technology was envisioned by Professors David Klenerman and Shankar Balasubramanian through a series of creative discussions in the lab and a meeting at a local pub during the summer of 1997. Together, they founded Solexa in 1998 which was purchased by Illumina in 2007 (http://www.illumina.com). Solexa was built on the approach of reversible dye termination (Bentley et al., 2008). Instead of emulsion PCR, DNA template cloning was performed by using bridge amplification. However, the most common error was substitution (Stranneheim & Lundeberg, 2012). The downstream processing of the sequencing reads was performed by using RDP datasets and RDP clustering algorithm (Degnan & Ochman, 2012). In 2007, a new sequencing approach named Sequencing by Oligonucleotide Ligation and Detection (SOLiD) was launched. SOLiD was based on the principle of emulsion PCR, where DNA was fragmented and attached to the beads (Valouev et al., 2008). These beads are then deposited to the glass slide and DNA ligase enzyme ligates the whole DNA strands by releasing the fluorescent dyes. These fluorescent dyes are then captured for imaging (Stranneheim & Lundeberg, 2012). Both, Solexa and SOLiD sequencing methods were based on different approaches altogether. Solexa's template preparation was based on solid-phase PCR, but SOLiD was based on emulsion PCR. Solexa provided with the longer reads whereas SOLiD could only produce short sequencing reads. Launch of these sequencing techniques substantially reduced the costs and allowed analysis of transcriptomes for qualitative as well as quantitative study of gene expression patterns (Bräutigam & Gowik, 2010).

Until 2010, all second-generation sequencing techniques were based on the optic detection to decode the DNA sequence. But in 2010, Ion torrent, a novel sequencing by synthesis platform was built (Rusk, 2011), which exploited the release of protons whenever a nucleotide gets added (Purushothaman et al., 2006). A sensor detects this proton shift converting it to an electric signal. Ion torrent had several advantages over other sequencing techniques as it did not require modified nucleotides, expensive optics and processing of large number of fluorescent images (Stranneheim & Lundeberg, 2012). For instance, Ion torrent was used to study the mitochondrial genome (mtGenome) in comparison with Sanger's sequencing (Parson et al., 2013). Ion torrent provided better insights on homopolymeric C tracks, deletions, substitution and point heteroplasmy. There was high level of consistency observed and the recorded difference was below 0.02%. However, the software used for analyzing Ion-torrent results made a large impact on results suggesting that the analyses pipelines are crucial for any downstream processing of the data.

Even though the second-generation sequencing technologies were efficient enough, they all were based on the principle of amplification (Liu et al., 2012). This initiated the development of third-generation sequencing technologies where single molecules were read in real-time (van Dijk et al., 2014), and offered faster and longer reads. While single-molecule sequencing was proposed as early as late 1900s (Jett et al., 1989), it took nearly two decades for it to be applied in practice to sequence the M13 bacteriophage genome (Harris et al., 2008). These third-generation sequencers are further classified into three categories (Bruijns et al., 2018): (1) Single-molecule synthesis (PacBio); (2) Nanopore based sequencing (Oxford Nanopore and NabSys); and (3) direct imaging of DNA molecules (VisiGen). While we discuss the first two here, direct imaging-based methods which use quantum dots (Beechem et al., 2011; Glenn, 2011) have not been described in this chapter as details on them are very limited since they have not yet been commercialized. However, the readers are directed to read through Schadt et al. (2010) for a review on the latter. As of latest information, VisiGen's Starlight technology (Ansorge, 2009; Perkel, 2010) has been deprioritized by Life Technologies which acquired the former in 2008 (https://allseq.com/knowledge-bank/ngs-necropolis/visigen/, accessed on December 10, 2020).

Single-molecule sequencing has developed significantly over the past decade. For starters, Helicos Biosynthesis commercially developed the first single-molecule sequencing system in 2008 and called it Heliscope. The method involves hybridization of individual molecules to a flow cell system containing covalently attached oligonucleotides (Harris et al., 2008). This is followed by a sequential addition of fluorescently labeled nucleotides and DNA polymerase, which is electronically monitored using laser excitation and a charge-coupled device (CCD) camera. The accuracy in incorporation is achieved by a Virtual Terminator nucleotide (Bowers et al., 2009). In 2009, Pacific Biosciences (PacBio) developed an advanced process that enabled real-time sequencing of single molecules (SMRT) (Korlach et al., 2008). In this process, the polymerase-bound DNA template is instead attached to a 50-nm-wide well called as zero-mode waveguide (ZMW) (Foquet et al., 2008; Levene et al., 2003). The polymerase uses fluorescently labeled nucleotides for the synthesis of the DNA strands. The ZMW does not allow the light to propagate, instead it transmits the energy which is captured by the fluorophores which are attached to the nucleotides. On addition of the correct base, a real-time detection of the fluorescence pulse is captured (Quail et al., 2012). Further, Life technologies made a third-generation sequencing platform which used Fluorescence Resonance Energy Transfer (FRET) principle. This platform uses immobilized DNA by ligating to a surface-attached oligonucleotide, modified nucleotides and engineered polymerase for single-molecule detection. Here, the polymerase is with the quantum dot (which serves as a FRET donor) and the nucleotides are with the emitting dye. This dye only gives fluorescence when in close proximity of quantum dots (Thompson & Milos, 2011). This platform allows base-specific and real-time detection of nucleotide incorporation events (Pettersson et al., 2009). All of these third-generation sequencing techniques are dependent on the modified polymerases or modified nucleotides. In contrast, Oxford Nanopore's nanopore sequencing technology reads the sequences as native DNA molecules are pulled through nanopores (Clarke et al., 2009; Rusk, 2009). Each passing nucleotide has a characteristic ionic current which is detected by the sensors. The nanopores are designed to allow only a single DNA molecule to pass through at a time (Dumschott et al., 2020; Thompson & Milos, 2011). A squiggle plot is used for visualization and base calling (Deamer et al., 2016). Ideally, sequencing can be performed till the end of DNA or until the pores are unusable due to physical blockage. Nanopore is very helpful in de novo genome assemblies and determining structural variations (Bolger et al., 2019; Branton et al., 2010). It has an added advantage of being portable and easily accessible to power and operate by using a standard laptop (Leggett & Clark, 2017). Another system based on the nanopore principle is the NabSys solid-state detector which can detect tags on single molecules of DNA (Oliver et al., 2017). The technology is very fast, accurate and can read 100 s of kilobases of DNA fragments at a rate greater than 1 Mbp/s. The accuracy of the NabSys HD-Mapping (NabSys 2.0 LLC, Providence, RI) system was useful in determining large chromosomal duplications in *Bordetella pertussis* genome (Weigand et al., 2018), which were not resolvable by high-throughput sequencing alone, but could be successfully resolved using the electronic detection of tagged single molecules suggesting that the NabSys technology has a great future. Recently, the system has also been successfully used for strain-level identification of microbial strains in complex microbial communities (Oliver et al., 2019). Overall, the third-generation sequencing methods have dramatically reduced the cost and time while improving the read lengths. Microbial diversity studies involving these technologies are likely to change the landscape of our understanding of complex communities in nature.

15.3 Development of databases

The advent in sequencing technology generated new information necessitating a central facility for its curation. In 1991, European Molecular Biology Laboratory (EMBL; http://www.embl.de), GenBank and DNA Data Bank of Japan (DDBJ; http://www.ddbj.nig.ac.jp), which used to maintain the taxonomic nomenclatures and classifications independently (Fig. 15.1), partnered and formed International Nucleotide Sequence Database Collection (INSDC). The data with the three partners was derived using common source, but the results were inconsistent. To bring at one platform, Entrez information retrieval system was initiated by National Center for Biotechnology Information (NCBI; http://www.ncbi.nlm.nih.gov) (see Table 15.2 for a timeline summarizing the developments in databases). Sequence entries were exchanged regularly and the source organism, nomenclature and classifications were updated continuously. During this period, protein sequences were maintained separately by Swiss-Prot and Protein Information Resource (PIR). These databases maintained their own taxonomies different from each other and whichever was closely related was used by INSDC. In 1997, INSDC came in agreement with NCBI that public taxonomy pages will have processed entries and will only display entries related to public sequences (Federhen, 2012). At the same time, Ribosomal Database Project (RDP; http://rdp.cme.msu.edu) was established as a curated database especially for ribosomal sequences (Maidak et al., 1994). RDP sequence alignments were drawn from GenBank and EMBL submissions and some direct submissions to RDP. RDP also offered analysis services like phylogenetically ordered alignment, derived phylogenetic trees and determination of rRNA secondary

TABLE 15.2 A timeline of the development of databases.

Year	Database and event	Reference
1991	NCBI—Inception	Smith (2013)
1994	RDP—as the basic database	Maidak et al. (1994)
2006	Greengenes	DeSantis et al. (2006)
2007	SILVA	Pruesse et al. (2007)
2007	EzTaxon	Chun et al. (2007)
2007	RDP II with alignment tools and better curation	Cole et al. (2007)
2009	RDP improved with more tools for analysis	Cole et al. (2009)
2017	EzBioCloud—a completely updated database for Bacteria and Archeae identification	Yoon et al. (2017)

structures. In addition, RDP also offered computer programs for checking similarity to possible probes for PCR amplification of target taxa, analysis of possible chimera and phylogenetic distance of user-submitted sequence on existing phylogenetic tree (Maidak et al., 1996). Given its wide range of tools, RDP became popular and was extensively used in nearly all microbial diversity and systematics studies at that time.

Comparative ribosomal studies had initiated till then. While most reports focused on the survey of microbial communities in different ecosystem, some studied bacterial and archaeal evolution of the development of microarrays. Given the futuristic implications of the outcomes, the quality of the sequences in various databases and their accurate curation came to the forefront. An analysis of GenBank revealed that over 40% of the full-length 16S rRNA gene sequences in the database were under pseudo divisions and were unclassified (DeSantis et al., 2006). For example, a sample with approx. 1400 sequence record from known phyla had 3% of chimeras. In 2006, Greengenes (https://greengenes.secondgenome.com) addressed these concerns by including only included chimera checked sequences and a standardized set of taxonomically accurate descriptive fields in the database with ARB project compatibility (DeSantis et al., 2006). Greengenes consisted of multiple-sequence alignment for archaeal and bacterial 16S rRNA genes, simplified taxonomic placement and tracked other independent curators, such as NCBI and RDP to promote awareness of other phylogenetic descents. This also helped in balancing the approach to unifying the accuracy of taxonomic nomenclature of sequences records in various databases.

As the knowledge increased, ARB project (http://www.arb-home.de) offered integrative knowledge of small and large subunits of rRNA genes. It was understood that highly variable regions of large subunit of rRNA in eukaryotes is the base for species discrimination. Triggered by rapid and cheap sequencing protocols, new sequences were added in the public domain every month and hence, the need for a new database emerged. In 2007, the SILVA database project was released (http://www.arb-silva.de). This comprehensive database contained 137,788 and 5660 high-quality SSU and LSU rRNA sequences, respectively. Since the database contained aligned rRNA gene datasets, it enabled reliable phylogenetic reconstructions and biodiversity analysis. It also allowed to assign a defined set of metadata to each sequence, while simultaneously flagging the potential problems with each sequence (Pruesse et al., 2007). The same year, RDP also made a major quality improvement by testing sequence and chimeric variances using Pintail (Ashelford et al., 2005). At the same time, RDP also introduced my*RDP* space as a new web component which allowed users to build their own databases. Additionally, the RDP Sample Comparison Tool and the RDP classifier within the RDP pipeline enabled users to perform quality checks, taxonomic assignments and further statistical tests on their raw datasets to compare the taxonomic compositions of samples (Cole et al., 2007). By 2009, RDP automated the process of data processing, further simplified the intensive analysis of large sequence libraries and added additional visualization tools for early detection of taxonomic differences (Cole et al., 2009). Over the years, both SILVA and RDP have maintained their pace with the changing scenario in the field of microbial diversity and systematics and are among the frontrunners for the most frequently used database for microbial diversity studies.

With multiple databases at ones' disposal, a systemic evaluation was required. An attempt was therefore made for comparing SILVA, RDP, Greengenes, and NCBI databases using shared taxonomic units (Balvočiute & Huson, 2017). Taxonomic ranks ranging from phylum to genus levels were compared using three different mapping procedures, viz; Strict, loose and path comparisons. A simple comparison of taxonomic classifications revealed a number of differences between the four databases, but of course with overlaps. Many taxa were unique to each database. For instance, while between 63% and 73% of phyla, class and order level sequences were unique to either SILVA, RDP, Greengenes, or NCBI, nearly 89%–90% of all families and genera were unique to one or the other database. Additionally, a higher number of sequences were shared between NCBI with SILVA than with RDP or Greengenes. This study recommended SILVA for 16S rRNA gene-based studies but also suggested to consider NCBI taxonomy in studies involving shotgun sequencing. In principle, SILVA offers easy flat file exports and constant updated releases (Quast et al., 2012). Due to these advantages, SILVA has now become one of the most popular reference dataset in nearly all analysis pipelines (described below).

Until 2007, all the databases included 16S rRNA gene sequences from both type and nontype strains. However, the coverage for type strains was lacking. This paved the way for EzTaxon (now ezBioCloud, http://www.ezbiocloud.net), a database with 16S rRNA gene sequences of all prokaryotic type strains and provided users with all other sequence comparison tools present in the established databases (Chun et al., 2007). EzTaxon soon became the most reliable source for taxonomically accurate identification of prokaryotes. In 2012, EzTaxon–e, was formally introduced as the next-generation database and contained potential species-level phylotypes with no cultured representatives (Kim et al., 2012). In addition, a new way of assessing the degree of completeness in sequencing was proposed, which became a boon for researchers involved in the isolation and identification of taxonomically novel species. By 2017, EzBioCloud was introduced as an integrated database of 16S rRNA gene and whole-genome sequences of Bacteria and Archaea

(Seok-Hwan et al., 2017). All sequences in the database are assigned phylogenetically based on the maximum-likelihood approach and OrthoANI values. Therefore the phylogenetic associations might vary with the current formal standing for certain taxa. Further, the database contains at least one 16S rRNA gene sequence per terminal taxa, which is completely and accurately assigned at all suprageneric ranks. Today, EzBioCloud is one of the most trusted databases for ascertaining the most reliable phylogenetic neighbor for both 16S rRNA gene sequences and whole genomes of Archaea and Bacteria.

15.4 Advancements in analyses tools

The progress in the sequencing technology and the database was hand-in-hand but the advancement in the field of Software packages and pipelines was never to suit database or technology. Software packages were developed to utilize the data outputs from the sequencing technologies and answer the curious questions in the field of science. With the shift to sequence-based microbial diversity studies, the need for quantitative comparison of sequence libraries was soon realized as an essential component of hypothesis-driven ecological research. The first step in this direction was taken by Whitman and colleagues at the University of Georgia with the development of LIBSHUFF, which was created to explain whether two or more communities have the same structure (Singleton et al., 2001). The methodology required that sequences in a given dataset be arranged in an ordered manner with sequences from one library be listed in the FASTA file together followed by those from the second library. This was followed by an ordered multiple-sequence alignment to convert their similarities into a distance matrix. LIBSHUFF was based on building coverage curves for each of the compared libraries (homologous coverage curves) and comparing them with those generated upon random shuffling between them (heterologous coverage curves) and calculating the Cramer-von Mises statistic to explain the difference between these curves. With the introduction of LIBSHUFF, statistical comparison of libraries were possible. The next in line was AMOVA, which used analysis of variance- type formulation to determine if the compared communities differed in their genetic diversity (Martin, 2002; Schloss & Handelsman, 2008). In 2004, -LIBSHUFF, a modification over LIBSHUFF was introduced, which calculated the exact and integral form of the Cramer-von Mises statistics (Schloss et al., 2004). Besides, -LIBSHUFF enabled analysis of more than two libraries with a single input file. It also estimated the probability of false identification with statistical difference and selected the number of randomization with accelerated execution.

Until this time, all analyses were limited to utilizing the distance matrix for all sequences without any clustering ability to assign sequences into OTUs. This lead to a substantial requirement on the computer hardware, especially for large sequence libraries. To rescue, a computer program named DOTUR was developed (Schloss et al., 2004), which used multiple clustering algorithms to group together sequences into OTUs. It allowed users to construct rarefaction and collector's curves on their datasets using the frequency distribution of each OTU. DOTUR enabled users to estimate the richness and evenness of microbial diversity in the sample. In 2005, a significant advancement was made toward a phylogenetic approach to compare microbial communities using UniFrac (Lozupone & Knight, 2005). This method calculated a metric based on the branch lengths in a phylogenetic tree between the compared communities. This phylogenetic distance was used as a proxy of the distance between the compared communities. As an extension, the program would also determine descendants from one or the other of these compared communities. Until now, researchers had decided the set protocols for the use of various diversity measurements tools depending on the required study (Table 15.3).

Even after advancement of the software packages, the ability to compare the membership and structure between communities was limited. The available tests did not indicate the similarities between the communities, instead they reported the probability of the hypothesis under

TABLE 15.3 Categories and use of various indices for diversity study.

	α-diversity (Intra-community)	**β-diversity (Inter-community)**
Based upon Richness (Qualitative)	**Species based:** Chao 1 ACE Rarefaction **Divergence based:** Phylogenetic Diversity (PD)	**Species based:** Sörensen index Jaccard index **Divergence based:** Unweighted UniFrac Taxonomic Similarity (Δ_S)
Based upon Richness and/or Evenness (Quantitative)	**Species based:** hannon's index Simpson's index **Divergence based:** Theta, Intra-*K*-function (I_{kf})	**Species based:** Sörensen quantitative index, Morisita-Horn measure **Divergence based:** Weighted UniFrac F_{ST} DPCoA, Cross-*K*-function (C_{kf})

Modified from Lozupone , C.A., & Knight, R. (2008). Species divergence and the measurement of microbial diversity. FEMS Microbiology Reviews, 32(4), 557–578; *Jangid, K., Kao, M. H., Lahamge, A., Williams, M. A., Rathbun, S. L., & Whitman, W. B. (2016).* K-*shuff: A novel algorithm for characterizing structural and compositional diversity in gene libraries.* PLoS One, 11*(12)*, e0167634.

investigation. This made way for SONS, a non-parametric estimator that designated OTUs for the compared communities and estimated the fraction and richness of OTUs shared between them (Schloss & Handelsman, 2006b). However, by now the sequencing methods had picked up and given way to high-throughput methods enabling comprehensive sampling of natural environments, therefore, the phylogenetic and ecological complexities of microbial communities required new methods for their comparisons. To address this shortcoming, TreeClimber was designed using Parsimony test and was effective with the one-tailed hypothesis for the soil community analysis (Schloss & Handelsman, 2006a). By 2009, MOTHUR (http://mothur.org/) was introduced as a comprehensive open-source software package to allow the analyses of sequence data from communities. It combined all previous smaller tools and provided a single platform for such analysis (Schloss et al., 2009). Soon after, a second open-source package was introduced which utilized much-advanced quality filtering of raw sequence data. The package, called Quantitative Insights Into Microbial Ecology (QIIME) (http://qiime.org/), was built using the PyCogent toolkit and enabled all standard analyses in addition to several visualization tools, including network analysis, histograms and analysis of the "core" microbiome (Caporaso et al., 2010). Recently, QIIME 2 has replaced the original version as a next-generation microbiome bioinformatics platform (Bolyen et al., 2019).

Most of these statistical tools which have been developed over the years are computationally expensive, require a large sample size for accurate calculations or have a low sensitivity for detecting ecological significant differences. These limitations paved way for the development of a new algorithm, called *K*-shuff (Jangid et al., 2016). This tool enables users to compare both the structural and the composition between communities and is based on Ripley's *K*-function (Ripley, 1981). Because the method uses all pairs of sequences to determine similarities between gene libraries, it is more reflective of the overall change. Briefly, the *K*-function or *K*(r) is the fraction of distances less than or equal to each distance (Fig. 15.2). For measurements of structural diversity or I_{kf}, *K*(r) is calculated within a single library. The I_{kf} is then represented by the area above the line described by *K*(r) and is a measure of alpha-diversity (Fig. 15.2). To compare the compositional diversity between libraries, C_{kf} is calculated from the *K*(r) of the sequence comparisons between two libraries. Graphically, it is the sum of the differences between the interlibrary and both intralibrary *K*(r)s. *K*-shuff has the advantages that it does not rely upon calculating and counting OTUs and it is highly sensitivity to small differences in the libraries. While its power is dependent on the sequencing method employed, *K*-shuff requires very few sequences per sample owing to its high

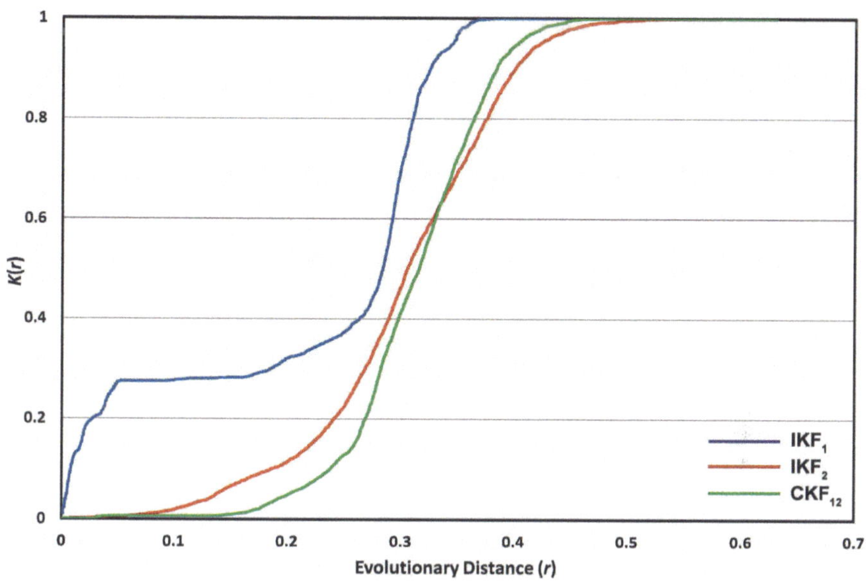

FIGURE 15.2 Measurement of structural and compositional differences between communities in *K*-shuff using intra-K-function (or I_{kf}) and cross-K-function (or C_{kf}), respectively. As illustrated, when the IKF is "steep" over a small distance (in library 1), the genetic diversity is small within the library since a large proportion of the gene sequences are closely related to each other. In contrast, when the genetic diversity measured between libraries (by CKFs) is "larger" than that measured within libraries (by IKFs), we tend to believe that the compositions of the libraries are different. A CKF, which is "flat" at the beginning, indicates the dissimilarity in the composition between the two corresponding libraries. While the area above the curves (*blue or green*) indicates the structural diversity (I_{kf}) for the corresponding library, the area between IKF1 (*blue curve*) and CKF (*green curve*) plus that between IKF2 (red curve) and CKF (*green curve*) represents the compositional difference (C_{kf}) between the two communities.

sensitivity. Based on mixing simulations, the differences in C_{kf} and I_{kf} are quasi-linear, whereas the distance estimates in UniFrac are dramatically nonlinear. This is why it was found to be more sensitive than UniFrac. *K*-shuff analyses can be carried out on a desktop PC and is available at https://research.franklin.uga.edu/whitman/k-shuff or https://microbial-lab.org/softwares/.

In the rapidly changing field of 16S rRNA gene sequence-based community analysis, bioinformatics pipelines have changed significantly. Today, the six most used pipelines utilize either the OTU-based approach used by QIIME-uclust, MOTHUR and USEARCH-UPARSE, or the amplicon sequence variant (ASV) approach used by DADA2 (Divisive Amplicon Denoising Algorithm 2), Qiime2-Deblur and USEARCH-UNOISE3. Even though OTU level perform well, they are said to have less sensitivity for the ASV—level pipelines. As far as

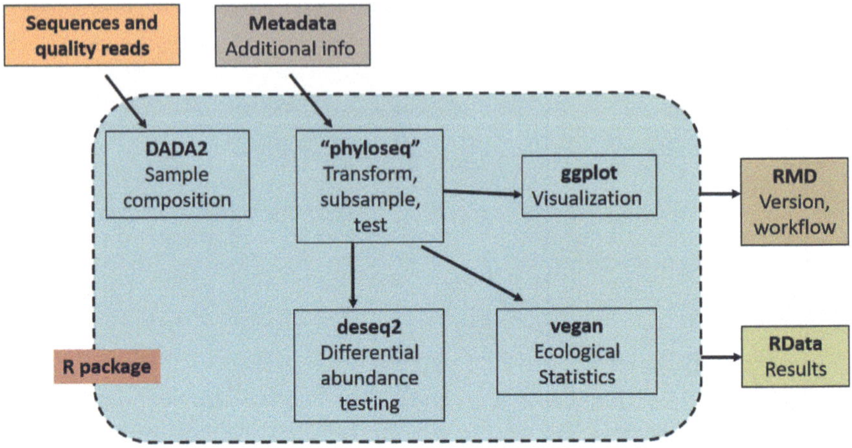

FIGURE 15.3 Overview of the data analysis performed in R using DADA2, knitr, and phyloseq. Raw sequences are denoised and used to create ASVs by merging forward and reverse sequence reads. ASVs are then used for further analysis wherein the metadata provides additional parameters for community analysis. While deseq2 identifies the outliners and removes them from the analysis, Vegan supports for multivariant analysis and ggplot helps in the formation of visualization graphs as the result of all the analysis. Source: *Modified from Callahan, B., Proctor, D., Relman, D., Fukuyama, J., & Holmes, S. (2016). Reproducible research workflow in R for the analysis of personalized human microbiome data. In* Pacific Symposium on Biocomputing *(pp. 183–194). World Scientific Publishing Co. Pte Ltd.*

sensitivity is considered, DADA2 has been the best (Prodan et al., 2020). DADA2 was launched in 2016 as an open-source software and it uses the R package (http://www.rstudio.com) for the pipeline implementation (Callahan, Proctor, et al., 2016). DADA2 offers a complete pipeline for the analysis of large datasets obtained using next-generation sequencing. It has enabled users to merge sequences from paired-end reads, perform quality checks and remove erroneous sequences, and perform specific hypothesis testing on these quality sequences on a regular desktop computer (Fig. 15.3).

15.5 How to study microbial diversity?

The extent of microbial diversity on our planet is enormous but we only know a little about the organisms from pure cultures in the laboratory. rRNA gene sequencing and shotgun metagenomics have both contributed significantly to our understanding of the ecological drivers of these communities and it's composition in various ecosystems like ocean, soil, freshwaters, etc. Microbial communities differ in their composition, the niches they occupy and their influence on the environment.

Due to the complex interaction between communities and environment, there is no fundamental definition to understand a particular site (Stubbendieck et al., 2016). To have a better understanding of the microbial communities in the environment, it is important to study them in the ecological context. Given the vastness, we will consider soil as our model ecosystem to study the microbial communities using rRNA gene sequencing to understand and derive the influence of various environmental parameters on the community structure and composition.

Soil microorganisms are the key components of the ecosystem and of utmost importance in maintaining soil processes like decomposition, nutrient cycling and plant and animal productivity. Even though soil is not an easy environment to survive, estimates show that there are thousands of microbial taxa in a gram of soil (Fierer, 2017). While we now understand about the most abundant microbial taxa present in the soil, very little information is available on the interactions they have within it and the changes in them due to climate change. Communities are often affected by the change in temperature and water content (Supramaniam et al., 2016), long term fertilization (Ling et al., 2016), altitudinal gradients (Ren et al., 2018), etc. However, a lot is yet to be unveiled about microbe-microbe interactions and plant/animal-microbe interactions in the soil. It is not necessary that every microbial taxa found in the sample interacts with the others or have a significant association with any other biotic organisms (Blanchet et al., 2020). This has proved to be a limiting step in predicting their changes in ecosystem functioning (Chu et al., 2020). After all these decades of soil microbial community analysis, ecologists have just started to understand the cooccurrence of the microbial species. Researchers are now involved in the study of cooccurrence and trying to understand if all the information of coexistence is proxy or not.

For any soil microbial diversity study to be undertaken, several key factors need to be considered. Given that soil is a dynamic ecosystem, there is no definition of the "ideal" soil microbiome. Studies have shown that soil samples collected even a few centimeters away exhibited variation in their microbial communities. These natural variations are inherent to the soil ecosystem and may be attributed to the microscopic soil structure within which the organisms thrive. Many-a-times, these differences can be attributed to the type of aboveground vegetation, interactions or just the spatial scale (O'Brien et al., 2016). Given the dynamic nature of the ecosystem, it is therefore imperative that any microbial diversity study undertaken must include as many replicate samples as possible to weed out this natural variation from the "real" effect of a particular treatment. Generally, three to five replicates per treatment plot with at least three plots per treatment condition are recommended. Another important consideration is the size of the plot from where the samples are being collected. Given the spatial variation, one must try to collect samples from plots with identical size as well as use identical strategy of collection of replicate samples within these plots. A very crude approach is to collect grab samples or use a more accurate core sampler to collect deeper samples from the soil. Regardless of the strategy, one must not pool replicates to a single bulk sample per treatment. Instead, each of the replicate samples may be prepared by pooling three to five grab samples or soil cores per treatment plot and processed accordingly as individual replicates. Given enough replication, one can ascertain community changes with decent accuracy.

For many years, the only way to learn about the microbial communities in a sample was the use of culture-dependent method. It is now well understood that Sanger's sequencing is the gold standard for the identification of microbial cultures. However, given that the vast majority (~99%) of the microbial species cannot be cultivated, rRNA gene-based studies are the only way to determine the extent of microbial diversity in a given ecosystem. These studies require the isolation of total community DNA from the soil followed by amplification and/or direct sequencing finally leading to sequence analysis. Each step is critical; however, the quality and quantity of isolated DNA is of utmost importance because it is dependent on the method of extraction and the type of soil itself. Having biological replicates (as stated above) and technical replicates during the extraction process is strongly recommended. At times, the quality of the isolated DNA is poor and loaded with inhibitors that do not allow successful downstream processing. During these times, additional purification is therefore required for such samples. However, this also leads to the loss of a portion of the original community DNA. Such samples, therefore require a conservative approach when being considered in the final analysis. Once good quality DNA is prepared, one may use the clone-library approach or the metagenomics way (see below).

Traditionally, microbial diversity studies based on Sanger sequencing follow the clone-library approach (Fig. 15.4). This is a laborious process involving multiple steps: (1) selective amplification of microbial rRNA genes using universal primers in a low-cycle PCR; (2) cloning of these amplicon products; (3) screening of clones; (4) plasmid isolation from positive clones; (5) sequencing of inserts by Sanger's sequencing; and (6) analyses of sequence data. Each step within the process has its own pitfalls and biases. However, the most influential is the generation of artificial amplicons during PCR, that is, synthetic chimeras. Although chimeric fragments are produced during the course of the PCR, their amplification is exponential during the later cycles in the reaction. Hence, a low-cycle PCR, typically 15–18 cycles, is used to reduce the formation of chimeric fragments (Jangid et al., 2008). At the same time,

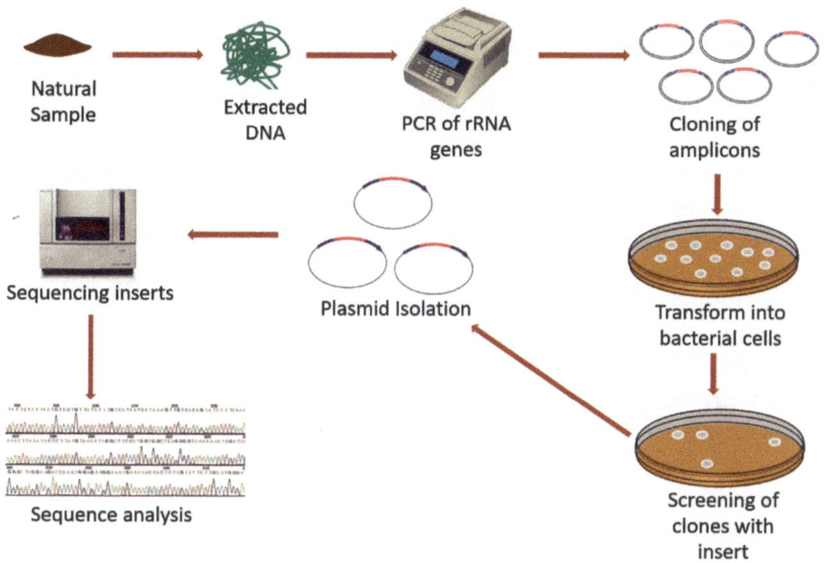

FIGURE 15.4 Overview of the steps involved in clone-library approach for microbial diversity studies.

it is also important to obtain a true representation of the relative abundance of the members within a community. For this, prior knowledge of the generation time of the recombinant host being used in heat transformation is crucial (Jangid et al., 2008). While this is very crucial, it is generally ignored in most clone-library based studies. This information allows to plate the clones as soon as the recombinant host has recovered from the heat shock and right before fission to avoid duplicate clones that might skew the relative abundance of the members in the library. In addition, the clone library approach has a preferential bias for cloning the most abundant fragments, generates a limited number of clones and the expensive sequencing cost associated with it. Regardless, the clone library approach generated some of the most crucial data, including the usage of functional gene markers, such as *mcrA* gene (Hales et al., 1996), and set the foundation for advanced research in microbial diversity and ecology.

Essentially, the shift from Sanger's clone-based library approach to high-throughput sequencing changed the way microbial diversity studies were conducted. Second-generation sequencing technologies gave high throughput because of the shift from single clone/sample to independent DNA fragments (Latorre-Pérez et al., 2020). This change enabled a large portion of the original community DNA to be sequenced in parallel (Thompson & Milos, 2011). Essentially, while all techniques involved library preparation, followed by clonal amplification of the library and sequencing of the amplicons; the sequencing technology used in the last step dictates the modifications in the first two steps. During library preparation, a major shift occurred in the way clonal fragments were generated. In clone-library approach, PCR amplicons are prepared and cloned separately for each sample, thereby allowing physical segregation of clones before sequencing. In contrast, the library preparation for high-throughput sequencing uses barcoded primers to amplify the target fragments from each sample, which are segregated into libraries postsequencing by bioinformatic analysis. Because the clonal amplification involves binding of individual DNA fragments to adapter-attached beads in emulsion PCR or glass slides with picolitre wells in bridge PCR, the target DNA fragment amplification may be achieved in two ways: using barcoded universal primers followed by adapter ligation, or using barcoded primers that already have the complimentary adapter sequences. Whereas emulsion PCR based clonal amplification fragments may then be sequenced using pyrosequencing or sequencing by ligation, those generated using bridge PCR may be sequencing by synthesis.

Today, soil samples are sequenced by either amplicon-metagenomics or shotgun sequencing. For amplicon-metagenomics, the choice of the primer will depend on the purpose of the study. If one is looking to understand the extent of diversity, a universal primer set for a taxonomic gene fragment, such as rRNA gene or ITS fragments will be helpful. However, for a functional survey of the ecosystem, a corresponding gene primer will help. Better yet, one should go for shotgun metagenomics or transcriptomics in this case. Regardless, massively parallel sequencing methods produce reads from the DNA fragments and bypasses the need for cloning these in a vector.

Whether amplicon or metagenome, as stated above, the choice of the sequencing technology will also dictate the method of library preparation for the study. For studies based on specific gene targets or diversity analysis, characterization of microbial diversity (Verma & Sharma, 2020), taxonomic profiling (Kenmotsu et al., 2020), CAG repeats (Rocca et al., 2020), and metagenomic functional profiling (Pacwa-Płociniczak et al., 2020), Illumina (MiSeq/HiSeq) is leading the way and contributes nearly 90% of the sequence data among all technologies. However, for studies like HLA typing (Cargou et al., 2020), mitochondrial genome sequencing (Harvey et al., 2019) and many others, Ion torrent is the preferred sequencing technology. For studying base modifications, such as methylation (Rand et al., 2017; Simpson et al., 2017), and direct RNA sequencing (Garalde et al., 2018), the third-generation Oxford Nanopore technology is widely used and has been supported by many algorithms and analytical tools that have been recently developed. The analysis involves multiple steps: base calling, mapping, sequence assembly and variant detection, the details of which have been discussed in detail by

Makałowski and Shabardina (2020). In addition, Nanopore has tremendously helped in translational applications in genomic medicine and clinical testing (Ameur et al., 2019) and metagenomic sequencing of viral genomes from clinical samples (Lewandowski et al., 2019). Hence, it is critical that the sequencing strategy be decided after careful consideration of the study objectives.

Once the sequencing of the sample is performed, the study turns toward bioinformatics pipelines. Most ecologists consider bioinformatic analyses as the "black box" of the study. The analyses pipeline consists of steps such as quality filtering and denoising, assembly of larger contigs and comparison with the database. While each step is critical, quality filtering and denoising are most crucial to the outcomes. While very strict filtering will lead to a substantial loss of usable data, very relaxed criteria will accumulate sequences with errors. Given this, we use a more conservative approach, one that does not dilute the quality criteria thresholds but uses nearly 75%–80% of the sequence raw data while ensuring that none of the quality passed data is erroneous. After all, the results are only as good as the data!

For marker genes like 16S rRNA, the most widely used pipelines are QIIME2 with Greengenes as the database (Nannipieri et al., 2020) and DADA2 using SILVA database (Weißbecker et al., 2020). With 16S rRNA amplicon-based sequencing, the coverage offered by public databases is extensive and only a minority of sequence reads remain unidentified. However, for metagenomes, only a smaller portion of the reads can be identified using such databases and therefore, most use database-independent approaches. As an example, Nonpareil calculates the extent of overlap among individual sequences within the metagenomic dataset and calculates the coverage for such samples (Vestergaard et al., 2017). To understand the functional capacity of the sampled community, the reads need to be assembled correctly using either of the two strategies commonly deployed in shotgun sequencing: Global assembly and local or targeted assembly (Guo et al., 2019). Among the two, Global assembly is a more widely used approach and attempts to recover most genomes from the metagenome dataset. For soils and other such highly diverse ecosystems, the presence of repeat sequences within the dataset, percolating sequencing errors due to bad quality filtering, an uneven coverage and the very large size of the datasets are a concern while using this approach. In case, the study focus is on the pathways or genes that are involved in community functions, the local or targeted assembly approach is used. In this approach, the aim is to assemble these target genes and not all of the metagenome data. Hence, the requirement on computational efforts is also reduced since only the assembly of alleles of specified genes is required.

15.5.1 Suggested data analysis pipeline(s)

Today, most microbial diversity studies use the amplicon sequencing approach and utilize multiple analysis pipelines to achieve the end result. While many labs have personal preferences into using one or the other, most published studies use DADA, MOTHUR, or QIIME either singly or in combination. Being open-source, importing analysis features into each other is easier and therefore we recommend using this strategy. Prodan and colleagues recently compared six of the most popular analysis pipelines discussed earlier (Prodan et al., 2020) and showed that DADA2 offered the best sensitivity, although with decreased specificity compared to USEARCH-UNOISE3 and Qiime2-Deblur. That said, most recent studies have utilized DADA2, which is an open-source R package-based analysis pipeline, is highly sensitive and also most appropriate to answer questions asked by the researchers (Fig. 15.3) (Callahan, McMurdie, et al., 2016). We, therefore, describe the basic functioning of this pipeline below based on these recommendations (Prodan et al., 2020).

DADA2 uses a parametric model to segregate true biological sequences from the total reads (Prodan et al., 2020). The model is dependent on input read abundances and distances. Input read abundances are calculated based on "true reads" that are considered to be present in abundance, and distances are calculated based on the less abundant reads that are separated by very few base differences and likely "error-derived" from a more abundant sequence. Using a substitution model generated from the Base Q-scores, DADA2 then decides to assign the "error-derived" sequence counts to the "true reads." DADA2 uses the "filterAndTrim" function for quality filtering (Fig. 15.5), followed by deduplicated and inferring ASVs. Another major shift in processing of the raw data in DADA2 is the independent denoising of forward and reverse reads followed by their merging. These merged ASVs are then

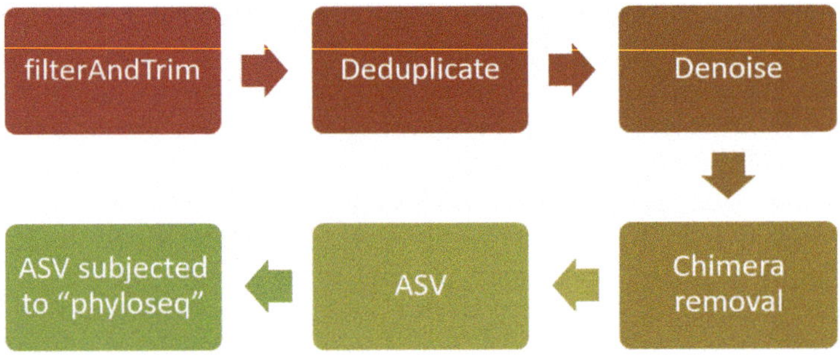

FIGURE 15.5 Workflow of DADA2 in R package suggested by Prodan et al. (2020). Only the main steps are highlighted.

subjected to "removerChimeradenova" to eliminate the chimeric ASVs. All ASVs are then converted to the phyloseq objects using "phyloseq" package. Alpha-diversity is estimated by "estimate_richness" function of "phyloseq." Plots are then designed in R using ggplot, corrplot, and VennDiagram packages.

As an extension of pure diversity studies based on 16S rRNA gene sequences, imputed metagenomics has enabled researchers to predict the most likely functions performed by different members of the community. Since this is an indirect approach, its results must be used very cautiously. However, very recently a strong correlation between functional prediction from 16S rRNA gene amplicons using Piphillin (http://piphillin.secondgenome.com/) and corresponding shotgun metagenomic data was shown (Narayan et al., 2020). In fact, Piphillin outperformed many of the comparable tools.

15.6 Conclusion

Microbial communities are diverse and their members have varied nutritional requirements which we have not been able to replicate in the laboratory even today. Today, we understand the extent of the microbial diversity on our planet and know that only a minuscule proportion of this enormous diversity has been cultivated and studied in the laboratory. If it was not for the ability to PCR amplify and sequence the rRNA gene from microbes, this vast majority would have remained obscured from mankind. Sequencing technology, databases and analyses tools for the assessment of microbial diversity have come a long way over the past half-century. The advances in all three have enabled researchers to answer questions beyond knowing who is present in the community. Overall, with the right balance of experimental design, sequencing strategy and analyses tools, microbial diversity studies have the potential to yield new insights into the roles and functioning of communities in any ecosystem and perform very specific hypothesis-driven research.

Acknowledgments

The authors thank Department of Biotechnology (DBT) for funding National Centre for Microbial Resource (NCMR) wide grant letter no. BT/COORD.II/01/03/2016.

References

Ahmadian, A., Gharizadeh, B., Gustafsson, A. C., Sterky, F., Nyrén, P., Uhlén, M., & Lundeberg, J. (2000). Single-nucleotide polymorphism analysis by pyrosequencing. *Analytical Biochemistry*, *280*(1), 103–110. Available from https://doi.org/10.1006/abio.2000.4493.

Ameur, A., Kloosterman, W. P., & Hestand, M. S. (2019). Single-molecule sequencing: Towards clinical applications. *Trends in Biotechnology, 37*(1), 72–85. Available from https://doi.org/10.1016/j.tibtech.2018.07.013.

Ansorge, W. J. (2009). Next-generation DNA sequencing techniques. *New Biotechnology, 25* (4), 195–203. Available from https://doi.org/10.1016/j.nbt.2008.12.009.

Ashelford, K. E., Chuzhanova, N. A., Fry, J. C., Jones, A. J., & Weightman, A. J. (2005). At least 1 in 20 16S rRNA sequence records currently held in the public repositories is estimated to contain substantial anomalies. *Applied and Environmental Microbiology, 71*(12).

Balvočiute, M., & Huson, D. H. (2017). SILVA, RDP, Greengenes, NCBI and OTT—How do these taxonomies compare? *BMC Genomics, 18*. Available from https://doi.org/10.1186/s12864-017-3501-4.

Beechem, J. M., Ulmanello, U., Wang, Y., Yue, M., Lafferty, M., & Sun, H. Y. (2011). *Single molecule real-time DNA sequencing using FRET-based reagents: Sequencing DNA on multiple size scales (from single bases to whole chromosomes) to resolve structural variation and enable de novo sequencing* (Poster). Genetic Systems.

Bent, S. J., & Forney, L. J. (2008). The tragedy of the uncommon: Understanding limitations in the analysis of microbial diversity. *ISME Journal, 2*(7), 689–695. Available from https://doi.org/10.1038/ismej.2008.44.

Bentley, D. R., Balasubramanian, S., Swerdlow, H. P., Smith, G. P., Milton, J., Brown, C. G., Hall, K. P., Evers, D. J., Barnes, C. L., Bignell, H. R., Boutell, J. M., Bryant, J., Carter, R. J., Keira Cheetham, R., Cox, A. J., Ellis, D. J., Flatbush, M. R., Gormley, N. A., Humphray, S. J., ... Smith, A. J. (2008). Accurate whole human genome sequencing using reversible terminator chemistry. *Nature, 456*(7218), 53–59. Available from https://doi.org/10.1038/nature07517.

Blanchet, F. G., Cazelles, K., & Gravel, D. (2020). Co-occurrence is not evidence of ecological interactions. *Ecology Letters, 23*(7), 1050–1063. Available from https://doi.org/10.1111/ele.13525.

Bolger, A. M., Poorter, H., Dumschott, K., Bolger, M. E., Arend, D., Osorio, S., Gundlach, H., Mayer, K. F. X., Lange, M., Scholz, U., & Usadel, B. (2019). Computational aspects underlying genome to phenome analysis in plants. *Plant Journal, 97*(1), 182–198. Available from https://doi.org/10.1111/tpj.14179.

Bolyen, E., Rideout, J. R., Dillon, M. R., Bokulich, N. A., Abnet, C. C., Al-Ghalith, G. A., Alexander, H., Alm, E. J., Arumugam, M., Asnicar, F., Bai, Y., Bisanz, J. E., Bittinger, K., Brejnrod, A., Brislawn, C. J., Brown, C. T., Callahan, B. J., Caraballo-Rodríguez, A. M., Chase, J., ... Caporaso, J. G. (2019). Reproducible, interactive, scalable and extensible microbiome data science using QIIME 2. *Nature Biotechnology, 37*(8), 852–857. Available from https://doi.org/10.1038/s41587-019-0209-9.

Bowers, J., Mitchell, J., Beer, E., Buzby, P. R., Causey, M., Efcavitch, J. W., Jarosz, M., Krzymanska-Olejnik, E., Kung, L., Lipson, D., Lowman, G. M., Marappan, S., McInerney, P., Platt, A., Roy, A., Siddiqi, S. M., Steinmann, K., & Thompson, J. F. (2009). Virtual terminator nucleotides for next-generation DNA sequencing. *Nature Methods, 6*(8), 593–595. Available from https://doi.org/10.1038/nmeth.1354.

Branton, D., Deamer, D. W., Marziali, A., Bayley, H., Benner, S. A., Butler, T., Ventra, M. D., Garaj, S., Hibbs, A., Huang, X., Jovanovich, S. B., Krstic, P. S., Lindsay, S., Ling, X. S., Mastrangelo, C. H., Meller, A., Oliver, J. S., Pershin, Y. V., Ramsey, J. M., ...Schloss, J. A. (2010). The potential and challenges of nanopore sequencing, *Nature Biotechnology* **26** (10), 2008, 1146–1153.https://doi.org/10.1038/nbt.1495.

Bräutigam, A., & Gowik, U. (2010). What can next generation sequencing do for you? Next generation sequencing as a valuable tool in plant research. *Plant Biology, 12*(6), 831–841. Available from https://doi.org/10.1111/j.1438-8677.2010.00373.x.

Bruijns, B., Tiggelaar, R., & Gardeniers, H. (2018). Massively parallel sequencing techniques for forensics: A review. *Electrophoresis, 39*(21), 2642–2654. Available from https://doi.org/10.1002/elps.201800082.

Callahan, B., Proctor, D., Relman, D., Fukuyama, J., & Holmes, S. (2016). *Reproducible research workflow in R for the analysis of personalized human microbiome data. Pacific Symposium on Biocomputing* (pp. 183–194). World Scientific Publishing Co. Pte Ltd. Available from https://doi.org/10.1142/9789814749411_0018.

Callahan, B. J., McMurdie, P. J., Rosen, M. J., Han, A. W., Johnson, A. J. A., & Holmes, S. P. (2016). DADA2: High-resolution sample inference from Illumina amplicon data. *Nature Methods, 13*(7), 581–583. Available from https://doi.org/10.1038/nmeth.3869.

Caporaso, J. G., Kuczynski, J., Stombaugh, J., Bittinger, K., Bushman, F. D., Costello, E. K., Fierer, N., Pẽa, A. G., Goodrich, J. K., Gordon, J. I., Huttley, G. A., Kelley, S. T., Knights, D., Koenig, J. E., Ley, R. E., Lozupone, C. A., McDonald, D., Muegge, B. D., Pirrung, M., ... Knight, R. (2010). QIIME allows analysis of high-throughput community sequencing data. *Nature Methods, 7*(5), 335–336. Available from https://doi.org/10.1038/nmeth.f.303.

Cargou, M., Ralazamahaleo, M., Blouin, L., Top, I., Elsermans, V., Andreani, M., Guidicelli, G., & Visentin, J. (2020). Evaluation of the AllType kit for HLA typing using the Ion Torrent S5 XL platform. *HLA, 95*(1), 30–39. Available from https://doi.org/10.1111/tan.13708.

Chu, H., Gao, G. F., Ma, Y., Fan, K., & Delgado-Baquerizo, M. (2020). Soil microbial biogeography in a changing world: Recent advances and future perspectives. *MSystems, 5*(2). Available from https://doi.org/10.1128/mSystems.00803-19.

Chun, J., Lee, J. H., Jung, Y., Kim, M., Kim, S., Kim, B. K., & Lim, Y. W. (2007). EzTaxon: A web-based tool for the identification of prokaryotes based on 16S ribosomal RNA gene sequences. *International Journal of Systematic and Evolutionary Microbiology, 57*(10), 2259–2261. Available from https://doi.org/10.1099/ijs.0.64915-0.

Clarke, J., Wu, H. C., Jayasinghe, L., Patel, A., Reid, S., & Bayley, H. (2009). Continuous base identification for single-molecule nanopore DNA sequencing. *Nature Nanotechnology, 4*(4), 265–270. Available from https://doi.org/10.1038/nnano.2009.12.

Cole, J. R., Chai, B., Farris, R. J., Wang, Q., Kulam-Syed-Mohideen, A. S., McGarrell, D. M., Bandela, A. M., Cardenas, E., Garrity, G. M., & Tiedje, J. M. (2007). The ribosomal database project (RDP-II): introducing myRDP space and quality controlled public data. *Nucleic Acids Research, 35*(suppl_1).

Cole, J. R., Wang, Q., Cardenas, E., Fish, J., Chai, B., Farris, R. J., Kulam-Syed-Mohideen, A. S., McGarrell, D. M., Marsh, T., Garrity, G. M., & Tiedje, J. M. (2009). The ribosomal database project: Improved alignments and new tools for rRNA analysis. *Nucleic Acids Research, 37*(1), D141–D145. Available from https://doi.org/10.1093/nar/gkn879.

Deamer, D., Akeson, M., & Branton, D. (2016). Three decades of nanopore sequencing. *Nature Biotechnology, 34*(5), 518–524. Available from https://doi.org/10.1038/nbt.3423.

Degnan, P. H., & Ochman, H. (2012). Illumina-based analysis of microbial community diversity. *ISME Journal, 6*(1), 183–194. Available from https://doi.org/10.1038/ismej.2011.74.

DeSantis, T. Z., Hugenholtz, P., Larsen, N., Rojas, M., Brodie, E. L., Keller, K., Huber, T., Dalevi, D., Hu, P., & Andersen, G. L. (2006). Greengenes, a chimera-checked 16S rRNA gene database and workbench compatible with ARB. *Applied and Environmental Microbiology, 72*(7), 5069–5072. Available from https://doi.org/10.1128/AEM.03006-05.

Dumschott, K., Schmidt, M. H. W., Chawla, H. S., Snowdon, R., & Usadel, B. (2020). Oxford nanopore sequencing: New opportunities for plant genomics? *Journal of Experimental Botany, 71*(18), 5313–5322. Available from https://doi.org/10.1093/jxb/eraa263.

Fakruddin., Mazumdar, R. M., Chowdhury, A., Hossain, N., Mahajan, S., & Islam, S. (2013). Pyrosequencing—A next generation sequencing technology. *World Applied*

Sciences Journal, 24(12), 1558–1571. Available from https://doi.org/10.5829/idosi.wasj.2013.24.12.2972.

Federhen, S. (2012). The NCBI taxonomy database. *Nucleic Acids Research, 40*(D1).

Fierer, N. (2017). Embracing the unknown: Disentangling the complexities of the soil microbiome. *Nature Reviews Microbiology, 15*(10), 579–590. Available from https://doi.org/10.1038/nrmicro.2017.87.

Foquet, M., Samiee, K. T., Kong, X., Chauduri, B. P., Lundquist, P. M., Turner, S. W., Freudenthal, J., & Roitman, D. B. (2008). Improved fabrication of zero-mode waveguides for single-molecule detection. *Journal of Applied Physics, 103*(3). Available from https://doi.org/10.1063/1.2831366.

Garalde, D. R., Snell, E. A., Jachimowicz, D., Sipos, B., Lloyd, J. H., Bruce, M., Pantic, N., Admassu, T., James, P., Warland, A., Jordan, M., Ciccone, J., Serra, S., Keenan, J., Martin, S., McNeill, L., Wallace, E. J., Jayasinghe, L., Wright, C., ... Turner, D. J. (2018). Highly parallel direct RNA sequencing on an array of nanopores. *Nature Methods, 15*(3), 201–206. Available from https://doi.org/10.1038/nmeth.4577.

Glenn, T. C. (2011). Field guide to next-generation DNA sequencers. *Molecular Ecology Resources, 11*(5), 759–769. Available from https://doi.org/10.1111/j.1755-0998.2011.03024.x.

Guo, J., Quensen, J., Sun, Y., Wang, Q., Brown, C. T., Cole, J. R., & Tiedje, J. M. (2019). Review, evaluation and directions for gene-targeted assembly for ecologic analyses of metagenomes. *Frontiers in Genetics, 10*.

Hales, B. A., Edwards, C., Ritchie, D. A., Hall, G., Pickup, R. W., & Saunders, J. R. (1996). Isolation and identification of methanogen-specific DNA from blanket bog peat by PCR amplification and sequence analysis. *Applied and Environmental Microbiology, 62*(2), 668–675. Available from https://doi.org/10.1128/aem.62.2.668-675.1996.

Harris, T. D., Buzby, P. R., Babcock, H., Beer, E., Bowers, J., Braslavsky, I., Causey, M., Colonell, J., DiMeo, J., Efcavitch, J. W., Giladi, E., Gill, J., Healy, J., Jarosz, M., Lapen, D., Moulton, K., Quake, S. R., Steinmann, K., Thayer, E., ... Xie, Z. (2008). Single-molecule DNA sequencing of a viral genome. *Science, 320*(5872), 106–109. Available from https://doi.org/10.1126/science.1150427.

Harvey, N. R., Albury, C. L., Stuart, S., Benton, M. C., Eccles, D. A., Connell, J. R., Sutherland, H. G., Allcock, R. J. N., Lea, R. A., Haupt, L. M., & Griffiths, L. R. (2019). Ion torrent high throughput mitochondrial genome sequencing (HTMGS). *PLoS One*, e0224847. Available from https://doi.org/10.1371/journal.pone.0224847.

Jangid, K., Kao, M. H., Lahamge, A., Williams, M. A., Rathbun, S. L., & Whitman, W. B. (2016). K-shuff: A novel algorithm for characterizing structural and compositional diversity in gene libraries. *PLoS One, 11*(12).

Jangid, K., Williams, M. A., Franzluebbers, A. J., Sanderlin, J. S., Reeves, J. H., Jenkins, M. B., Endale, D. M., Coleman, D. C., & Whitman, W. B. (2008). Relative impacts of land-use, management intensity and fertilization upon soil microbial community structure in agricultural systems. *Soil Biology and Biochemistry, 40*(11), 2843–2853. Available from https://doi.org/10.1016/j.soilbio.2008.07.030.

Jett, J. H., Keller, R. A., Martin, J. C., Marrone, B. L., Moyzis, R. K., Ratliff, R. L., Seitzinger, N. K., Shera, E. B., & Stewart, C. C. (1989). High-speed dna sequencing: An approach based upon fluorescence detection of single molecules. *Journal of Biomolecular Structure and Dynamics, 7*(2), 301–309. Available from https://doi.org/10.1080/07391102.1989.10507773.

Kenmotsu, H., Uchida, K., Hirose, Y., & Eki, T. (2020). Taxonomic profiling of individual nematodes isolated from copse soils using deep amplicon sequencing of four distinct regions of the 18S ribosomal RNA gene. *PLoS One, 15*(10). Available from https://doi.org/10.1371/journal.pone.0240336.

Kim, O. S., Cho, Y. J., Lee, K., Yoon, S. H., Kim, M., Na, H., Park, S. C., Jeon, Y. S., Lee, J. H., Yi, H., Won, S., & Chun, J. (2012). Introducing EzTaxon-e: A prokaryotic 16s rRNA gene sequence database with phylotypes that represent uncultured species.

International Journal of Systematic and Evolutionary Microbiology, 62(3), 716–721. Available from https://doi.org/10.1099/ijs.0.038075-0.

Korlach, J., Marks, P. J., Cicero, R. L., Gray, J. J., Murphy, D. L., Roitman, D. B., Pham, T. T., Otto, G. A., Foquet, M., & Turner, S. W. (2008). Selective aluminum passivation for targeted immobilization of single DNA polymerase molecules in zero-mode waveguide nanostructures. *Proceedings of the National Academy of Sciences*, 1176–1181. Available from https://doi.org/10.1073/pnas.0710982105.

Latorre-Pérez, A., Pascual, J., Porcar, M., & Vilanova, C. (2020). A lab in the field: Applications of real-time, in situ metagenomic sequencing. *Biology Methods and Protocols, 5*(1).

Leggett, R. M., & Clark, M. D. (2017). A world of opportunities with nanopore sequencing. *Journal of Experimental Botany, 68*(20), 5419–5429. Available from https://doi.org/10.1093/jxb/erx289.

Lemos, L. N., Fulthorpe, R. R., Triplett, E. W., & Roesch, L. F. W. (2011). Rethinking microbial diversity analysis in the high throughput sequencing era. *Journal of Microbiological Methods, 86*(1), 42–51. Available from https://doi.org/10.1016/j.mimet.2011.03.014.

Levene, H. J., Korlach, J., Turner, S. W., Foquet, M., Craighead, H. G., & Webb, W. W. (2003). Zero-mode waveguides for single-molecule analysis at high concentrations. *Science, 299*(5607), 682–686. Available from https://doi.org/10.1126/science.1079700.

Lewandowski, K., Xu, Y., Pullan, S. T., Lumley, S. F., Foster, D., Sanderson, N., Vaughan, A., Morgan, M., Bright, N., Kavanagh, J., Vipond, R., Carroll, M., Marriott, A. C., Gooch, K. E., Andersson, M., Jeffery, K., Peto, T. E. A., Crook, D. W., Sarah Walker, A., & Matthews, P. C. (2019). Metagenomic nanopore sequencing of influenza virus direct from clinical respiratory samples. *BioRxiv, 23*. Available from https://doi.org/10.1101/676155.

Ling, N., Zhu, C., Xue, C., Chen, H., Duan, Y., Peng, C., Guo, S., & Shen, Q. (2016). Insight into how organic amendments can shape the soil microbiome in long-term field experiments as revealed by network analysis. *Soil Biology and Biochemistry, 99*, 137–149. Available from https://doi.org/10.1016/j.soilbio.2016.05.005.

Liu, L., Li, Y., Li, S., Hu, N., He, Y., Pong, R., Lin, D., Lu, L., & Law, M. (2012). Comparison of next-generation sequencing systems. *Journal of Biomedicine and Biotechnology, 2012*. Available from https://doi.org/10.1155/2012/251364.

Lozupone, C., & Knight, R. (2005). UniFrac: A new phylogenetic method for comparing microbial communities. *Applied and Environmental Microbiology, 71*(12), 8228–8235. Available from https://doi.org/10.1128/AEM.71.12.8228-8235.2005.

Lozupone C.A. and Knight R., Species divergence and the measurement of microbial diversity. *FEMS Microbiology Reviews*, **32**(4), 2008, 557–578.

Luckey, J. A., Drossman, H., Kostichka, A. J., Mead, D. A., D'cunha, J., Norris, T. B., & Smith, L. M. (1990). High speed DNA sequencing by capillary electrophoresis. *Nucleic Acids Research, 18*(15), 4417–4421. Available from https://doi.org/10.1093/nar/18.15.4417.

Mahato, N. K., Gupta, V., Singh, P., Kumari, R., Verma, H., Tripathi, C., Rani, P., Sharma, A., Singhvi, N., Sood, U., Hira, P., Kohli, P., Nayyar, N., Puri, A., Bajaj, A., Kumar, R., Negi, V., Talwar, C., Khurana, H., … Lal, R. (2017). Microbial taxonomy in the era of OMICS: Application of DNA sequences, computational tools and techniques. *Antonie van Leeuwenhoek, International Journal of General and Molecular Microbiology, 110*(10), 1357–1371. Available from https://doi.org/10.1007/s10482-017-0928-1.

Maidak, B. L., Larsen, N., McCaughey, M. J., Overbeek, R., Olsen, G. J., Fogel, K., Blandy, J., & Woese, C. R. (1994). The ribosomal database project. *Nucleic Acids Research, 22*(17), 3485–3487. Available from https://doi.org/10.1093/nar/22.17.3485.

Maidak, B. L., Olsen, G. J., Larsen, N., Overbeek, R., McCaughey, M. J., & Woese, C. R. (1996). The ribosomal database project (RDP). *Nucleic Acids Research, 24*(1), 82–85. Available from https://doi.org/10.1093/nar/24.1.82.

Makałowski, W., & Shabardina, V. (2020). Bioinformatics of nanopore sequencing. *Journal of Human Genetics, 65*(1), 61–67. Available from https://doi.org/10.1038/s10038-019-0659-4.

Margulies, M., Egholm, M., Altman, W. E., Attiya, S., Bader, J. S., Bemben, L. A., Berka, J., Braverman, M. S., Chen, Y. J., Chen, Z., Dewell, S. B., Du, L., Fierro, J. M., Gomes, X. V., Godwin, B. C., He, W., Helgesen, S., Ho, C. H., Irzyk, G. P., ... Rothberg, J. M. (2005). Genome sequencing in microfabricated high-density picolitre reactors. *Nature, 437*(7057), 376–380. Available from https://doi.org/10.1038/nature03959.

Martin, A. P. (2002). Phylogenetic approaches for describing and comparing the diversity of microbial communities. *Applied and Environmental Microbiology, 68*(8), 3673–3682. Available from https://doi.org/10.1128/AEM.68.8.3673-3682.2002.

Md, F., & Bin, M. K. S. (2013). Methods for analyzing diversity of microbial communities in natural environments. *Ceylon Journal of Science (Biological Sciences), 19*. Available from https://doi.org/10.4038/cjsbs.v42i1.5896.

Moorthie, S., Mattocks, C. J., & Wright, C. F. (2011). Review of massively parallel DNA sequencing technologies. *HUGO Journal, 5*(1–4), 1–12. Available from https://doi.org/10.1007/s11568-011-9156-3.

Nannipieri, P., Ascher-Jenull, J., Ceccherini, M. T., Pietramellara, G., Renella, G., & Schloter, M. (2020). Beyond microbial diversity for predicting soil functions: A mini review. *Pedosphere, 30*(1), 5–17. Available from https://doi.org/10.1016/S1002-0160(19)60824-6.

Narayan N.R., Weinmaier T., Laserna-Mendieta E.J., Claesson M.J., Shanahan F., Dabbagh K., Iwai S. and DeSantis T.D. (2020). Piphillin predicts metagenomic composition and dynamics from DADA2-corrected 16S rDNA sequences. *BMC Genomics*. Available from https://doi.org/10.1186/s12864-019-6427-1.

Nyrén, P., & Lundin, A. (1985). Enzymatic method for continuous monitoring of inorganic pyrophosphate synthesis. *Analytical Biochemistry, 151*(2), 504–509. Available from https://doi.org/10.1016/0003-2697(85)90211-8.

Oliver, J. S., Bready, B., Catalano, A. P., Davis, J. R., Kaiser, M. D., & Sage, J. M. (2019). Analysis of complex microbial samples using high definition mapping. *Journal of Biomolecular Techniques: JBT, 30*, S5.

Oliver, J. S., Catalano, A., Davis, J. R., Grinberg, B. S., Hutchins, T. E., Kaiser, M. D., Nurnberg, S., Sage, J. M., Seward, L., Simelgor, G., Weiner, N. K., & Bready, B. (2017). High-definition electronic genome maps from single molecule data. *BioRxiv*. Available from https://doi.org/10.1101/139840.

O'Brien, S. L., Gibbons, S. M., Owens, S. M., Hampton-Marcell, J., Johnston, E. R., Jastrow, J. D., Gilbert, J. A., Meyer, F., & Antonopoulos, D. A. (2016). Spatial scale drives patterns in soil bacterial diversity. *Environmental Microbiology, 18*(6), 2039–2051. Available from https://doi.org/10.1111/1462-2920.13231.

Pacwa-Płociniczak, M., Biniecka, P., Bondarczuk, K., & Piotrowska-Seget, Z. (2020). Metagenomic functional profiling reveals differences in bacterial composition and function during bioaugmentation of aged petroleum-contaminated soil. *Frontiers in Microbiology, 11*. Available from https://doi.org/10.3389/fmicb.2020.02106.

Parson, W., Strobl, C., Huber, G., Zimmermann, B., Gomes, S. M., Souto, L., Fendt, L., Delport, R., Langit, R., Wootton, S., Lagacé, R., & Irwin, J. (2013). Evaluation of next generation mtGenome sequencing using the Ion Torrent Personal Genome Machine (PGM). *Forensic Science International: Genetics, 7*(5), 543–549. Available from https://doi.org/10.1016/j.fsigen.2013.06.003.

Perkel, J. M. (2010). Bright lights, single molecules. *BioTechniques, 49*(6), 875–878. Available from https://doi.org/10.2144/000113564.

Pettersson, E., Lundeberg, J., & Ahmadian, A. (2009). Generations of sequencing technologies. *Genomics, 93*(2), 105–111. Available from https://doi.org/10.1016/j.ygeno.2008.10.003.

Prodan, A., Tremaroli, V., Brolin, H., Zwinderman, A. H., Nieuwdorp, M., & Levin, E. (2020). Comparing bioinformatic pipelines for microbial 16S rRNA amplicon sequencing. *PLoS One, 15*(1). Available from https://doi.org/10.1371/journal.pone.0227434.

Pruesse, E., Quast, C., Knittel, K., Fuchs, B. M., Ludwig, W., Peplies, J., & Glöckner, F. O. (2007). SILVA: A comprehensive online resource for quality checked and aligned ribosomal RNA sequence data compatible with ARB. *Nucleic Acids Research, 35*(21), 7188–7196. Available from https://doi.org/10.1093/nar/gkm864.

Purushothaman, S., Toumazou, C., & Ou, C. P. (2006). Protons and single nucleotide polymorphism detection: A simple use for the ion sensitive field effect transistor. *Sensors and Actuators, B: Chemical, 114*(2), 964–968. Available from https://doi.org/10.1016/j.snb.2005.06.069.

Quail, M. A., Smith, M., Coupland, P., Otto, T. D., Harris, S. R., Connor, T. R., Bertoni, A., Swerdlow, H. P., & Gu, Y. (2012). A tale of three next generation sequencing platforms: Comparison of Ion Torrent, Pacific Biosciences and Illumina MiSeq sequencers. *BMC Genomics, 13*(1), 341. Available from https://doi.org/10.1186/1471-2164-13-341.

Quast, C., Pruesse, E., Yilmaz, P., Gerken, J., Schweer, T., Yarza, P., Peplies, J., & Glöckner, F. O. (2012). The SILVA ribosomal RNA gene database project: Improved data processing and web-based tools. *Nucleic Acids Research, 41*(D1).

Rand, A. C., Jain, M., Eizenga, J. M., Musselman-Brown, A., Olsen, H. E., Akeson, M., & Paten, B. (2017). Mapping DNA methylation with high-throughput nanopore sequencing. *Nature Methods, 14*(4), 411–413. Available from https://doi.org/10.1038/nmeth.4189.

Ren, C., Zhang, W., Zhong, Z. K., Han, X., Yang, G., Feng, Y., & Ren, G. (2018). Differential responses of soil microbial biomass, diversity, and compositions to altitudinal gradients depend on plant and soil characteristics. *Science of the Total Environment, 610–611*, 750–758. Available from https://doi.org/10.1016/j.scitotenv.2017.08.110.

Ripley, B. D. (1981). Spatial statistics.New York: Wiley.

Rocca, M. S., Ferrarini, M., Msaki, A., Vinanzi, C., Ghezzi, M., De Rocco Ponce, M., Foresta, C., & Ferlin, A. (2020). Comparison of NGS panel and Sanger sequencing for genotyping CAG repeats in the AR gene. *Molecular Genetics and Genomic Medicine, 8*(6). Available from https://doi.org/10.1002/mgg3.1207.

Ronaghi, M., Uhlén, M., & Nyrén, P. (1998). A sequencing method based on real-time pyrophosphate. *Science, 281*(5375), 363–365. Available from https://doi.org/10.1126/science.281.5375.363.

Rusk, N. (2009). Cheap third-generation sequencing. *Nature Methods, 6*(4), 244–245. Available from https://doi.org/10.1038/nmeth0409-244a.

Rusk, N. (2011). Torrents of sequence. *Nature Methods, 8*(1), 44. Available from https://doi.org/10.1038/nmeth.f.330.

Sanger, F., Nicklen, S., & Coulson, A. R. (1977). DNA sequencing with chain-terminating inhibitors. *Proceedings of the National Academy of Sciences*, 5463–5467. Available from https://doi.org/10.1073/pnas.74.12.5463.

Schadt, E. E., Turner, S., & Kasarskis, A. (2010). A window into third-generation sequencing. *Human Molecular Genetics, 19*(2), R227–R240. Available from https://doi.org/10.1093/hmg/ddq416.

Schloss, P. D. (2008). Evaluating different approaches that test whether microbial communities have the same structure. *ISME Journal, 2*(3), 265–275. Available from https://doi.org/10.1038/ismej.2008.5.

Schloss, P. D., & Handelsman, J. (2006a). Introducing SONS, a tool for operational taxonomic unit-based comparisons of microbial community memberships and structures. *Applied and Environmental Microbiology, 72*(10), 6773–6779. Available from https://doi.org/10.1128/AEM.00474-06.

Schloss, P. D., & Handelsman, J. (2006b). Introducing TreeClimber, a test to compare microbial community structures. *Applied and Environmental Microbiology, 72*(4), 2379–2384. Available from https://doi.org/10.1128/AEM.72.4.2379-2384.2006.

Schloss, P. D., & Handelsman, J. (2008). A statistical toolbox for metagenomics: Assessing functional diversity in microbial communities. *BMC Bioinformatics, 9*(1).

Schloss, P. D., Larget, B. R., & Handelsman, J. (2004). Integration of microbial ecology and statistics: A test to compare gene libraries. *Applied and Environmental Microbiology, 70*(9), 5485–5492. Available from https://doi.org/10.1128/AEM.70.9.5485-5492.2004.

Schloss, P. D., Westcott, S. L., Ryabin, T., Hall, J. R., Hartmann, M., Hollister, E. B., Lesniewski, R. A., Oakley, B. B., Parks, D. H., Robinson, C. J., Sahl, J. W., Stres, B., Thallinger, G. G., Van Horn, D. J., & Weber, C. F. (2009). Introducing mothur: Open-source, platform-independent, community-supported software for describing and comparing microbial communities. *Applied and Environmental Microbiology, 75*(23), 7537–7541. Available from https://doi.org/10.1128/AEM.01541-09.

Seok-Hwan, Y., Sung-Min, H., Soonjae, K., Jeongmin, L., Yeseul, K., Hyungseok, S., & Jongsik, C. (2017). Introducing EzBioCloud: A taxonomically united database of 16S rRNA gene sequences and whole-genome assemblies. *International Journal of Systematic and Evolutionary Microbiology*, 1613–1617. Available from https://doi.org/10.1099/ijsem.0.001755.

Simpson, J. T., Workman, R. E., Zuzarte, P. C., David, M., Dursi, L. J., & Timp, W. (2017). Detecting DNA cytosine methylation using nanopore sequencing. *Nature Methods, 14*(4), 407–410. Available from https://doi.org/10.1038/nmeth.4184.

Singleton, D. R., Furlong, M. A., Rathbun, S. L., & Whitman, W. B. (2001). Quantitative comparisons of 16S rRNA gene sequence libraries from environmental samples. *Applied and Environmental Microbiology, 67*(9), 4374–4376. Available from https://doi.org/10.1128/AEM.67.9.4374-4376.2001.

Smith K., A brief history of NCBI's formation and growth, *The NCBI Handbook*, 2013.

Stranneheim, H., & Lundeberg, J. (2012). Stepping stones in DNA sequencing. *Biotechnology Journal, 7*(9), 1063–1073. Available from https://doi.org/10.1002/biot.201200153.

Stubbendieck, R. M., Vargas-Bautista, C., & Straight, P. D. (2016). Bacterial communities: Interactions to scale. *Frontiers in Microbiology, 7*. Available from https://doi.org/10.3389/fmicb.2016.01234.

Supramaniam, Y., Chong, C. W., Silvaraj, S., & Tan, I. K. P. (2016). Effect of short term variation in temperature and water content on the bacterial community in a tropical soil. *Applied Soil Ecology, 107*, 279–289. Available from https://doi.org/10.1016/j.apsoil.2016.07.003.

Thompson, J. F., & Milos, P. M. (2011). The properties and applications of single-molecule DNA sequencing. *Genome Biology, 12*(2).

Valouev, A., Ichikawa, J., Tonthat, T., Stuart, J., Ranade, S., Peckham, H., Zeng, K., Malek, J. A., Costa, G., McKernan, K., Sidow, A., Fire, A., & Johnson, S. M. (2008). A high-resolution, nucleosome position map of C. elegans reveals a lack of universal sequence-dictated positioning. *Genome Research, 18*(7), 1051–1063. Available from https://doi.org/10.1101/gr.076463.108.

van Dijk E.L., Auger H., Jaszczyszyn Y. and Thermes C. Ten years of next-generation sequencing technology. (2014). *Trends in Genetics*. (2014), 418–426. Available from https://doi.org/10.1016/j.tig.2014.07.001.

Verma, S. K., & Sharma, P. C. (2020). NGS-based characterization of microbial diversity and functional profiling of solid tannery waste metagenomes. *Genomics, 112*(4), 2903–2913. Available from https://doi.org/10.1016/j.ygeno.2020.04.002.

Vestergaard, G., Schulz, S., Schöler, A., & Schloter, M. (2017). Making big data smart—How to use metagenomics to understand soil quality. *Biology and Fertility of Soils, 53*(5), 479–484. Available from https://doi.org/10.1007/s00374-017-1191-3.

Wadapurkar, R. M., & Vyas, R. (2018). Computational analysis of next generation sequencing data and its applications in clinical oncology. *Informatics in Medicine Unlocked, 11*, 75–82. Available from https://doi.org/10.1016/j.imu.2018.05.003.

Weigand, M. R., Pawloski, L. C., Peng, Y., Ju, H., Burroughs, M., Cassiday, P. K., Davis, J. K., DuVall, M., Johnson, T., Juieng, P., Knipe, K., Loparev, V. N., Mathis, M. H.,

Rowe, L. A., Sheth, M., Williams, M. M., & Tondella, M. L. (2018). Screening and genomic characterization of filamentous hemagglutinin-deficient *Bordetella pertussis*. *Infection and Immunity*, *86*(4). Available from https://doi.org/10.1128/IAI.00869-17.

Weißbecker, C., Schnabel, B., & Heintz-Buschart, A. (2020). Dadasnake, a Snakemake implementation of DADA2 to process amplicon sequencing data for microbial ecology. *GigaScience*, *9*(12). Available from https://doi.org/10.1093/gigascience/giaa135.

Yarus M. and Barrell B.G., The sequence of nucleotides in $tRNA^{Ile}$ from *E. coli* B, *Biochemical and Biophysical Research Communications* **43** (4), 1971, 729–734.

Yoon S.H., Ha S.M., Kwon S., Lim J., Kim Y., Seo H. and Chun J., Introducing EzBioCloud: A taxonomically united database of 16S rRNA gene sequences and whole-genome assemblies, *International Journal of Systematic and Evolutionary Microbiology* **67** (5), 2017, 1613.

CHAPTER

16

Legal protection of microbial biodiversity

Rashmi Arya[1] and Sonali Shinde[2]

[1]CSIR-Unit for Research & Development of Information Products, Pune, India, [2]Annasaheb Kulkarni Department of Biodiversity, MES Abasaheb Garware College, Pune, India

16.1 Introduction

Norman Myers, an environmentalist, has rightly said, "Without knowing it, we utilize hundreds of products each day that owe their origin to wild animals and plants. Indeed, our welfare is intimately tied up with the welfare of wildlife. Well may conservationists proclaim that by saving the lives of wild species, we may be saving our own" (Myers, 1983). This idea can further be extended for the inclusion of microbiomes and their largely unexplored world. Apart from the sustainable exploitation of microorganisms in obtaining valuable secondary metabolites, their key role in the ecosystem resides in the fact that they are the ultimate decomposers in the energy pyramid. According to the study (Myers, 1983) published by researchers at the Indiana University in 2016, an estimated one trillion microbial species exist on our planet (Locey & Lennon, 2016), out of which only 0.001% of them have been discovered as of yet. This provides an insight toward the enormous scope in exploring a great domain altogether (Locey & Lennon, 2016).

The enormous potential that resides in the diverse microflora has to be explored for the various benefits. Efforts made by the researchers in exploring the microbial diversity is currently supported by various International Conventions and Agreements along with the National Legislations of the respective states on several key points involving the scope of research, taxonomic nomenclature system, infrastructural and

Microbial Diversity and Ecology in Hotspots
DOI: https://doi.org/10.1016/B978-0-323-90148-2.00004-3

logistical capacity for research, etc. An example for such a system in place can be considered to be that formed by the Indian Government, in which, the onus of conservation of the biodiversity of a particular region, is distributed evenly with different authorities and stakeholders. With the National Biodiversity Authority being the nodal body for such purposes, the powers and responsibilities involved in the decision-making processes are vested equitably with the State Biodiversity Authorities (SBA). The SBA advise the respective State Governments upon matters related to the sustainable use of the elements of biodiversity and the benefit-sharing processes. At the grassroots levels, the regional Biodiversity Management Committees function in a harmonious way to ensure the conservation work and the active involvement of the stakeholders along with the participating nongovernmental organizations, as well (Perinchery, 2020). Of them, the aspect of carrying out research in the areas inhabited by Indigenous Tribes and Communities and considering the sentiments of those people, sharing of the resources and information obtained by such an exploration, and benefit-sharing mechanisms are among the prime areas of focus, especially in recent years where cases of biopiracy and thefts have undermined the principle of fair and equitable scientific exploration in the International Community. For example, within the Indian context, the concept of benefit-sharing and legitimate access to the genetic resources in the Biodiversity Conservation regions remains to be limited in scope, despite the effective directions of the National Green Tribunal in place. This can be attributed to the fact that most of the companies using such resources are unaware of these procedural rules and/or hesitant to comply with them (Perinchery, 2020).

This chapter aims at providing insights into the current international regime supporting the scientific exploration and protection associated with the microbial life with a special emphasis on the discovery and research of microorganisms in Biodiversity Hotspots across the world.

16.2 Biodiversity Hotspots: Introduction

Norman Myers, in the late 1980s through his published works, conceptualized the idea of "biodiversity hotspot" regions across the world. In its current form, Conservation International (CI), a nonprofit American organization, is actively involved and contributes majorly toward the protection of such identified regions. According to the CI (https://www.conservation.org/priorities/biodiversity-hotspots), if a region has a minimum of 1500 vascular plants native to a specific area with either a high percentage of them being unique plants or its original natural vegetation being 30% or less, then it will be declared as a

Biodiversity Hotspot (Biodiversity Hotspots, n.d.). The Biodiversity Hotspot must be irreplaceable and must have threatened plant vegetation.

As of October 2020, a total of 36 hotspots have been identified for protection. There is a dire need to protect such areas as these regions despite being a mere 2.4% of the total land area of the Earth, constitute more than 35% of the "ecosystem services" that vulnerable human population depend directly upon.

16.3 Protection of microbial diversity in hotspot regions

Apart from the soil microbiota, which is primarily responsible for maintaining ecological processes, energy flows, degradation of toxic materials and other vital functions, microbes form considerable relationships with various other organisms in the niche as well, which may be commensal or parasitic or symbiotic (Frey-Klett et al., 2011). A few prominent examples from the hotspot regions are of considerable interest in the above context. *Alternaria* sp., *Cladosporium* sp., *Aspergillus* sp., *Bipolaris* sp., *Fusarium* sp., and *Talaromyces* sp. have been isolated from the seeds of common mesembs in Namaqualand, in Succulent Karoo, which is a Biodiversity Hotspot region located in South Africa (Pieterse et al., 2018). These fungal species are purported to support the growth of *Aizoaceae* (ice plants) even in the hot and dry weather of Africa as they are specialized in the formation of spores with thick walls that remain viable and can grow slowly in extreme arid conditions. Another example can be accounted for the discovery of a novel species of green phototrophic bacterium from the genus *Rhodobacter* isolated from a mud-stream of the Western Ghat Biodiversity Hotspot region of the Indian Subcontinent (Nampoothiri et al., 2013). However, apart from flora and fauna of the hotspot regions, the microbiota associated with it is still largely unexplored. Hence, it is important to include the exploration of microbiome from the Biodiversity Hotspot regions across the world along with the primary aim of its protection and conservation. There are different acts and policies which govern the protection of microorganisms which are discussed below.

16.4 Existing framework for the research and exploration of microbiome

Access, collection, deposition, and utilization of samples for the isolation of microbiome and its associated products is currently based on the guidelines and rules governing them. "Intellectual Property Rights"

(IPR) give protection to any new creation due to the intellectual capacity of man. These rights include different types such as patents, trademarks, copyright, designs, geographical indications, and trade secrets. The new varieties of plants are protected by Plant Breeders' Rights. The primary objective is to avoid misappropriation of one's rights over the invention. A good amount of time, money, and effort is invested in research and technology development. The returns of this investment can be maximized if the technology is successfully commercialized. IPR play a pivotal role in the protection of different products and processes leading to economic growth and technology diffusion (Falvey & Foster-McGregor, 2006). Securing IPR in a given technology area, helps to gain a competitive advantage and the same holds true for microbial inventions just like any other invention. The microflora from a Biodiversity Hotspot help can be explored for useful products. The protection of the products from diverse microorganisms using genetic engineering or recent technologies such as genome engineering can be eligible for patent protection. Thus patents are key to the protection of modified microbes or products from the microbes. Additionally, the "Convention on Biological Diversity" (CBD) and "Trade-Related Aspects of Intellectual Property Rights" (TRIPS) Agreement are two important international treaties that have implications on the protection of biodiversity.

A patent is an exclusive right granted to an applicant by the state for a new and useful invention, for a limited duration of time, and restricted to the geographical territory. In order to seek protection in different countries, separate applications have to be made in each country of interest. The patent rights are valid for a limited time duration, that is, 20 years from the filing date of the patent application. The applicant or the patent owner is required to pay an annual maintenance fee to the patent office to keep the patent valid. A patent grants the exclusionary right that does not allow others to practice, make, use, or sell the invention. In case any patent-protected invention is required for an invention to be practiced, prior permission from the patent owner is essential and licenses can be sought. If anybody practices any protected invention without prior permission, they can be sued for infringement. An invention qualifies for patent protection if it satisfies the three criteria of novelty, inventive step (being nonobvious), and industrial application/utility. (https://www.wipo.int/ip-outreach/en/ipday/2017/innovation_and_intellectual_property.html) (Innovation and Intellectual Property, n.d.).

Patents provide the exclusive property right in exchange for disclosing the technical information to the public. The inventor has to describe the mode to carry out the invention in detail and also ensure that his invention can be enabled. The rationale of all patents is to stimulate investment in inventions and innovation through a temporary monopoly. The inventor, during the period of his patent, can market and commercialize the invention to recover the research and development costs. On the other hand, society at large benefits from new products and from the disclosure of the invention, which can support technological development. Patents are therefore incentives for innovation, research, and development as well as further investment in the development of new and improved products and processes.

The patenting of a novel and modified microorganism is not very different from patenting other forms of inventions in any field of technology. According to the TRIPS agreement, patents for microorganisms should be granted by all World Trade Organization (WTO) members (TRIPS: REVIEWS, ARTICLE 27.3(B) AND RELATED ISSUES, n.d.).

16.5 Patents and microorganisms

Streptomyces venezuelae culture was deposited in the American Type Culture Collection (ATCC) by Parke Davis Co. on July 8, 1949. The culture, assigned ATCC number 10712, was disclosed in the patent United States2483892, claiming "a process for the manufacture of chloramphenicol." Similarly, American Cyanamid Company deposited a culture of *Streptomyces aureofaciens* strain in August 1949 in the Agricultural Research Service Culture Collection, which is now named as Northern Regional Research Laboratory (NRRL), and assigned the NRRL number 2209. This species was disclosed in the patent United States2482055, "claiming the production of aureomycin." The abovementioned two deposits are among the forerunners to file patent applications with a deposit in a recognized patent depositary (United States Congress, Office of Technology Assessment, New Developments in Biotechnology: Patenting Life–Special Report, OTA-BA-370, 1989).

The microorganism, to be patentable, should be new and should involve human intervention. The microorganism that is merely isolated from nature is not patentable. The microbiological inventions that involve novel strains of microorganisms, including a step of human intervention so that they can either produce a new compound or a compound that is already known more efficaciously with better yield or purity are considered for patent protection. However, it is very difficult to completely describe a microbial strain in writing in the patent disclosure. A microbial strain is a living system, having variation in structure and characteristics, and capable of altering its behavior in response to changes in its environment. Even if a complete description would be possible, still, the public would not have the invention on expiry of the patent. Just as, in any mechanical invention, the drawing is important, similarly, for any microbiological invention, the microorganism is important. When an invention is based on a microorganism or any other biological material or its use in various industries such as food, pharmaceutical, etc., but is not easily accessible to the public, then the description of such an invention is not sufficiently disclosed. Thus, in almost all countries, it is required to file the patent application giving the written description along with a deposit of the sample of the microorganism. Thus the World Intellectual Property Organization brought in force the "Budapest Treaty on the International Recognition of the Deposit of Microorganisms for the Purposes of Patent Procedure" in 1980 (Budapest Treaty on the International Recognition of the Deposit of Microorganisms for the Purposes of Patent Procedure, 2020). It puts forth the requirement of deposit of a microorganism as a part of the procedure in processing the grant of Patent Rights for microbes (Goel & Parashar, 2013). A single deposit made at any of the international depositary authorities listed will suffice for all signatory states. As defined in the Budapest Treaty, an "International Depositary Authority (IDA) is a scientific institution which is capable of storing microorganisms." If a sample of the microorganism deposited is no more viable, a new deposit should be made. If the IDA ceases to function, the deposit should be transferred to another IDA confirming the viability of the deposited organism. The IDA is allowed to charge a fee for the deposit in which the deposited organism should be stored for a minimum period of 30 years (*Article 9 of Budapest Treaty*). Anybody can get the sample of the microorganisms deposited from the IDA. Thus the "Budapest Treaty on the International Recognition of the Deposit of Microorganisms for the Purposes of Patent Procedure" needs the applicants with microbial inventions to have to deposit the microorganism to satisfy the requirement of complete disclosure and being reproducible for biological materials.

For any invention involving the microorganisms from diverse environments, the opportunity to file for patent protection can be possible if the inventions of these microorganisms satisfy the criteria of the invention being novel, having an inventive step and an industrial application, provided it is deposited at anyone of the International Depositary Authorities. These days it is also required to give the source from where the biological material is taken along with the deposition details.

The TRIPS Agreement includes other forms of protection of biodiversity, other than patents, such as the "sui generis system" for the protection of new plant varieties.

The basic legislation framework related to biodiversity conservation, its supporting guidelines and rules, is the CBD which was brought into force on December 29, 1993. This treaty aims at the protection of the bioresources and prevents unfair exploitation of these resources in various countries (Convention on Biological Diversity, n.d.). The treaty provides for the "conservation of biological diversity, the sustainable use of its components and the fair and equitable sharing of the benefits arising out of the utilization of genetic resources, including by appropriate access to genetic resources and by appropriate transfer of relevant technologies, taking into account all rights over those resources and to technologies, and by appropriate funding" (*Article 1 of CBD*). In its primary meaning, the CBD includes variability in the living organisms (including microbes) from all sources, as an element of biodiversity, as well (*Article 2 of CBD*). "States have the sovereign right to exploit their resources under their environmental policies, and the responsibility to ensure that activities within their jurisdiction or control do not cause damage to the environment of other States or of areas beyond the limits of national jurisdiction" (*Article 3 of CBD*). Thus the CBD ensures that the natural environment is not disturbed due to exploration and exploitation of the flora present. The CBD also entrusts the countries to adopt strategies that will rely on conservation and the sustainable use of biodiversity. The CBD also states that if any bioresources from foreign countries are to be accessed, prior consent from the country of origin should be taken. It is also essential to share the benefits that come up from the technology transfer, if any, in a fair and equitable manner with the country of origin of bioresource. The IPR law should not hinder the objectives of the CBD. Hence, if patents are being filed, the requisite permissions sought for using bioresources also have to be provided. The CBD is also committed to recognizing and continuously monitoring the important elements of the biodiversity that should be conserved and ensure their sustainable use. Moreover, CBD is entrusted to set up preserved areas for conserving the diversity and reinstating the damaged ecosystems in order to encourage the revival of the endangered species. CBD also is committed to respect, preserve, and conserve the traditional knowledge of indigenous people and local communities for the sustainable use of the biodiversity. CBD further aims to prevent the appearance of, restrict, and eliminate species that have come from outside and promote participation from the public to assess the projects that threaten the biodiversity. However, CBD faces challenges due to the significant growth in population which has resulted in the increased demand for bioresources. CBD also has to cope up with the ability to document, understand, realize the value, and identify the threats to the biodiversity. It is also needed to build requisite expertise for policy making, crafting legislation, and devising measures for governing the effective utilization of biodiversity.

On the similar lines, another international agreement, "The Cartagena Protocol on Biosafety to the Convention on Biological Diversity," came into force in 2003 (The Cartagena Protocol on Biosafety to the Convention on Biological Diversity, 2020). "The Cartagena Protocol on Biosafety to the CBD ensures an adequate level of protection in safe transfer, handling and use of living modified organisms resulting from modern biotechnology that may have adverse effects on the conservation and sustainable use of biological diversity" (*Article 1 of Cartagena Protocol*). The Protocol states that the countries "should take necessary and appropriate legal, administrative, and other measures to implement its obligations under this protocol and also ensure that the development, handling, transport, use, transfer, and release of any living modified organisms are undertaken in a manner that prevents or reduces the risks to biological diversity, taking also into account risks to human health" (*Article 2 of Cartagena Protocol*). This Protocol also takes into account the transboundary movements of the collected samples.

Further, "The Nagoya Protocol on Access and Benefit Sharing" another international agreement was brought into force in 2014 that aimed at fair and equitable sharing of the benefits which arise from the exploitation of the genetic resources (The Nagoya Protocol on Access and Benefit-Sharing, 2020). It thus further strengthens the CBD concerning its benefit-sharing terms. The "ABS Clearing-House is a key tool for facilitating the implementation of the Nagoya Protocol, by enhancing legal certainty and transparency on procedures for access and benefit-sharing, and for monitoring the utilization of genetic resources along the value chain, including through the internationally recognized certificate of compliance" (*Article 14 of the Protocol*).

The "Microorganisms Sustainable Use and Access Regulation International Code of Conduct" (MOSAICC) was launched by the "Belgian Coordinated Collections of Microorganisms." The MOSAICC provides the international framework to ensure ethical and legitimate access to the microbiome associated with a region. It is a "Voluntary Code of Conduct developed to facilitate access to microbial genetic resources (MGRs) and to help partners to make appropriate agreements when transferring MGRs, in the framework of the Convention on Biological Diversity (CBD) and other applicable rules of international and national laws" (MOSAICC Micro-Organisms Sustainable Use and Access Regulation International Code of Conduct, n.d.). It supports the execution of the CBD for microbial and other genetic resources. The Bonn Guidelines complements the Nagoya Protocol and the MOSAICC as it provides for the benefit-sharing terms as well as the procedural aspect for accessing the microbes from their niche. "The Bonn Guidelines on Access to Genetic Resources and Fair and Equitable Sharing of the Benefits Arising out of their Utilization" were adopted in 2002 "to assist Parties when establishing administrative, legislative or policy measures on access and benefit-sharing and/or when negotiating contractual arrangements for access to genetic resources and benefit-sharing" (The Bonn Guidelines, n.d.).

16.6 Microbial access and research: pipeline

The CBD assures the sovereignty of Contracting Parties by mandating the jurisdictional scope of the Parties within their respective national boundaries (*Articles 3 and 4 of CBD*). Further, it also provides for the Contracting Parties to formulate respective National Policies and mechanisms for ensuring the sustainable use and protection of its BioDiversity (*Article 6 of CBD*). The CBD also encourages the Parties to carry out scientific research and development for exploration and protection of their respective biodiversity aspect (*Article 12 of CBD*). For research on the microbiome of a particular region or for obtaining the already isolated microbes endemic to a specific area, a researcher has to proceed in a certain way. This is schematically represented in Fig. 16.1.

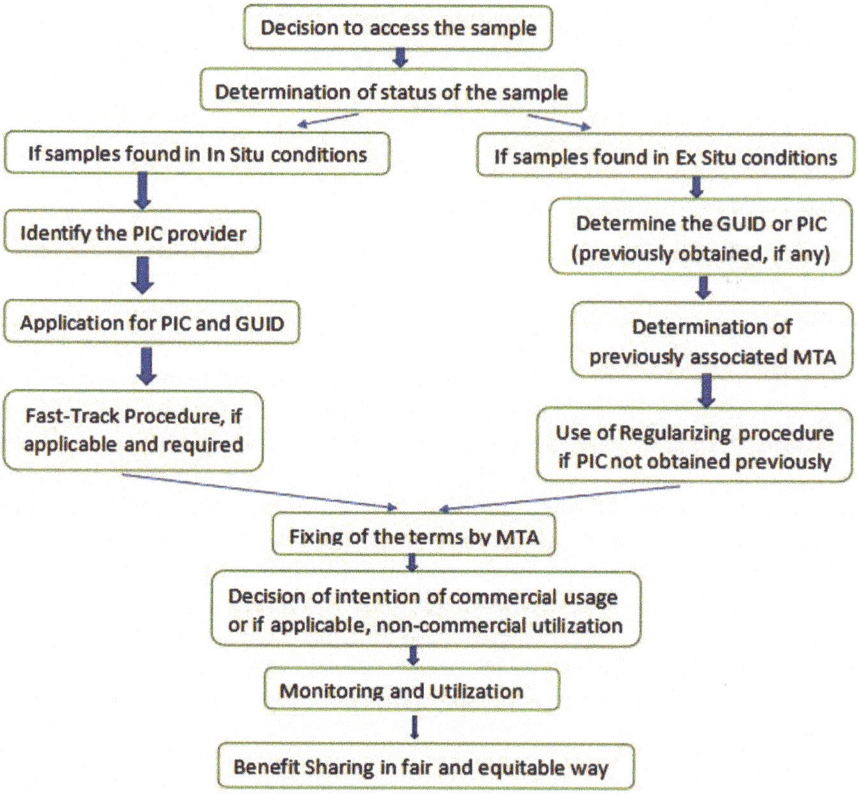

FIGURE 16.1 Procedural Pipeline for microbial research and utilization (derived partially from MOSAICC code of conduct brochure).

16.6.1 Decision to access the sample

For an individual researcher or an organization endeavoring toward exploring into microbiome or its associated metabolic products of commercial interests from a particular area, it is important to consider the feasibility aspect for such a venture in the initial stage itself. A tentative strategy can be developed considering the aspects related to the purpose of research (whether commercial exploitation intended or not), procedure for accessing the samples in a legitimate way, infrastructural and technological capacity, and other things. Recently, owing to the explicit development of state-of-the-art technologies for data mining and handling, researchers can make use of the processing capacity of computers for planning their exploration before initiating the research. As an example, the United States Environmental Protection Agency has devised an online software-based tool called MicroSAP (Silvestri et al., 2018) for conducting the sampling and planning strategies before conducting research. The Preamble to CBD encourages the scientific community to venture into the domain especially involving the protection and scientific exploration of biodiversity by gaining knowledge and information about it and striving toward capacity-building for scientific, technical, and institutional phenomena.

16.6.2 Determination of status of the sample

Biological samples consisting of microbes can be in either of the states of origin and/or conservation for accessing them—in situ or ex situ (Carlos et al., 2020). Article 2 of the CBD defines the scope and the meaning of the above terms, where "*in situ* conservation" means protecting the elements of biodiversity (implying microbes as well) within their natural habitats by certain conservative measures and "*ex situ* conservation" includes protecting these elements outside their natural habitats (which includes collection centers, cell-culture laboratories, etc. for microbes). It is to be remembered that in any of the below circumstances, the handling of microbes, that is, their safe-handling and transport, is to be carried out under the rules stipulated by the National Authority concerned in conservation of biodiversity of the respective region. The guidelines and general rules for the same are described under the Cartagena Protocol to CBD (*Article 1 of the Cartagena Protocol*).

For a microbial sample accessible within a region where in situ conservation has been undertaken (guidelines for in situ conservation are defined within the aegis of *Article 8 of the CBD*), the first step for a researcher (or the organization) is to identify all the stakeholders pertaining the conservation and sustainable utilization of the hotspot region. Section 1.2 of the MOSAICC guidelines, 2011 mandates for obtaining the Prior Informed Consent (PIC) from the country of origin of the hotspot region. The obligation for obtaining a PIC before starting the acquisition of samples is based on the concept of fair and equitable access to genetic resources as per Article 15 (5) of the CBD. The researcher further has to identify the competent authorities having the legitimate capacity to grant the PIC for the region in contention. If there is no such system in place, a template for the same has been described in the MOSAICC guidelines, which would potentially serve the purpose on an acceptable scale globally. The authority concerned with providing the PIC must take into account the welfare and sentiments of the Indigenous Communities associated with that region, as mentioned in Articles 6 and 7 of the Nagoya Protocol on the CBD. The PIC application can be followed by assigning a Global Unique Identifier (GUID) to the sample so that there is an assurance of the legitimate obtaining of the sample from its in situ conservation region. As per the MOSAICC guidelines, the entire process described here can be fast-tracked or is speed up in case of medical emergencies or other considerable situations. This is in line with the implied meaning of Article 14 of the CBD, where the Contracting Parties are under a moral obligation to fast-track such procedure in case of emergencies.

The microbial sample can also be obtained from an ex situ conservation facility which may include (but not be limited in the scope) cell-culture laboratories, microbial collection centers, etc. The associated scope and measures to be undertaken by the Contracting Parties for the purpose of setting up such a facility is described in Article 9 of the CBD. A microbial sample deposited in an ex situ conservation facility is usually associated with the PIC and the GUID, as well, when it is obtained previously from its in situ conservation region. The accession form, filled at the time of depositing the sample for ex situ conservation, serves the purpose of correctly identifying the microbial strain and the culture-type throughout the lifespan of its intended conservation. If in case the PIC is not obtained previously during the transfer of the MGRs or simply the samples from in situ conservation regions to the ex situ facilities, MOSAICC guidelines do suggest for a regularizing procedure that can be followed for ensuring the compatibility of the legitimacy of the samples, obtained in such a manner. A researcher must also consider understanding the terms associated with a previously signed Material Transfer Agreement (MTA), if any, for the sample available in an ex situ facility.

A distinct scenario is to be considered when dealing with the secondary products associated with the microbes and the flora and fauna of a biodiversity conservation region. As mentioned earlier, microbes can be commensal, parasitic, or symbiotic in relationship with their host in the wild. The result of such a sophisticated relation between them can yield valuable secondary metabolites which mostly, do possess immense potential for commercialization. The trade and international exchange of endangered species and their derived products takes place through the law enforced by "Convention on International Trade in Endangered Species of Wild Fauna and Flora" (CITES). The metabolic products, as discussed above, are to be obtained or traded by adhering to the CITES standard procedure for operation.

16.6.3 Fixing of the terms by MTA and the decision for commercial exploitation or noncommercial usage

The CBD mandates, through firm assurance in Article 15 (4), about granting access to the genetic resources (in this case the microbial samples) in a fair and legitimate way through a mutual understanding and an agreement for the same considering the dignity and sentiments of all the stakeholders involved within the in situ conservation region of the sample. These mutually agreed terms are decided before granting the possession of the sample through an agreement, called as the Material Transfer Agreement (MTA). Section 1.4 of the MOSAICC guidelines describe the potential contents of a MTA, which includes the agreed terms documented with the consent of the stakeholders on subjects like the purpose of usage of the sample (commercial or noncommercial), ownership of the IPRs emerging out of such a venture, benefit-sharing from the usage of the sample, and other things. The terms in the MTA should be devised along with the PIC application, as it would meet the purpose of satisfying the concept of benefit-sharing between the stakeholders and fair and equitable access to the samples, with the obligation of sustainable usage of it.

The scientific exploration and study arising out of the sustainable usage of the sample can potentially lead to the development of a commercial model. Article 41 of the TRIPS Agreement envisages the concept of benefit-sharing in a fair and equitable way associated with all forms of IPRs. So, if the samples are intended to be used for a potentially commercial exploitation, such terms are to be decided mutually in the MTA. Also, the scope for redistribution of the samples is to be decided in the MTA itself. If the microbes are to be subjected to legitimate ownership under the IPR, they are first required to be deposited as genuine samples to the Microbial Collection centers, designated as "International Depositary Authorities" according to the Budapest Treaty.

16.6.4 Monitoring and utilization

After granting the access of its genetic resources (microbial samples in this case), it is important for the supplying authority as a Contracting

Party of the CBD to effectively monitor the transfer of such genetic resources and the study arising out from it, as well, so that the benefit-sharing arising from its utilization can be done in a fair and equitable way. This is emphasized in Article 7 of the CBD where Contracting Parties are under an obligation to monitor their resources effectively, to counter any form of misconduct, or mitigate any adverse effects arising from unwarranted utilization. Section 1.5 of MOSAICC suggests such a system which can be effectively applied on a global scale as well.

16.6.5 Fair and equitable benefit-sharing

It is morally intuitive for a researcher (or an organization) utilizing the genetic resources associated with a conservation region, especially such as a Biodiversity Hotspot, to distribute the benefits coming out of its use to the stakeholders involved in the conservation of that region. This has been made legally binding for all through the adaption of CBD and several other protocols, thereafter. Benefit-sharing in a fair and equitable manner, especially with the Indigenous Communities associated with the Conservation regions, is the essence of the CBD, as mentioned in its Article 1. Identification of the rights of such people pertaining to the conservation of biological diversity of the region is described in Article 8 of the CBD, through means of in situ conservation measures. A standout example is that of the Indian Traditional Knowledge Digital Library which ensures that the IPRs of Indigenous Communities, arising out from the traditional knowledge that they possess, are not captured by others, that is, to avoid biopiracy. The Nagoya Protocol on CBD further strengthens the goal aimed in CBD about the aspect on sharing of benefits. The Annex to the Nagoya Protocol describes the types of monetary and nonmonetary benefits coming out of the use of the samples. An example of effective execution of the concept of benefit-sharing can be considered through the case of the traditional knowledge of the Kani tribe in Southern Kerala on the antifatigue properties of the herb arogyapacha (*Trichopus zeylanicus*) that grows in their forests. The success reflected in this case is still considered a benchmark for strengthening the idea of benefit-sharing in a fair and equitable way, along with the protection of biodiversity of the region.

Even as already being susceptible and vulnerable to an irreversible consequence of drastic climate change globally (Reinold et al., 2019) and the added miseries being caused by the excess usage of artificial chemicals as fertilizers (Souza et al., 2016), microbial communities, especially in the biodiversity hotspot regions, are also under the increased threat to becoming more prone to fragmented habitation and disturbed ecological interactions with the surroundings due to unaccounted human activities and interference (Donachie et al., 2004). Being fragile, the microbial communities will only be subjected to increased imbalance in their invaluable diversity and as important component of the ecosystem, as well, if there are no concrete measures in place for sampling or utilization of the microbial samples, especially in the hotspot regions. Though there is a promising commitment at the international stage, under the CBD agreement, toward the concerted efforts in conserving and protecting the biological diversity in general, the ecological interests and expertise required for protecting the microbial communities, and especially in the hotspot regions, is an intricate domain in itself as microorganisms are of quintessential importance in maintaining a healthy ecosystem overall (Cavicchioli et al., 2019). The in situ and ex situ conservation strategies projected for protecting the biodiversity in general as affirmed through the CBD, needs to be strengthened and sophisticated for microbial subjects. This is evident from the fact that the microbiota associated with animals in the wild, have been affected through habitat destruction, as of recent times (Trevelline et al., 2019). In upcoming times, this trend can continue with the hotspot regions as well and so, there is an urgent need to have effective and strict protocols in place for microbial handling and utilization in these regions. The onus would completely be vested with the local governmental authorities to make strict provisions and enforcement as well, for acts of entering and/or carrying out certain activities that would threaten the microbial communities. For instance, for adopting a sustainable conservation strategy for biodiversity in the Western Ghats Biodiversity Hotspot region of India, the Western Ghats Ecology Expert Panel headed by the renowned ecologist Prof. Madhav Gadgil, designated the entire hill range as an Ecologically Sensitive Area (ESA). Also, the report proposed to categorize the region into three zones—ESA1, ESA2, and ESA3. According to it, the ESA1 zone has restricted developmental activities such as mining, thermal, power plants, etc. (Report of the Western Ghats Ecology Expert Panel Part I, 2011) and there would be minimum human interference in those regions, so as not to disturb the ecosystem of that region. Such strategies need to be strictly applied, emphasizing more on the conservation and protection of microbial communities. However, implementation remains to be a prime concern, as most of these strategies are prone to get stuck in the red-tape.

References

Biodiversity Hotspots. (n.d.). Retrieved November 18, 2020, available from https://www.conservation.org/priorities/biodiversity-hotspots

Budapest Treaty on the International Recognition of the Deposit of Microorganisms for the Purposes of Patent Procedure. (2020). http://www.wipo.int/treaties/en/registration/budapest/index.html

Carlos, M.-R., M., H. S., Paola, V. D., Yadira, V. T., Maritza, S. C., Diana, C. M., Mirian, J. G., Manuel, C. G., & Paúl, M. R. (2020). In-situ and ex-situ biodiversity conservation in ecuador: A review of policies, actions and challenges. *Diversity, 315*. Available from https://doi.org/10.3390/d12080315.

Cavicchioli, R., Ripple, W. J., Timmis, K. N., Azam, F., Bakken, L. R., Baylis, M., Behrenfeld, M. J., Boetius, A., Boyd, P. W., Classen, A. T., Crowther, T. W., Danovaro, R., Foreman, C. M., Huisman, J., Hutchins, D. A., Jansson, J. K., Karl, D. M., Koskella, B., Mark Welch, D. B., & Webster, N. S. (2019). Scientists' warning to humanity: Microorganisms and climate change. *Nature Reviews Microbiology, 17*(9), 569–586. Available from https://doi.org/10.1038/s41579-019-0222-5.

Convention on Biological Diversity. (n.d.). Retrieved November 18, 2020, available from https://www.cbd.int/

Donachie, S. P., Hou, S., Lee, K. S., Riley, C. W., Pikina, A., Belisle, C., Kempe, S., Gregory, T. S., Bossuyt, A., Boerema, J., Liu, J., Freitas, T. A., Malahoff, A., & Alam, M. (2004). The Hawaiian archipelago: A microbial diversity hotspot. *Microbial Ecology, 48*, 509–520. Available from https://doi.org/10.1007/s00248-004-0217-1.

Falvey, R., & Foster-McGregor, N. (2006). *The role of intellectual property rights in technology transfer and economic growth: Theory and evidence*. https://www.unido.org/sites/default/files/2009-04/Role_of_intellectual_property_rights_in_technology_transfer_and_economic_growth_0.pdf

Frey-Klett, P., Burlinson, P., Deveau, A., Barret, M., Tarkka, M., & Sarniguet, A. (2011). Bacterial-fungal interactions: hyphens between agricultural, clinical, environmental, and food microbiologists. *Microbiology and Molecular Biology Reviews: MMBR, 75*(4), 583–609. Available from https://doi.org/10.1128/MMBR.00020-11.

Goel, D., & Parashar, S. (2013). *IPR, biosafety and bioethics*. Pearson Education India.

Innovation and Intellectual Property. (n.d.). Retrieved November 18, 2020, available from https://www.wipo.int/ip-outreach/en/ipday/2017/innovation_and_intellectual_property.html

Locey, K. J., & Lennon, J. T. (2016). Scaling laws predict global microbial diversity. *Proceedings of the National Academy of Sciences of the United States of America, 113*(21), 5970–5975. Available from https://doi.org/10.1073/pnas.1521291113.

MOSAICC Micro-Organisms Sustainable use and Access Regulation International Code of Conduct. (n.d.). Retrieved November 20, 2020, available from https://www.wipo.int/tk/en/databases/contracts/texts/bccm.html

Myers, N. (1983). *A wealth of wild species: Storehouse for human welfare*. Boulder, Colo: Westview Press.

Nampoothiri, K. M., Ramkumar, B., & Pandey, A. (2013). Western ghats of india: Rich source of microbial biodiversity. *Journal of Scientific and Industrial Research, 72*(9–10), 617–623. Available from http://nopr.niscair.res.in/bitstream/123456789/20956/1/JSIR%2072%289-10%29%20617-623.pdf.

Perinchery, A. (2020). *Bioresource access and benefit-sharing: How far have we come in India?* https://india.mongabay.com/2020/04/india-bioresource-access-and-benefit-sharing-how-far-have-we-come/

Pieterse, Z., Aveling, T. A. S., & Jacobs, A. (2018). Fungi associated with Aizoaceae seed in the Succulent Karoo. *Acta Horticulturae, 1204*, 177–185. Available from https://doi.org/10.17660/ActaHortic.2018.1204.23.

Reinold, M., Wong, H. L., MacLeod, F. I., Meltzer, J., Thompson, A., & Burns, B. P. (2019). The vulnerability of microbial ecosystems in a changing climate: Potential impact in shark bay. *Life, 9*(3). Available from https://doi.org/10.3390/life9030071.

Report of the Western Ghats Ecology Expert Panel Part I. (2011). Submitted to The Ministry of Environment and Forests, Government of India. Available from https://www.cppr.in/wp-content/uploads/2013/03/Gadgil-report.pdf

Silvestri, E., Hall, K., Chambers-Velarde, Y., Chandler, J., Cuddeback, J., & Jones, K. (2018). *Sampling, laboratory and data considerations for microbial data collected in the field*. Report number: EPA 600/R-18/164. United States Environmental Protection Agency.

Souza, R. C., Mendes, I. C., Reis-Junior, F. B., Carvalho, F. M., Carvalho, F. M., Vasconcelos, A. T. R., Vicente, V. A., & Hungria, M. (2016). Shifts in taxonomic and functional microbial diversity with agriculture: How fragile is the Brazilian Cerrado? *BMC Microbiology, 16*(41). Available from https://doi.org/10.1186/s12866-016-0657-z.

The Bonn Guidelines. (n.d.). Retrieved November 20, 2020, available from https://www.cbd.int/abs/infokit/revised/web/factsheet-bonn-en.pdf

The Cartagena Protocol on Biosafety to the Convention on Biological Diversity. (2020). http://bch.cbd.int/protocol

The Nagoya Protocol on Access and Benefit-sharing. (2020). https://www.cbd.int/abs/

Trevelline, B. K., Fontaine, S. S., Hartup, B. K., & Kohl, K. D. (2019). Conservation biology needs a microbial renaissance: A call for the consideration of host-associated microbiota in wildlife management practices. *Proceedings of the Royal Society B: Biological Sciences, 286*(1895). Available from https://doi.org/10.1098/rspb.2018.2448.

TRIPS: REVIEWS, ARTICLE 27.3(B) AND RELATED ISSUES. (n.d.). Retrieved November 18, 2020, available from https://www.wto.org/english/tratop_e/trips_e/art27_3b_-background_e.htm

U.S. Congress, Office of Technology Assessment, New Developments in Biotechnology: Patenting Life–Special Report, OTA-BA-370. (1989). Washington, DC: United States Government Printing Office. Available from https://ota.fas.org/reports/8924.pdf.

Index

Note: Page numbers followed by "*f*" and "*t*" refer to figures and tables, respectively.

Q

R

S

Kolibri 360 GmbH
Ferdinand-Jühlke-Straße 7
99095 Erfurt, Deutschland
produktsicherheit@kolibri360.de